DATE			

The Coastline

The Coastline

A contribution to our understanding of its ecology and physiography in relation to land-use and management and the pressures to which it is subject

Edited by

R. S. K. Barnes

St. Catharine's College and Department of Zoology, Cambridge

A Wiley–Interscience Publication

JOHN WILEY & SONS

London · New York · Sydney · Toronto

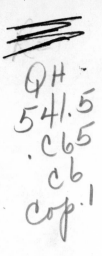

Copyright © 1977 by John Wiley & Sons Ltd.

Library of Congress Cataloging in Publication Data:

Main entry under title:

The Coastline.

  "A Wiley–Interscience publication."
    Bibliography: p.
  Includes index.
  1. Coastal ecology. I. Barnes, Richard Stephen Kent.
QH541.5.C65C6 574.5'26 76-51343

ISBN 0 471 99470 7

Printed in England by J. W. Arrowsmith Ltd, Bristol BS3 2NT

For

Hilary, David and Andrew

and in reply to a gadoid
who said:

'What matters it how far we go? . . .
There is another shore, you know,
upon the other side.'

(L. Carroll, 1865)

List of contributors

Dr R. S. K. Barnes Department of Zoology, University of Cambridge, Downing Street, Cambridge CB2 3EJ, U.K.

Dr Ir W. G. Beeftink Delta Instituut voor Hydrobiologisch Onderzoek, Vierstraat 28, Yerseke, The Netherlands.

Dr L. A. Boorman Colney Research Station (Institute of Terrestrial Ecology), Colney Lane, Norwich NR4 7UD, U.K.

Prof. G. Colombo Istituto di Zoologia e Biologia Generale, Università di Ferrara, Via Previati 24, Ferrara 44100, Italy.

Dr G. Dickinson Department of Geography, University of Glasgow, Glasgow W.2, U.K.

Dr F. B. Goldsmith Department of Botany and Microbiology, University College London, Gower Street, London WC1E 6BT, U.K.

Dr A. J. Gray Colney Research Station (Institute of Terrestrial Ecology), Colney Lane, Norwich NR4 7UD, U.K. (Present address: Furzebrook Research Station (Institute of Terrestrial Ecology), Near Wareham, Dorset BH20 5AS, U.K.)

Dr J. R. Lewis Wellcome Marine Laboratory (University of Leeds), Robin Hood's Bay, Yorkshire YO22 4SL, U.K.

V. J. May, Esq. Dorset Institute of Higher Education, Wallisdown Road, Wallisdown, Poole, Dorset, U.K.

A. D. McIntyre, Esq. The Marine Laboratory (Department of Agriculture and Fisheries for Scotland), Victoria Road, Aberdeen AB9 8DB, U.K.

Dr A. Nelson Smith Department of Zoology, University College of Swansea, Singleton Park, Swansea SA2 8PP, U.K.

Dr R. E. Randall Board of Extra-Mural Studies, University of Cambridge, Madingley Hall, Madingley, Cambridge, U.K.

C. R. Tubbs, Esq. The Nature Conservancy Council, Shrubbs Hill Road, Lyndhurst, Hampshire SO4 7DJ, U.K.

Dr M. B. Usher Department of Biology, University of York, Heslington, York YO1 5DD, U.K.

Contents

PART IV: CONCLUSION

Preface

Descriptive terms such as 'heathland', 'moorland', 'grassland', and 'coniferous forest' undoubtedly conceal considerable variety, but nevertheless they do realistically indicate the large measures of physiographic and vegetational unity displayed by these habitats. Such cannot be said of the 'coastline'. One has only to think of the mud-flats of the Wadden Zee, the cliffs of St Kilda, the marshes of the Rhône delta, the dunes of the Pte d'Arcachon, the polders of the IJsselmeer, and the shingle of Dungeness (to consider only north-western Europe) to realize that, their intimate association with the sea notwithstanding, coastal environments share few common features. Certainly the vegetation frequenting the coast is often halophytic or at least salt-tolerant, and coastal regions are usually in a state of flux—salt-marshes may be accreting sediment, for example, whilst cliffs are being eroded. But the gulf between the rate of erosion of Land's End in Cornwall and that of sedimentation in the vicinity of Mont-Saint-Michel in Normandy is as vast as the phylogenetic distance separating the wracks on a rocky shore and the spring squills upon a cliff-top,

This diversity has dictated the form of this book. Rather than adopt a 'process approach', with chapters treating such subjects as the effects of trampling or ecological monitoring, I have divided the book into a series of more or less separate sections, each covering one fairly uniform group of habitats—sandy foreshores, rocky cliffs, salt-marshes, and the like. These divisions closely follow those proposed by the then Nature Conservancy in Volume 2 of the Countryside Commission's Report on Coastal Preservation and Development (H.M.S.O., London, 1969). They will, I trust, permit seekers of specific information to turn directly to the habitat of concern. This system has also enabled me to inveigle a number of acknowledged authorities on the individual coastal habitats into contributing chapters on their particular specialities.

Although I have endeavoured to encourage uniformity of approach wherever possible—so that individual chapters are self-contained yet comparable—the unequal state of ecological knowledge of the various coastal habitats and the differing degrees of management required by these environments militate against the rigid application of such a plan. This, of course, is not without its advantages: clearly, it is important to realize that the problems to be faced in different environments are of differing magnitude, and that our knowledge of some is more complete than of others. Where the details of applied ecology and management have not yet been worked out, I have not hesitated to include the basic ecological information that will help eventually to formulate appropriate policies. In order to make each chapter as self-contained as possible, I have also permitted a certain amount of duplication: I trust that this will not offend those who seek to read the tome 'from cover to cover'!

The common structural plan of the chapters is: first, a summary of the fundamental ecological and geomorphological processes operating in the environment in question; secondly, the special features of the environment in terms of ecology and characteristic pressures; thirdly, methods of study; fourthly, educational, scientific, recreational or

economic uses of the environment, and fifthly, recommended conservation and management policies, including international case-histories of 'success or failure stories'. General
introductory and concluding chapters are included to place the individual habitats in a
wider perspective.

We are by no means in a position yet to produce a true handbook, in which the answers
to any question and the appropriate policies for any situation are clearly set out; I hope,
however, that this book is a step in the right direction, and that those concerned with the
future of our coastlines may find in the following pages material of interest, value and
enlightenment. The need for some such manual is evident:

> Those who know the coast well realize how good it is, and how fast some parts
> of it are being spoiled, . . .

> J. A. Steers, 1969
> (*Coasts and beaches*, Oliver & Boyd, Edinburgh)

This spoliation has frequently been brought about more by ignorance and thoughtlessness
than by economic necessity: if this handbook can partly rectify the situation it will have
achieved much.

This brings me to the subject of the audience for whom the book has been written. It is
becoming increasingly more obvious that communication between the producers and the
users of scientific information is in need of improvement. Ecological surveys, for example,
frequently do not provide the data required by engineers or planning authorities, or
conceal their findings from non-specialist comprehension by flagrant use of esoteric
terminology. This book (mainly written by the producers of scientific data) is an attempt to
tell those responsible for the initiation and operation of coastal policies what the ecological
factors of particular importance in the various coastal habitats are, and to provide an
ecological basis against which management decisions aimed at minimizing pressures whilst
permitting necessary utilization can be judged. It is thus aimed at the professionally-
concerned, but not necessarily ecologically-trained planner with coastal responsibility.
This is not to say that other potential audiences have been ignored: whilst editing, I have
tried to bear in mind all those professional and amateur people with an interest in the
coastal fringe.

I would like here to acknowledge the help and assistance given to me by many people
during the preparation of this book, not least by the individual contributors whose labours
form the substance of the work. In particular, I would like to express my gratitude to
Michael Packard and Derek Ranwell who provided much support and encouragement
during the book's early stages. I would also like to record my thanks to my wife, without
whose forbearance and understanding under difficult circumstances it could not have been
completed. It only remains to hope that their trouble has not been wasted.

Cambridge, 1976   R. S. K. Barnes

POSTSCRIPT

Owing to circumstances beyond the control of the editor or publishers—circumstances
which will readily be appreciated by all those who have endeavoured to produce multi-
authored works—it is regrettably also necessary to apologize to the contributors for the
delay in publication of *The Coastline*. The editor would particularly like to thank Drs L. A.
Boorman and F. B. Goldsmith for rescuing him from severe difficulties.

R.S.K.B.

PART I

INTRODUCTION

Chapter 1

The coastline

R. S. K. BARNES

Man is a terrestrial organism, and therefore the coastline forms a very real boundary for us—beyond it lies an essentially alien world on or in which we can survive for extremely short periods without the aid of technology. The fringe of this alien world, however, serves as a lure both to large numbers of people seeking recreation and relaxation (how many people in Britain have not been to the 'seaside' for a holiday?) and to industry dependent on the import of materials from overseas, the export of home produce, or the extraction of resources from the sea. Such industrial use creates the need for residential accommodation and so, as I have noted elsewhere (Barnes, 1974), in the last one and a half centuries the coastal fringe has become an area with a very high concentration of human habitation and employment.

To all (or most) of these permanent or temporary coastal inhabitants it is clearly of importance that this boundary zone should remain (a) aesthetically pleasing (or at worst tolerable), and (b) in the same place, save only for the rise and fall of the tides. Some countries suffering from a shortage of agricultural or building land would presumably take issue with the second requirement, rewriting it so that rather than maintain the *status quo*, the proportion of land to sea should rise wherever possible. These desired ends will not be achieved without some form of awareness of the problems involved and without management programmes designed to limit the power of the sea to erode or to deposit sediment, and to reconcile the often conflicting demands of educational, scientific, recreational, agricultural, industrial, fishery, and coastal-protection interests. One also needs the means to recognize the pressures that are altering given areas, to predict what effects other potential pressures might have, were they to be introduced or to evolve, to counteract undesirable influences, and to evalute different control measures. Such matters form the underlying theme of this book.

In this introductory chapter, I would like to state some of the problems in general terms and attempt to create some form of perspective.

MAN, THE WITNESS OF CHANGE

At any one time, the position of a coastline can be considered as the result of the operation of three independent factors (excluding man's influence, for the moment): continental drift, relative land–sea level, and the local characteristics of the rate of erosion or accretion resulting from subaerial, fluviatile or marine processes (including those of biological

3

origin). Since the first two factors operate in the long term, they are beyond man's control (and are likely to remain so for some considerable time), and are therefore outside the scope of this book. However, in a book concerned with what man can do, it is a salutary experience to reflect that the plates of the earth's crust, on which we live, are mobile, so that, for example, the northern shore of India is now concealed somewhere in the Himalayas; and that there is no reason to suppose that the Pleistocene ice-ages have really finished (although we frequently seem to assume that they have done so—witness the division into Pleistocene and Holocene). According to a number of eminent climatologists and students of the Quaternary, we are already past the half-way mark of the present interglacial period.

Although, therefore, the East African rift valley may one day be a pair of coastlines and the North Sea may once again be dry land (albeit partly ice-covered) as a result of these global fluctuations, more local but still dramatic alterations to a coastline can also result from erosive and accretive forces acting over much shorter periods of time. The processes continuously moulding the shape of a coastline have been reviewed in detail by such authors as Guilcher (1954), King (1959), Williams (1960), Steers (1953, 1964, 1973), Zenkovich (1967), Bird (1969), and Davies (1972): what I would like to do here is to look at some of the effects of these processes.

The night of 31 January 1953 in the North Sea area was characterized by (a) high northerly winds, of up to 160 km h^{-1}, veering round later to south-westerly, (b) low atmospheric pressure (down to 723 mmHg or 964 mbar), and (c) spring tides, although by no means extreme ones. Most people resident in the North Sea area, however, will remember that night and the following day by a different set of statistics: in Britain, 81,000 ha were flooded by seawater, 307 people (not to mention 3,000 pigs, 9,000 sheep, and 35,000 poultry) died, and 24,000 houses were destroyed; in the Netherlands, 162,000 ha were flooded, 18,000 people (and some 50,000 sheep and 100,000 poultry) died, and 50,000 houses were destroyed or severely damaged. Sea level passed all previous records, by 30 cm in The Wash and by 50 cm in the Thames Estuary. And the tide could have been 1 m higher, the storm could have coincided with the point of high water, and the rivers could have been in flood after previous heavy rain!

To many Europeans, although not to the Dutch, this was their first real experience of the erosive power of the sea—waves poured over natural and artificial coastal-defence systems and removed an incalculable amount of soil, huge concrete blocks were, in the words of one eyewitness, 'tossed like dandelion seeds in the wind', and the cliffs at Covehithe on the Suffolk coast were cut back 10·7 m in two hours. On a global scale, however, such events are not infrequent: hurricanes or cyclones cause frequent inundation of land in several tropical and subtropical regions, and sandy beaches are removed overnight. In the Low Countries of Europe, serious floods have been reported for the years 1014, 1099, 1170, 1175, 1225, 1277, 1285, 1288, 1323, 1337, 1357, 1377, 1404, 1421 (when from 16 to 72 villages were destroyed, killing 10,000–100,000 people), 1468, 1526, 1530, 1532, 1551, 1570, . . . , etc. (Wagret, 1959). Similar figures could be produced for the fens of East Anglia and Lincolnshire (Darby, 1940, 1956), and for the Thames Estuary.

Erosion

Important an erosive force as such floods may be, many areas of the coastline are being continually eroded in a much less spectacular manner and the loss of material resulting from these chronic causes in, say, a century is much larger than can be attributed to the sporadic storm surges in the same period. The loss of 10·7 m in two hours from the

Covehithe cliffs has been mentioned above, but, discounting that loss, these cliffs retreated at a rate of 0·5 m per month between August 1951 and March 1954—a total loss of 15·5 m through causes other than those of 31 January 1953 (Williams, 1960). The Suffolk, Norfolk, Lincolnshire, and Holderness coasts of eastern England supply many other examples of such chronic erosion: that of the Holderness region has been investigated by Valentin (1954).

He studied a 61·5 km stretch of boulder-clay cliffs, making measurements every 200 m, and was able to document the changes that had occurred since 1852 by reference to known buildings and field boundaries. Figure 1 shows the area in question, with the site of lost

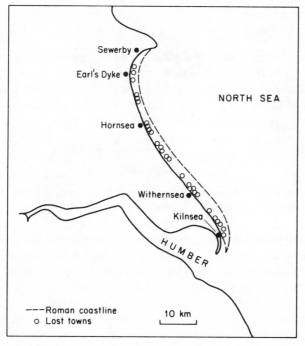

Figure 1. Erosion in Holderness (after Steers, 1964, reproduced by permission of Cambridge University Press—see text)

towns and villages indicated by a circle, and with the position of the coastline in Roman times marked approximately. Historical evidence suggests that erosion at a rate of 1–6 m per annum has been a feature of that strip of coastline for more than a thousand years, and Valentin's measurements indicate that since 1852 one million cubic metres of cliff have been lost annually. Valentin's results (as analysed by Steers, 1964) are set out in Table 1—the places named are marked on Figure 1.

Taking a coastline such as that of Britain as a whole, erosion is occurring along a much greater length than is accretion, although, of course, most cliffs will be more resistant than those described above. Accretion mainly occurs in the comparatively quiet waters of estuaries, semi-enclosed bays and natural harbours, but as the Royal Commission on Coastal Erosion (set up in the early years of the present century) pointed out, 'The evidence as regards the total superficial area gained and lost in recent years on the coasts

Table 1   Cliff retreat in Holderness between 1852 and 1952
(adapted with permission from Steers, 1964)

	Yearly retreat (m)	Length of section (m)	Yearly surface loss (m²)	Average cliff height (m)	Total yearly loss (m³)
Sewerby–Earl's Dyke	0·29	8,100	2,357	11·0	25,927
Earl's Dyke–Hornsea	1·10	13,650	15,015	11·8	177,177
Hornsea–Withernsea	1·12	24,250	27,160	16·2	439,992
Withernsea–Kilnsea Warren	1·75	15,525	27,200	13·2	359,040
Whole coast (averaged)	1·20	61,500	72,000	14·0	1,000,000

and in the tidal rivers of the United Kingdom shows that far larger areas have been gained by accretion and artificial reclamation than have been lost by erosion' (Royal Commission on Coastal Erosion and Afforestation, Final Report, 1911). The Commission quoted figures to the effect that in the previous 35 years over seven times more land was gained than was lost. Let us now, therefore, examine accretion rates.

Accretion

Although organisms (other than man) do play a role in speeding the process of erosion, except in certain tropical areas their effect is comparatively minor (Davies, 1972). The effect of organisms (particularly plants) in promoting accretion, however, is overwhelming: without salt-marsh plants there would be very little natural land-reclamation; without sand-dune grasses (especially *Ammophila*) there would be few coastal sand-dunes protecting the hinterland from tidal inundation (here one may point out that many coastal sand-dunes stood up remarkably well to the 1953 storm surge—much better than many concrete structures). A plant cover reduces the velocity of tidal water, or air stream in the case of sand-dunes, thereby allowing suspended particles to sediment out, and it acts as a filter retaining many of the particles which might otherwise drain from the marsh in the ebb tide. The root systems of plants also help to bind the surface sediment and render it less mobile: this fact is extremely relevant to the theme of coastal (and terrestrial) management—allow the plant cover to be denuded by various pressures and the accreted sediment will be removed by wind or water. I will return to the use of plants as a management tool later in the chapter.

Accretion, then, is a feature of sheltered areas of the coastline and is largely effected by maritime plants. Much of the gain in land thereby achieved is, at least in one sense, not really reclaimed from the sea. The sediment trapped by salt-marsh plants, etc., is only partly derived from the adjacent marine environment; indeed much of it has never been in the sea. The regions experiencing the greatest accretion rates are those such as estuaries, where rivers discharge considerable quantities of sediment recently eroded from the land—this then sediments in the estuarine zone. So we have here a redistribution of terrestrial soils: the land so to speak loses in mean height but gains in area, admittedly at the expense of the sea. Artificial land-reclamation, as we shall see, is a real winning of land from the sea.

Accretion can be analysed into two components—increase in height and increase in area. The effects of salt-marsh plants in accelerating the vertical component of accretion has been investigated at a number of sites. To do this, a layer of distinctive particles (iron filings or coloured sand) is spread on to a marsh surface just before the tide floods over it, and thereafter one returns at periodic intervals, cuts sections through the substratum and measures the accumulation of sediment above the marked layer. Some results from such studies at Scolt Head Island (Norfolk) are given in Table 2.

Table 2   Accretion (cm), averaged from 17 stations on Missel Marsh, above a sand layer introduced in 1935 (from data in Steers, 1960)

	1937	1947	1957	Yearly rate
Mean	1·6	10·8	18·8	0·9
Range	0·8–2·5	7·5–14·0	14·0–23·0	0·6–1·0

Change in the vertical height of a marsh surface (brought about by accretion) will affect the frequency of tidal cover and therefore the accretion rate (as the sediment is transported on to the marsh by the tides). As a marsh increases in height, the accretion rate will first accelerate as the plant cover becomes more complete and comparatively bushy species succeed smaller ones, but it will decrease when tidal cover becomes infrequent. As noted above, the species present will also change, as, apart from other reasons, different species require longer periods of aerial exposure for successful germination and can therefore invade as the height of the marsh increases. At Scolt, the following temporal sequence is found: Salicornietum → Asteretum → late Asteretum → Limonietum → Armerietum → Plantaginetum → Juncetum—by which time the number of submergences per annum will have dropped from about 450 to less than 100. (An '. . . etum' is a community of plants dominated by the genus stated to which has been added the suffix 'etum', e.g. a Salicornietum is dominated by *Salicornia*.) (Chapman, 1960a.)

If these different successional stages can be related to the sedimentation rate, some impression of the time taken for the marsh to approach a terrestrial condition can be gained. One such analysis is given in Table 3 (after Steers, 1964) based on the above Scolt Head Island sequence. Some two centuries are required at Scolt to convert a mud-flat into marshy land (assuming no intervening erosive phase). In many areas a salt-marsh may reclaim land very much more quickly: accretion rates of up to 20 cm per annum are known from the vicinity of Mont-Saint-Michel and in Zeeland. (Further discussion of salt-marshes and their ecology can be found in Chapman, 1960b, 1964; Ranwell, 1972a; Waisel, 1972; etc., and see Chapter 6.)

As a salt-marsh increases in vertical height it also extends in area, and this, of course, is the main agency reclaiming land. Man therefore invariably encourages the horizontal spread of salt-marsh by management techniques, and so it is comparatively difficult to obtain figures for the natural rate of spread. Such figures are available for a number of mangrove zones in the tropics, however. Macnae (1968), for example, cites the case of Palembang in south-eastern Sumatra which was a coastal port 400 years ago but is now 50 km inland—a seaward extension of the mangroves of 125 m per annum. Near Indramaju in north-western Java and Semarang in the eastern part of that island, rates of 108 m and 200 m per annum respectively have been recorded. The seaward extension of salt-marshes, sometimes aided inadvertently by man, along the southern shores of The

Table 3   Accretion in a salt-marsh at Scolt Head Island
(adapted with permission from Steers, 1964)

	Maximum depth of silt at conclusion of successional stage (ave.) (cm)	Rate of accretion per annum (ave.) (cm)	Time required for accumulation of that depth at those rates (ave.) (years)
Salicornietum	39·6	0·68	58
Asteretum	67·1	0·98	86
Late Asteretum	85·3	0·90	106
Limonietum	97·5	0·80	121
Armerietum	115·8	0·36	172
Plantaginetum	128·0	0·42	201

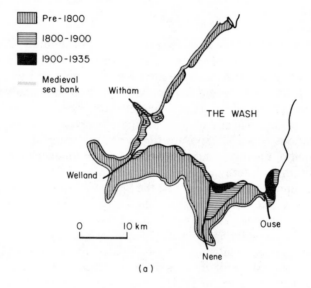

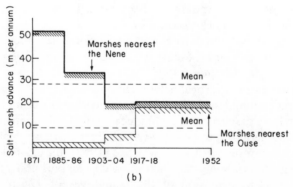

Figure 2. Reclamation of salt-marshes bordering The Wash. (a) Extent (after Darby, 1956, reproduced by permission of Cambridge University Press); (b) Speed (from data in Inglis and Kestner, 1958)

Wash has been investigated by Inglis and Kestner (1958). The fastest growing series of saltings increased at an average rate of 28·5 m per annum over an 80-year period, with a maximum advance of 513 m in a decade (Figure 2). Since the seventeenth century, some 30,000 ha of The Wash have been reclaimed by salt-marsh.

Longshore movements of material

Up to this point, I have been considering erosion and accretion as if they were two totally separate processes occurring in different and isolated parts of a given coastline. This is seldom the case: material is frequently eroded from place A and then deposited at place B a few kilometres farther along the coast. This can result in another form of coastal change—large structures can move slowly along a shore. Many a prosperous port has sunk into oblivion because a shingle spit, once providing a modicum of shelter, lengthened beyond the port and marsh developing in the lee effectively severed connections with the sea for all but the smallest boats. This state of affairs can be seen at Scolt and Blakeney (Norfolk), Orford Ness (Suffolk) and at Dungeness (Kent). Orford Ness has elongated southwards at an average rate in the region of 12 m per annum during the last eight centuries or so. In so doing, it has markedly deflected the mouth of the River Alde and has deprived the one-time port of Orford of easy access to the sea (Figure 3).

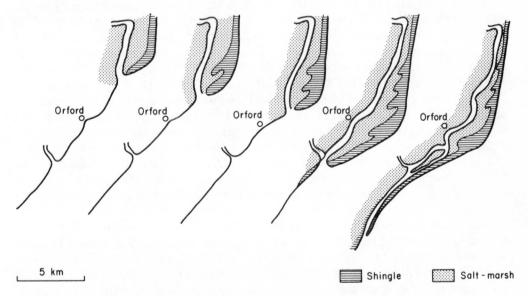

5 km   ▦ Shingle   ▨ Salt - marsh

Figure 3. Sketch map showing the growth of Orford Ness (reconstructed from the position of individual shingle ridges)—see text and Figure 66

The process resulting in the southward movement of shingle along Orford Ness and similar structures is basically longshore drift: pebbles are moved up a beach along an oblique path by waves striking the shore at an angle, but both they and the swash return perpendicularly to the water-line. Hence if a shingle structure is aligned north–south and the predominant direction from which waves approach is north-east or north-west, pebbles will tend to move southwards. The rate at which this occurs has been measured at, for example, Scolt, along a shore running approximately east–west (Table 4).

Table 4   Movement of brick fragments along Scolt Head Island under different weather conditions (adapted with permission from Steers, 1960)

	Direction of movement	Ave. daily movement (m)	Max. movement (m)	Wind direction	Ave. wind velocity (km h^{-1})	Direction of wave approach	
Day 1	W	5–9	58 (W)	ENE	8	NE	(waves small)
Day 2	W	5	73 (W)	NE	8	ENE	(waves small)
Day 3	W	9	183 (W)	NE	16	ENE	⎫
Day 4	E	5–9	110 (W)	NNW	32–40	N by W	(waves
Day 5	E	4	110 (W)	N	24	onshore	increasing
Day 6	0	0	110 (W)	N	29	onshore	in size)
Day 7	E	5–9	110 (W)	NNW	32–40	N by W ⎭	

MAN, THE BUILDER

By now it should be apparent that coastlines formed by soft sediments (sand or mud), shingle or comparatively soft rock (boulder-clay, chalk, some sandstones, etc.) are by nature in a state of flux. The other major activity determining the shape and position of the land–water boundary is the behaviour of man, particularly in his roles as a reclaimer of land and a protector of coastlines. At the time of writing, plans and proposals exist in Britain for the construction of barrages across The Wash, Morecambe Bay, Solway Firth and the Cheshire Dee in order to create freshwater coastal reservoirs, for reclaiming the Maplin–Foulness flats for a docks complex and London's third airport, and for reclaiming parts of a number of estuaries to site industrial and housing developments, e.g. Southampton Water and the Humber—Figure 4. In the Netherlands, there are the continuing Zuider Zee and Delta projects, and the proposed Wadden Zee Reclamation Scheme (Vaas, 1966) and so on. New, strengthened, coastal-protection walls have been built around most of the low-lying North Sea coasts after their 1953 battering. I would therefore now like to consider the constructional activities of man in connection with coastal reclamation and protection.

Before considering wholly artificial land-reclamation, however, I must briefly return to the role of salt-marsh plants. *Spartina* is a genus of halophytic grasses; some 14 species occur naturally around the shores of the North and South Atlantic Oceans. An American species, *S. alterniflora*, was imported into Europe in the nineteenth century, and in Southampton Water it crossed with the native European *S. maritima* to form a sterile hybrid, *S. townsendii*, first noticed in 1870. This doubled its chromosome number shortly after 'formation', thereby giving rise to a fertile species, *S. anglica*, which proved to have marked powers of spread, particularly vegetatively and by vegetative fragments (*S. anglica* is the '*S. townsendii*' of much of the literature—the sterile hybrid being in fact still restricted to a small area around Southampton Water). It is also able to act as a pioneer colonist, being capable of withstanding 24 h continual submersion in the field (and over 4 months submergence in the laboratory).

Once its potential for land-reclamation was realized, it was soon introduced to several parts of Europe, Australia, New Zealand, etc. Plants introduced to Zeeland in the

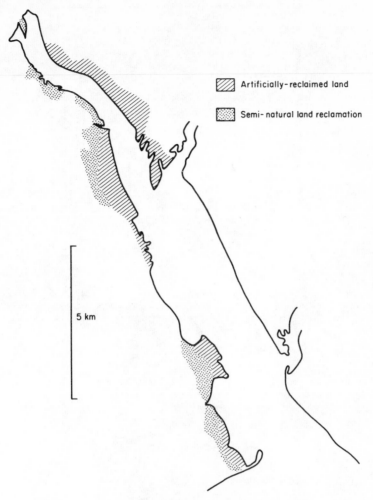

Figure 4. Land reclamation around Southampton Water (after
Barnes and Jones, 1974)

Netherlands caused the accretion rate to increase by such an extent that only five years
after its introduction marsh levels were raised by 1 m. Its abilities to reclaim land at speeds
beyond those achieved by other salt-marsh plants certainly led to large areas of mud-flat
being converted into land in several regions of the world, but *S. anglica* must be considered
a mixed blessing. The rate at which it can invade new areas, and spread once established,
have given rise to alarm (in parts of France it is known as *le péril vert*); it is invading and
replacing native salt-marsh communities; and perhaps more importantly for the would-be
reclaimer, some *Spartina* marshes are now manifesting die-back and the accreted material
is being eroded again.

   Besides deliberately introducing salt-marsh plants to virgin mud-flats, man can encourage natural sedimentation by building appropriately sited fences and ditches, and by
embanking saltings (Wagret, 1959). The two most outstanding and ambitious examples of
anthropogenic land-reclamation, however, are undoubtedly those of the Zuider Zee and
the Rhine–Maas estuary in the Netherlands. The Dutch have for a long time been masters

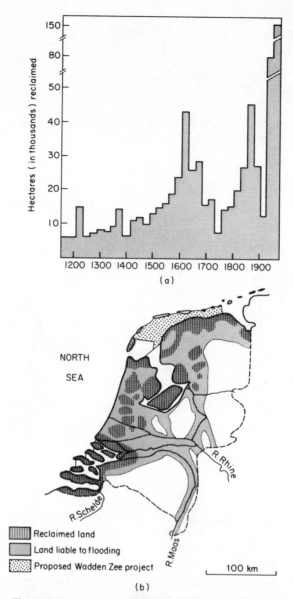

Figure 5. Land reclamation in the Netherlands (after Wagret, 1968, with additions, reproduced by permission of Dunod). (a) Land reclaimed during different quarter centuries from the twelfth century to the present day; (b) Extent of land reclamation

of the art of land-reclamation (Figure 5)—as indicated by their proverb *Deus mare Batavus littora fecit*—and they exported experts in this field to all parts of Europe in the seventeenth and eighteenth centuries. Some 18% of the land area of the Netherlands is now land reclaimed from the sea, and a further 4% has been reclaimed from coastal lakes and marshes.

The first plan to isolate the Zuider Zee from the Wadden Zee (and hence the North Sea) was put forward in 1667 and other plans followed at regular intervals thereafter, but it was

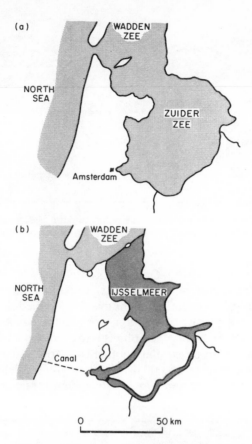

Figure 6. Reclamation of the Zuider Zee
(after Wagret, 1968, reproduced by permission of Dunod). (a) Before reclamation; (b)
After reclamation

not until 1917 that a definite proposal was accepted by the Dutch Government and work
started two years later (Wagret, 1959). A barrage was first constructed between the small
island of Wieringen and the adjacent mainland, and then the main barrage was built across
the 32 km stretch from Wieringen to Zurich via Breezand (a sand-bank upon which an
artificial island was formed) (Figure 6). The outer walls of the barrage were mainly
constructed using boulder-clay dredged from the Zuider Zee itself, and a 30 m-thick layer
of sand (dredged from the sea) was then pumped into the space between the two walls and
allowed to consolidate (Figure 7). Finally, in May 1932, the separation from the Wadden
Zee was effected and the IJsselmeer created, the waters of which gradually became fresh.

The creation of a 350,000 ha brackish inland sea was, of course, only the first step
towards reclamation: there remained the task (still in progress) of creating 220,000 ha of
agricultural land in five large polders—Wieringermeer (completed 1930), North East
(1942), East Flevoland (1956), Markenwaard (1964) and South Flevoland (as yet uncom-
pleted). Reclamation of these areas proceeded by isolating each region of proposed new
land with further barrages, and then by pumping out the impounded water to leave a wet
muddy surface (muddy because only regions with a clay substratum were reclaimed).
Before draining, channels were dredged to serve as thoroughfares for barges, as obviously
the new land surface (some of which is as much as 5 m below sea level) would be incapable

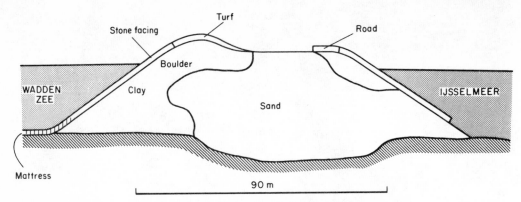

Figure 7. Simplified cross-section of the main dike separating the IJsselmeer from the Wadden Zee (after Wagret, 1968, reproduced by permission of Dunod)

of supporting wheeled traffic until dry. Once dry, however, roads, villages, schools, and all the paraphernalia of an agricultural community were built extremely rapidly.

The Delta Plan shows many similarities to the events described above, although whereas the IJsselmeer polders are nearing completion, work on the Rhine–Maas estuary started comparatively recently (in the 1950s). The object is to construct a series of barrages across the mouths of the estuary, thereby creating another large freshwater lake, which will make possible the reclamation of some 16,000 ha of potentially fertile land and considerably shorten the length of sea-walls required for coastal defence (Figure 8). Allowance will be

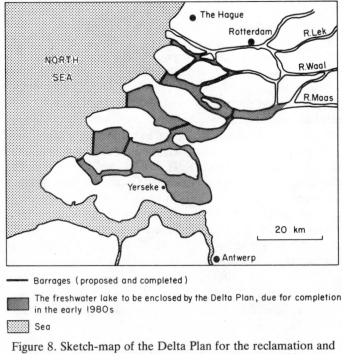

——— Barrages (proposed and completed)

▓▓ The freshwater lake to be enclosed by the Delta Plan, due for completion in the early 1980s

░░ Sea

Figure 8. Sketch-map of the Delta Plan for the reclamation and protection of the Rhine–Maas estuary (after Wagret, 1968, reproduced by permission of Dunod)

made for the outflow of freshwater by installing numerous locks and sluices, the whole system being capable of storing water for several days in the event of protracted periods of high water in the North Sea. Similarly, two sets of sluices discharge water from the IJsselmeer into the Wadden Zee during periods of low tide.

In the Netherlands, the problem of coastal-protection is inseparable from that of land-reclamation—less than half of the country is free from the danger of coastal flooding or of river flooding induced by high sea levels (Figure 5). Other countries bordering the North Sea have a qualitatively, though not quantitatively, similar problem. The best natural form of coastal-protection is a wide shelving beach, against which waves can dissipate their energy, and a supralittoral or submaritime zone stabilized by vegetation (considering here only those areas in need of coastal-protection). However, as we have seen, erosion can bring the problem of coastal-protection farther and farther inland, and normal natural defences may succumb to storm surges such as that of 1953.

Vegetation can be further stabilized by judicious planting (e.g. of *Hippophae* or conifers on coastal dunes) and, as Randall suggests later in this book (see Chapter 3), shingle can be

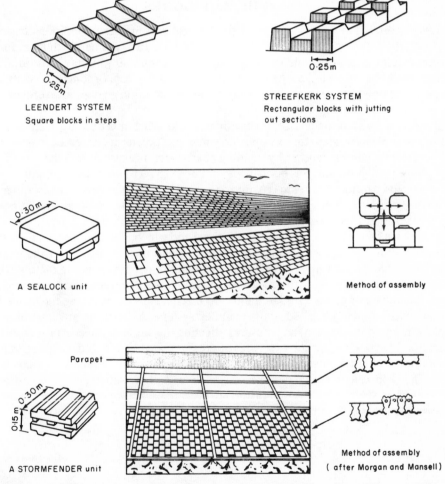

Figure 9. The nature of the facing of sea-walls and dikes (from Wagret, 1968, reproduced by permission of Dunod)

manoeuvred back into position after displacement by stormy seas; but by far the most commonly adopted coastal-protection measure is the construction of a sea-wall. In essence, sea-walls are identical to the barrages built around polders in the Netherlands and to those isolating areas like the IJsselmeer from the adjacent sea: both are designed to keep the sea from invading that which is being protected. Modern sea-walls have a foundation of clay and a facing of polygonal or rectangular concrete slabs (or stone or large sand-bags). Two fundamental requirements are imperative: (a) that the structures should be sufficiently rigid to serve as a wall; and (b) that it should be sufficiently elastic to dissipate as much of the wave energy as possible to avoid collapse under strain. An interlocking series of preformed concrete slabs will fulfil both requirements, and if individual slabs bear stout projections, wave energy will be further dissipated (Figure 9). The wall may be completed by a parapet designed to keep out the surge from breaking waves—as most sea-walls are not symmetrical (the side facing the land being comparatively weak), there is a danger of undermining from behind should large quantities of water overtop the wall. Finally, one can note that just as a shelving beach absorbs more wave energy than a vertical one, so a sea-wall should be at a small angle to the horizontal (*c.* 20°) in exposed conditions.

MAN, THE AMBIGUOUS

Having now completed a brief survey of the natural and artificial conditions basically determining the stability and position of the coastline, we can summarize thus far in this discussion as follows. Marine (and subaerial) influences continually modify coastlines: any given region will be subjected to the effects of erosion, accretion, or longshore drift, and over a period of time the dominant effect may change—accretion giving way to erosion, for example. The action of these three processes can therefore be considered to form the natural basis for short-term change of coastlines, on to which must be superimposed the effects of man. We have seen, for example, how man can accelerate or retard their actions by deliberately reclaiming land or by building coastal-defence systems. If these were the only effects of man, however, this book would not need to be written. Unfortunately, man unwittingly causes other changes in coastal zones, and hence we must turn to consider the many ways in which our species uses the coast and the pressures which this use puts upon it.

The pressures can be separated into two types: those which have a bearing on the subjects with which we have been concerned above and which, if unchecked, could lead to increased erosion or could weaken coastal defences; and those which affect the general amenity and educational value of the coast, including under this heading factors which influence those features making this zone into a favoured area for recreation, teaching and research (including, of course, research into the mechanisms determining the shape and evolution of coastlines described above). Other systems of analysis are possible—for example, one could investigate the pressures affecting the value of the coast for industrial or agricultural purposes—but the two types of pressure singled out above appear, at least to this author, to be the most important ones and those in urgent need of management policies. The economic potential of some coastal areas and the subject of conservation as an end in itself must also receive mention. From an ecological point of view, one can simply summarize the pressures and their effects as in the diagram.

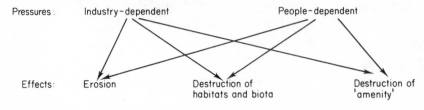

Pressures effecting erosion

Mining of a coastal cliff to gain stone or clay is obviously artificial erosion, but erosion can also be caused in a less direct manner. I have noted above that one of the best forms of coastal defence is a good beach, and if beach material is removed it can have unexpected consequences on local cliffs. A good example of such a process is provided by events at Hallsands in southern Devon. The village of Hallsands was built upon a raised-beach platform at the foot of a cliff in Start Bay. The shore was formed by shingle, and in retrospect this served to prevent erosion of the cliff-base. But in 1887, the beach was mined to provide shingle for the construction of dockyards at Plymouth, some 660,000 tonnes being removed from the intertidal zone. Although shingle is a common beach material in Start Bay, there is no sublittoral shingle reservoir, the bay being floored by sand. How the shingle was originally deposited on the beaches is unknown, but it is evident that the supply had ceased and that the material removed from the Hallsands beach was not naturally replaced. In consequence, the shore level was reduced by about 4 m and the loss of the protective shingle soon resulted in erosion of the cliffs—between 1907 and 1957 the cliffs retreated 6 m. The village was severely attacked by wave action and is now in ruins (Robinson, 1961).

As a broad beach serves to protect the base of a cliff, one means of cliff protection is to encourage beach formation by the use of groynes, etc., and this method can be used to prevent the erosion of sandy beaches near tourist resorts, and to maintain beaches to seaward of sea-walls (thereby helping to reduce the collossal maintenance expenditure frequently required on such structures). But the construction of groynes may merely displace the erosion (in an even more marked form) farther along the coast. If the line A–B in Figure 10 represents a shore subject to long-shore drift in the direction A → B, then the

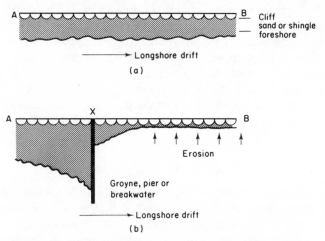

Figure 10. Diagrammatic representation of erosion resulting from interruption of longshore drift—see text

construction of a groyne (or a structure having a similar effect) at X would serve to maintain a wide beach along the section A–X, but it would result in the loss of beach material from the section X–B as material moved in the direction X → B could no longer be replaced. If the cliffs to landward are 'soft', considerable erosion could be expected along X–B.

This has indeed occurred at several places on the Sussex and Suffolk coasts—near Newhaven and Lowestoft, for example. In the former case, material drifting from the west collects at the Newhaven breakwater, causing loss of material from the beaches of Seaford and the erosion of the chalk cliffs of Seaford Head. In the Lowestoft example, material drifting southwards collects behind piers at Gorleston, starving the region to the south (Corton and Lowestoft) and causing erosion of the sand and gravel cliffs. Both Seaford and Lowestoft have built sea-walls to protect against this erosion.

The erosion at Hallsands was a consequence of exploitation of the coast's resources without sufficient prior investigation of the possible effects of this action, whilst that caused by the erection of piers, breakwaters, etc. can result from unco-ordinated coastal or harbour protection schemes, or from the provision of facilities for tourists. The former is industry-dependent and the latter is both industry- and people-dependent in the classification given above. These types of erosion are potentially under the control of local authorities (or the equivalent administrative body), provided that different authorities co-ordinate their policies to avoid piecemeal treatment of a stretch of coastline subject to the same physical processes. Permission to erect a structure capable of obstructing the flow of beach materials, or to remove the material itself can and should be refused when erosion, either locally or farther along the coast, is considered a possible consequence. A further cause of erosion is less easy to control—the people-dependent pressure exerted by human feet.

The effects of trampling are most evident on sand-dunes and on 'soft' cliffs, i.e. on slopes including those partially stabilized by vegetation. Many such areas receive a large influx of visitors in the summer-holiday season, and a comparison of a given area before (e.g. April) and after (e.g. October) that season frequently shows only too clearly the damage that can be caused. The vegetation cover is damaged and in severe cases removed, allowing the underlying material to be eroded by wind or by gravity, and this material may also be displaced downslope by people climbing on dunes and cliffs. Complete prevention of this pressure, short of denying or severely limiting human access, is probably impossible, but areas showing the worst effects can be rested through the use of fencing and can be reseeded, and the problem can be greatly reduced by the construction of footpaths. Paths channel access to the adjacent beach (which is where most people want to be) and can be floored by concrete slabs (as at Morfa Harlech), wooden boards (as at Gibraltar Point, Lincolnshire, etc.), or by resistant turf. Should the area through which the path runs be a nature reserve or be particularly sensitive to erosion, it may be necessary for the path to be fenced, as the provision of a path does not always guarantee that most people will in fact use it.

Value of the coast for teaching and research

We must now turn from a consideration of the protection of land for its own sake and look at the value of the coastline for education and research; later we will investigate the pressures which can affect this value. That the coast possesses great value for the teaching of geomorphology and physiography should have been obvious from the preceding pages: it also has great importance for other geological studies as all the major subdivisions of geological time can be seen in section at the coast, even a comparatively small country like Britain showing most of them. So, to quote the Countryside Commission (1969, pp. 4–5):

> For geologists, the coast provides unique facilities for research and demonstration. Coastal exposures are characterized by their length and continuity, in contrast to the much smaller and more intermittent outcrops inland. Further, they are the only exposures seen in three dimensions with the gently sloping shore platform . . . complementing the steep . . . rocky slopes behind. Where

soft rocks adjoin the sea are found the only naturally permanent exposures of such strata, as the natural cleaning process of the sea prevents the accumulation of fallen material and the development of vegetation cover which occurs inland.

Consequently the coast of Britain is studded with geological sections of national importance, easily available for research and field education [unless obscured by sea-walls, etc.]. As much of the early development of the science took place in Britain, many of these are type sections of international importance. In addition, the coastline provides dramatic exposures of structure . . . many of which . . . are seldom exposed inland

Coastal districts thus attract the major portion of geological field teaching by universities and other educational bodies. The Dorset coast alone is the site of 11 per cent of British field instruction in geology.

The coastline is no less important for biological teaching and research due to the combined effect of many factors. In the Preface, I stressed the great variety of coastal habitats and the fact that they show few common features. This broad spectrum of environmental diversity ensures that very many different communities of animals and plants are present, often within a limited area. Within a 10 km radius of the town of Pembroke, for example, one can find good examples of a wide range of submaritime habitats, rocky cliffs, sand-dunes, salt-marshes, lagoons, estuaries, sandy foreshores, mud-flats, rocky beaches, etc. Secondly, although the sea possesses comparatively few individual species of animals, those species which do occur are representatives of a much wider range of different types than can be found on land—of the 69 classes of living animals listed by Clark and Panchen (1971) which are not comprised solely by endoparasites, 90% are represented in the sea, 41% in freshwater, and only 23% on land (excluding from the category 'terrestrial' those essentially freshwater animals which on land can only inhabit films of water). The greatest variety of animal types is to be found on the shore and thus it is no coincidence that the university field courses designed to acquaint students with the diversity of animals are held at the coast. Thirdly, the coastal environment is subject to considerable fluctuations—littoral organisms have to withstand temperature, desiccation, and salinity extremes—and further there are many problems to be faced in connection with settlement and reproduction by a marine animal regularly exposed to the air. Hence, the coast is a particularly good site to demonstrate the physiological, anatomical and behavioural adaptations shown by organisms to their environment. Fourthly, a number of fundamental ecological processes can be demonstrated more clearly and more commonly on the coast than elsewhere: one can study the manner in which plants colonize new surfaces exposed by erosion or created by accretion, the succession from pioneer to climax communities (e.g. in sand-dune or salt-marsh systems), zonation along environmental gradients (e.g. along the salinity gradients of estuaries or the vertical gradients of rocky shores), competition between different species (e.g. between barnacles and mussels, and between different species of barnacles), etc. These fields of study or demonstration are some of what can be termed the specialities of the coastal zone—all can be more easily seen here than elsewhere—but, of course, the coast also possesses all the other advantages of natural or semi-natural environments for teaching and investigating the nature of the basic ecological properties of our planet. In Britain, the coast is in fact one of the very few places where one can still find truly 'natural' communities of organisms. The ecology of the coastline is described in detail in such works as Yonge (1949), Hepburn (1952), Schäfer (1962), Lewis (1964), Friedrich (1965), Green (1968), Eltringham (1971), Stephenson and Stephenson (1972), and Barnes (1974), besides those listed earlier in connection with salt-marshes.

A number of food-webs can also be easily demonstrated along the coast, some of them of considerable commercial import. The top carnivores of coastal regions are birds and fish. Very few commercially-important fish are restricted to coastal regions (using the term 'coast' in the sense of the littoral and immediately sublittoral regions), but many are dependent on it. The productivity of coastal areas is very high, in part because of the abundant nutrients derived from land drainage, and much of the energy fixed is exported, either in the form of detritus or as foodstuffs taken by migratory species. Many fish either spawn in such areas as mud-flats or estuaries, or else spawn in regions from which their developing eggs and larvae will be carried to these areas. Plaice, for example, spawn in the southern North Sea and their larvae are carried by currents to the coastal Waddens of the Netherlands, etc. By these means, the stages in the life history which grow most rapidly find themselves in areas where an abundant food supply can most easily be obtained. Having grown in shallow coastal waters, the young fish may then migrate elsewhere. So although man may not catch many fish along coastlines, he is nevertheless dependent on such areas for many of the fish which he wishes to catch in the adjacent seas. The shellfish industries of several European countries also earn considerable revenue.

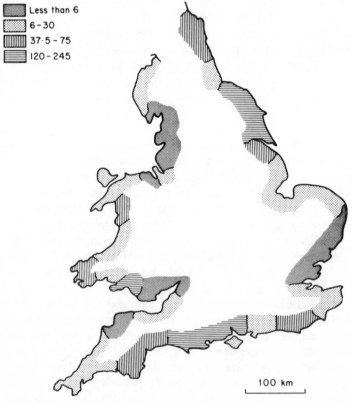

Figure 11. Use of the English and Welsh coastlines for teaching purposes in 1967: number of student-days per year per km (after Countryside Commission, 1969, with the permission of the Controller of Her Majesty's Stationery Office)

These facilities for teaching and research are not just hypothetical. They are indeed used by many universities, technical colleges, schools, and amateurs. In Britain, centres of higher education maintain more than a dozen coastal laboratories and many more centres are already situated near the sea; local education authorities are increasingly building residential centres on the coast (of which there are now more than 30); seven of the nine centres run by the Field Studies Council make use of the coast for teaching purposes; and more than 20 marine laboratories are run by other organizations concerned with coastal research. Figure 11 shows the use made of various stretches of the British coastline by students in 1967. The over-all research importance of the coast can be gauged from the fact that of the 6,200 km of the English and Welsh coastlines, 3,600 km are classified as areas of special or outstanding scientific importance (although only 1,800 km are in any sense protected against 'development', and of these the protection afforded to 1,500 km is much less than total).

Pressures affecting habitats of scientific and educational importance

I have listed above some of the non-quantifiable resources of the coastline: unfortunately the pressures adversely affecting these resources are sometimes only too quantifiable. Perhaps the greatest threat to several habitats is that posed by land-reclamation; other threats including the dumping of mining spoil, pollution, the requirements for more land for building houses and factories, trampling and over-collecting.

Conversion of mud-flats or salt-marshes into land or freshwater lakes obviously destroys the habitat of many resident animals and plants, and local populations will be extinguished. As many of these organisms have wide distributions, however, this has not given rise to fears that particular species may become extinct, although if encroachment is continued into the indefinite future, this situation could well arise. The main effects of reclamation are probably felt by the migratory waders and wildfowl which use mud-flats and salt-marshes as wintering and feeding grounds, and as roots.

> The areas in Britain already undergoing or scheduled for possible reclamation are all particularly important sites for waders and wildfowl: Morecambe Bay supports the largest numbers of waders and the second-largest numbers of wildfowl of any British estuary; the Dee supports the second-largest numbers of waders; the Wash is the fourth most important site for wildfowl and the fifth for waders; etc. In more concrete terms, Morecambe Bay probably supports 15% of the populations of knot wintering in Europe and North Africa, whilst the Wash, Dee, and Solway Firth also contribute markedly to the British total of 40% of these populations. Foulness supports 40–50% of the British populations of wintering [dark-bellied Brent Geese], which themselves constitute some 55% of the world population . . . Removal of many of these areas will therefore have definite repercussions on the ecology of the estuarine avifauna. If any one habitat was destroyed, the other existing sites could probably absorb the displaced individuals. But as reclamation is a general feature of much of the coastline of north-western Europe (and of the eastern seaboard of the U.S.A.), the number of suitable feeding and wintering grounds is becoming seriously limited. If this process continues, considerable diminution in numbers of several species must be expected. (Barnes, 1974)

Reclamation is generally an all-or-none phenomenon: one cannot recommend management policies for a habitat in the process of destruction, although such policies may be required for the new environment, be it land or freshwater. It is possible, however, to create artificial habitats to compensate partially for losses, but one can usually only recommend that the effects of reclamation be thoroughly investigated before permission for a specific project is given—it is not only birds and the mud-flat invertebrates that will be affected. Zijlstra (1972) has drawn attention to the consequences of the proposed Wadden Zee Plan on the fisheries of the southern North Sea.

The sea is used as a dumping ground for many substances, ranging from poison-gas containers and radioactive materials to sewage sludge, motor cars and bottles. Most of the dumped materials are deposited in deep water offshore, but 2,550,000 tonnes of colliery waste are tipped each year from cliff-tops along the Durham coast on to the foreshore, and almost 1,000,000 tonnes of china-clay waste per annum pass down the rivers draining into St. Austell and Mevagissey Bays in Cornwall (Royal Commission on Environmental Pollution, 1972). This material smothers the foreshore and adjacent sublittoral regions, and not only renders it unsightly but also unsuitable for many of the organisms which would originally have been present (and with effects, therefore, on its use as a nursery area for fish). One sometimes wonders why this waste could not either be more frequently used for constructional purposes or be put back into the ground in disused mine shafts, etc. The use of the coastline as a rubbish bin cannot be condoned when other methods of disposal are available.

The discharge of pollutants into coastal seas has now an extensive literature (see, e.g., Cole, 1971; Russell and Gilson, 1972; Goldberg, 1972; and Figure 12) and it is unnecessary here to review once again the effects of pollution on coastal communities. Suffice it to say that, apart from the localized effects of reduced oxygen tensions in estuaries and semi-enclosed bays, the main type of pollution forming a threat to coastal organisms in general is oil, together with oil-dispersing detergents. The acute form of oil pollution resulting from major oil-spills from tankers makes the newspaper headlines at irregular intervals, but although this is the 'best known' way in which oil is deposited on the coast, it only accounts for some 20% of the oil discharged into the world's oceans. The remaining 80% is derived from effluents and from discharges of unknown origin (oil tipped into the public drains, etc.) and this frequently gives rise to more serious, chronic forms of pollution. Refinery effluents, for example, may contain from 5 to 50 p.p.m. of oil, and one large refinery can therefore discharge some 4,500 l of oil per day. The problems associated with the prevention of oiling and the dispersal of oil which has come ashore have been admirably summarized by Nelson-Smith (1972a) (see also Cowell, 1971), and a case history of an oil spill, from the treatment of which many lessons can be learned, is contained in Smith (1968). Treatments of areas affected by oil vary with the nature of the environment and with the uses to which that environment is put. We can note, however, that as far as the effects on coastal organisms are concerned, chronic pollution is worse than the acute form, areas subject to little wave action (e.g. mud-flats) are affected more seriously than those in more exposed situations (e.g. many rocky shores), and that in many cases no action at all is better than dosing the environment with detergent.

These factors affecting coastal habitats and their organisms have all been industry-dependent. As a bridge between industry- and people-dependence, we can consider the destruction of the more terrestrial habitats for agricultural, industrial or housing developments. As Dickinson describes later in this book (Chapter 14), those coastal industries which are not sited on reclaimed land occupy areas of the submaritime fringe. In several areas of Dorset, etc., agriculture extends to within 2 m of cliff faces. Many, if not most, of Britain's sand-dunes now bear golf courses. Resort towns along the coastline are still expanding and in several areas of the south coast of Britain it is very difficult to gain access

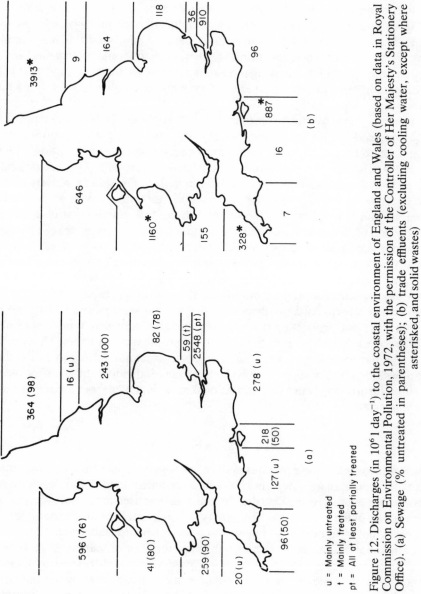

Figure 12. Discharges (in 10^6 l day^{-1}) to the coastal environment of England and Wales (based on data in Royal Commission on Environmental Pollution, 1972, with the permission of the Controller of Her Majesty's Stationery Office). (a) Sewage (% untreated in parentheses); (b) trade effluents (excluding cooling water, except where asterisked, and solid wastes)

u = Mainly untreated
t = Mainly treated
pt = All at least partially treated

to the shore because of a barrier of houses, bungalows, and hotels: this barrier is particularly well developed, and is still growing, in those coastal towns which form part of a 'retirement belt'. And when one adds Ministry of Defence properties, such as those in Pembrokeshire, Dorset, Devon, and Suffolk, to the list, it is not difficult to appreciate that the submaritime fringe is under pressure. A country such as Britain would not be reclaiming land unless this commodity was in short supply (even if only locally), so it is evident that industry and housing will continue to require more coastal land, but, as outlined in connection with land-reclamation, authorities granting permission for the development of an area of coastal fringe should bear in mind that this is a diminishing resource of great scientific and educational importance, and should consider as a matter of some urgency the possibility of leaving representative and particularly important habitats free of concrete in otherwise built-up areas. Cliff walks can be a great asset to a holiday town!

Lastly in this section, we must consider the two people-dependent pressures of trampling and over-collecting. Trampling has already been discussed in connection with erosion and, of course, destruction of the local vegetation cover will not only facilitate erosion, but will affect the communities of organisms present. The educational and scientific communities themselves are not without guilt in this sphere: a party of school-children or students under ineffective supervision can cause havoc on a salt-marsh, sand-dune, or mud-flat. But besides the greater self-control and 'thoughtfulness' required from that section of the population which seeks to safeguard the coast for the purposes of teaching and research, the remedies for the effects of trampling pressure are as indicated earlier. Over-collecting is a feature of a number of beaches situated either near teaching or research establishments, or near holiday resorts (where the numbers of edible molluscs like winkles may become severely reduced—as along several regions of the north French coast). With the exception of a number of species, such as sea-urchins, which have a high market value, the over-collecting is not severe and the increasing tendency to observe instead of to collect should do much to diminish the problem even further. Those in charge of educational parties could, however, do more to correct any false impression that the purpose of visiting a rocky shore is to collect as many individuals of as many species as possible, and, further, to ensure that educational material taken from a beach is returned to it in a living state.

MAN, THE AESTHETE

We must now investigate those features of the coast which account for its high amenity and aesthetic value, and here we enter a field fraught with difficulty, as all too clearly one man's beauty is another man's eyesore. Further 'amenity' can be used in several different senses: most dictionaries define amenity as 'the quality of being agreeable or pleasant' and thus a pleasant rocky cove would be an amenity to a tourist board; but when a local council provides 'amenities', it is usually erecting public lavatories or car parks and these can only barely be described as agreeable or pleasant, particularly with respect to their visual appearance. A caravan site can therefore either create or totally destroy the amenity of an area depending on one's preferred definition or standpoint, and so my remarks under this head must be considered to be, by nature, essentially personal.

Several areas of the British coastline are classified as 'Areas of Outstanding Natural Beauty' and large stretches of unclassified coast have considerable appeal. Clearly, the building of an oil refinery on Scolt Head Island would be a desecration, but here further difficulties start. We all want electricity in our homes, but no one wants the power station providing that electricity sited near them. Someone who spends his time on the south-west coast of Britain may not object to a power station sited near Bootle or Scunthorpe; alternatively a coastline buff would prefer it to be inland, say at Huddersfield or Birmingham; and so on. Of course, the problem is not always in the form 'shall we build the next steelworks on the Chesil, at Robin Hood's Bay, or on the Mawddach?'. Amenity is doubtless considered, but other factors such as economics, rates of unemployment, etc., have to be taken into account. Unfortunately, some installations, such as nuclear reactors, are deliberately sited in areas of minimum habitation, and although this may be for a very good reason, a large power station can stick out like the proverbial sore thumb in an hitherto 'unspoilt' region like Dungeness, and can scarcely be encouraged on amenity grounds. The argument in Britain over the site for London's third airport displays many of the problems associated with 'amenity'.

Perhaps the major problem is that one just cannot quantify aesthetic or amenity values (and neither can one quantify the values of many different habitats): is the destruction of 200 ha of moorland equivalent to that of a similar area of mud-flat (or cliff-top); and is that of one million ants equivalent to that of one million sea-anemones (or mallard)? And if not, why not? Some views are unsurpassed, others are downright squalid, and some habitats are nationally or internationally rare and others are still comparatively wide-spread; here the decisions may seem deceptively straightforward, but what if these decisions clash with the number of unemployed who would doubtless consider a factory to be an amenity?

Such difficulties notwithstanding, the fact that it is still possible to obtain views of unspoilt scenery along many coastlines is undoubtedly something to be treasured beyond price. It should therefore be obvious that the construction of an eyesore in an area of natural beauty can only be deplored. But whilst the construction of a blast furnace in St James's Park would be almost universally condemned (except by those people who fear that it would otherwise be sited near them), people flock to an area of beauty because (presumably) it is beautiful, and then in the name of amenity paradoxically demand additions to the environment which effectively destroy its aesthetic appeal. A blast furnace in St James's Park could be far less of an eyesore than many of the caravan sites which litter parts of the south-west of England. Yet many hideous public toilets, cafeterias, car parks and fun fairs continue to detract from areas of one-time beauty. This is not to say that public toilets are necessarily hideous. Some local authorities have managed to screen such 'amenities' very well by taking advantage of the natural topography or by planting trees (Dickinson describes one such example in Chapter 14). It is to be hoped that more authorities will emulate these enlightened ones and not destroy amenity in an attempt to create it. Car parks etc. are certainly necessary, as uncontrolled car parking can cause serious disturbance and erosion, but in most areas it is usually possible to hide the park in such a manner that it does not form a blot on the landscape.

MAN, THE MANAGER

Having now ridden one of my favourite hobby-horses, it is time to effect a summary of the uses and pressures peculiar to the coastline and to draw some general conclusions. As we have seen, the coastline is an area of high aesthetic value and a zone of the highest importance to education and science. It is also a region subject to many pressures, particularly those resulting from the high population densities and the industries attracted to it. Further, this boundary between land and sea is mobile: considerable effort is devoted to preventing increase of the sea relative to the land and to increasing the area of land at the expense of the sea. To ensure that most of these factors can coexist requires management plans at both the national and regional levels (Wastes Management Concepts for the Coastal Zone, 1970; Brahtz, 1972; Ketchum, 1972; Usher, 1973; Perkins, 1974; etc.). Note that 'most' and not 'all' figures in the sentence above: although many of the pressures are legitimate in the sense that a country's prosperity depends upon them, some can and should be removed by management. Although, for example, use of the coast for residential, industrial, agricultural and recreational purposes is obviously necessary and must inevitably increase; trampling-pressure, many forms of pollution, etc. are avoidable and therefore management plans should seek to eliminate them, rather than incorporate them into a multi-use scheme.

Perhaps few of the contrasting uses to which the coast is put can be permitted at any one point—one should not mine coastal-defence systems for shingle or bedrock, neither should one allow high population pressures in areas where erosion would be the result—and so a system of zoning must operate in which certain areas are set aside for one major purpose (together with any compatible uses), and others are given over to different activities. However, one must be extremely careful in deciding which areas should be devoted to which purpose! One particular example of such zoning is in areas where conservation is required as a matter of fundamental principle. Conservation of certain areas is not advocated solely because they are of great aesthetic, scientific or educational importance and because 'development' would interfere with this status, nor because of 'destruction of our natural heritage' arguments, although both of these are clearly important: there can be practical reasons for conservation. Many of our domesticated vegetables, for example, have been derived from coastal ancestors—beetroot (and its sugar-beet, spinach-beet and mangel-wurzel varieties), cabbage (together with broccoli, Brussels sprouts, cauliflower, kale, kohlrabi, savoy, etc.), sea kale, and asparagus all have this origin—and the coast also forms the habitat of wild species of carrot, lettuce, radish, orache, fennel, etc. Even today, the various wild 'samphires' (*Salicornia* spp, *Crithmum maritimum* and *Inula crithmoides*) are gathered for human consumption. The tropics provide similar examples (sweet potato, etc.). When the world as a whole is suffering food shortage, it would clearly be disastrous to destroy the natural genetic stocks of these and other potential food-plants, from which new varieties can be bred and which serve as a reservoir guarding against extinction of the largely inbred cultivars through epidemic disease. Other coastal plants can be viewed as valuable genetic resources; for example, grasses which stabilize coastal habitats may be used to cover spoil tips and inland habitats subject to erosion, and several coastal nitrogen-fixing legumes have been used with great effect inland. Conservation is emphatically not just a matter of preserving 'pretty flowers, birds and animals (meaning mammals)', although this too can be worthwhile.

Although, therefore, a strong case can be made for maintaining some stretches of the coastline in a natural or semi-natural state, other regions must be subject to a multitude of uses under strict management and control—the authors who follow describe the priorities and pressures in the different coastal habitats. Indeed *laissez-faire* is seldom possible at the coast—management is required even in areas set aside as nature reserves, for otherwise successional stages in the vegetation may convert the habitat for which the reserve was established into, for example, scrubland, and the fauna will also change accordingly. Throughout the coastal zone, management must seek to prevent and to regulate.

It is appropriate to conclude this introductory chapter by briefly considering which of the different coastal habitats are under the greatest pressure and are therefore in particular need of conservation. The habitat probably experiencing least pressure is that provided by rocky cliffs. Although some are mined commercially, the difficulty of this terrain is sufficient to exclude most people and most industries, leaving the cliff face to nesting sea-birds. Similarly, as emphasized by Lewis (Chapter 8), rocky foreshores are in a comparatively 'safe' position, pollution by oil and over-collecting or disturbances consequent on collecting being the only pressures seriously affecting this habitat (and the pressures resulting from the collection of biological specimens are very local phenomena). Intertidal expanses of shingle possess a very impoverished fauna and flora for a variety of reasons—the pebbles are easily rotated one against the other, their water-retaining capacity is low, etc.—and, apart from their use as a source of building materials which can give rise to problems (see the Hallsands history above), pressure on this habitat is generally low. All these habitats are also comparatively frequent along most coastlines.

The next two environments, semi-terrestrial shingle formations and sandy foreshores, experience more pressure but are still by no means severely affected. Shingle formations are mined locally, but although this pressure may not be too serious in itself, it becomes more so when one considers the limited extent of this habitat, and the peripheral effects of the mining (cart tracks, etc.). Sandy foreshores are mainly considered as amenity beaches, and therefore oil-spills which have coated them have in the past been dispersed as quickly as possible by methods incidentally inflicting great biological damage. The effects of recreational pressure and of various types of pollution are marked locally, but fortunately the frequency of occurrence of this habitat, and its areal extent at low tide mitigates these pressures when viewed at a national scale.

The remaining habitats are all subject to greater pressures. Lagoons, mud-flats and salt-marshes are threatened by pollutants; mud-flats and salt-marshes by reclamation; earth cliffs and sand-dunes by trampling-pressure; salt-marshes by the spread of *Spartina anglica*; and earth cliffs, lagoons, the submaritime fringe and sand-dunes by 'development'—in several parts of Britain, coastal lagoons are particularly exposed to deliberate development, the original reed-fringed pond being converted into a concrete-walled boating lake, complete with artificial islands, 'picturesque' bridges, gnomes, pagodas, and storks. Whilst admitting, as stated earlier, that one man's meat is another man's poison, it must be stressed that in Britain coastal lagoons, together with several other of these habitats, are not only under great pressure but are also comparatively rare: few British counties possess lagoons, for example, and of those that do several no longer possess any in a natural state. The environment experiencing the greatest pressure, however, is undoubtedly the estuary.

Many estuaries are heavily polluted, some like the Tees and (until recently) the Thames being anaerobic over considerable stretches, and land-reclamation and 'development' generally are increasingly converting the regions both above and below the high-water mark into concrete. Yet estuaries, together with the subsidiary mud-flat and salt-marsh habitats, are one of the most productive ecosystems known to man, and they help to maintain the productivity of the adjacent coastal seas by exporting organic matter in the form of detritus, young fish, etc.

Naturally, specialists on habitats which I have considered to be relatively free from pressure are liable to object strongly to this analysis. We all consider our own particular preferred habitat to be under the greatest pressure—consequently it should not be a surprise to learn that mine are lagoons and estuaries! As with so many of the decisions which have to be taken in regard to conservation, in reality all one can do to protect a given area from what one considers to be undue pressure is to state one's own genuinely-felt (but inevitably biased) views as forcibly as one can, and be thankful that the consequences of the decision finally taken rarely lie upon one's own head.

PART II

PREDOMINANTLY AQUATIC
ENVIRONMENTS

Chapter 2

Sandy foreshores

A. D. McIntyre

THE NATURE OF SANDY FORESHORES

The basic nature of a shore is determined by the geology and topography of the coastline and adjacent land, and by the physical processes operating on them, particularly the actions of rivers and waves which create, supply, and distribute sedimentary material. When the slope is too small for the available wave energy to carry away this material, a beach is formed. The type and configuration of the beach will again depend on the geomorphology of the adjacent land and the prevailing waves, currents and tides, and also on the nature, size, type and quantity of available beach material.

The presence of sand as the dominant constituent of a beach indicates exposure to wave action (which may range from slight to severe), and the geological description of sand—sedimentary particles of rock between about 0·05 and 2·0 mm in diameter—provides an objective means of distinguishing sandy foreshores from coarser beaches of gravel and finer ones of mud. Sand-beaches are much flatter than those of shingle. They are seldom steeper than 1 in 25, and gradients may be as small as 1 in 200 or even occasionally 1 in 1,000. The precise nature of sand particles on any given beach will depend on the type of parent rock, which may be limestone (calcium carbonate), felspar (aluminium silicate) or quartz (silicon dioxide). To these will be added varying quantities of material of recent biological origin—shells of molluscs, plates and spines of echinoderms, and other fragments including organic matter from organisms which lived in the vicinity. In some areas these recent biogenic particles may predominate, and beaches composed largely of fragments of lamellibranch or even barnacle shells are known. Around the coast of western Europe, however, the commonest sand-beach material is quartz. While calcareous fragments tend to break up or dissolve fairly quickly, quartz grains, protected by a water film, are very much more resistant, and once deposited tend to retain their original shape, varying from angular to rounded, which may be characteristic of a particular locality. A size description of the sedimentary particles should include not only the range but also the distribution of particle size. A widely used classification, the Wentworth scale, relating descriptive terms to size, is set out in Table 5. Since this is a geometric scale it provides closer grade intervals at the smaller end of the spectrum and wider ones at the coarser end where a narrow definition is not required. A logarithmic transformation (the phi notation where ϕ equals $-\log_2$ of the particle size in mm) applied to the Wentworth scale produces an arithmetic series of integers which is useful in graphical and statistical treatment. Details of analysis and presentation are set out on page 40.

Table 5   The Wentworth scale and the ϕ notation

Type	Grade limits	
	mm	ϕ units
Boulder	above 265	above −8·0
Cobble	265–64	−8·0 to −6.0
Pebble	64–4	−6·0 to −2·0
Granule	4–2	−2·0 to −1·0
Very coarse sand	2–1	−1·0 to 0
Coarse sand	1−0·5	0 to 1·0
Medium sand	0·5–0·25	1·0 to 2·0
Fine sand	0·25–0·125	2·0 to 3·0
Very fine sand	0·125–0·062	3·0 to 4·0
Silt	0·062–0·004	4·0 to 8·0
Clay	less than 0·004	beyond 8·0

Sandy foreshores range in size from small pocket beaches often only a few hundred metres from side to side and less than 100 m between tide marks, to vast stretches of sand extending uninterrupted for many kilometres along the coast, and with intertidal distances of several hundred metres. Within the constraints of over-all topography, the width of a sand-beach will depend on the quantity of material available, which is supplied ultimately by rivers and streams, and by coastal erosion. Thus the sediment budget of the adjacent coast is of considerable importance, but so also is the movement of material from offshore. The zone just below the low-water mark acts as both a source and a trap for sediment supplied to the beach and removed from it. Since the detailed configuration of the sediment is determined by the interaction of waves, currents, and tides with the beach, the direction and varying strength of the winds which control the length, frequency and steepness of the waves are of major importance.

When waves approach shallow water they begin to 'feel' the bottom (this point is sometimes taken as the seaward limit of the beach) and will break when the depth of water is about 1·3× that of the wave height at breakpoint. The steepness of the waves determines to an important extent the effect they will have on a beach and on its profile. With steep waves, sand is transported landward outside the breakpoint, but seaward inside it. On striking a sloping beach, such waves tend to have a plunging action so that much of their energy is dissipated at the point of impact and they cut into the beach, eroding it and carrying material offshore. With flat waves on the other hand, sand is transported towards the land at all depths, resulting in deposition on the beach, usually building up the berm or flat platform above mean high water. The beach profile, in turn, affects the character of the waves, so that the interactions are highly complex. In general the swash (or wave surge up

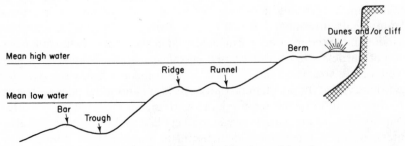

Figure 13. Generalized profile of a sand-beach

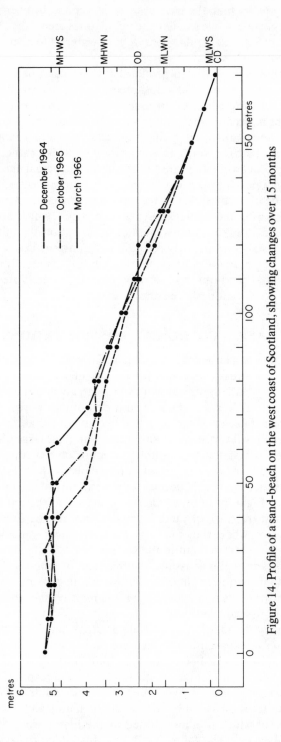

Figure 14. Profile of a sand-beach on the west coast of Scotland, showing changes over 15 months

the shore) of a breaking wave slows down due to gravity, friction or percolation, but it tends to have more energy than the backwash; so larger and heavier particles are carried up the beach and deposited, while smaller and lighter ones are washed back. There is thus a continuous sorting process depending on varying water velocities, and exposed sand-beaches are often comparatively well sorted.

A sand-beach is thus never static. Its profile at any particular time represents a dynamic equilibrium of offshore, onshore, and longshore sediment movements. The scale of these movements varies from a few hours during severe storms, to weeks between spring and neap tides, or months between summer and winter. Although deposition or erosion may be cumulative, these effects are often cyclic, producing periodic changes on the shore— erosion during times of strong winds and deposition in calm weather. Figure 13 shows a generalized beach profile, while Figure 14 illustrates, from actual measurements, changes which took place on a beach over several seasons. In extreme cases, such as Village Bay Beach on the Atlantic island of St. Kilda, the sand may be almost completely removed and deposited offshore during the winter, leaving bare rocks, which are covered again by deposition in summer. Longshore movement of sediment on the other hand, caused by oblique waves or coastal currents, may result in net transport in one direction only, and unless the supply of new material is adequate, a beach may eventually be eliminated. It has been demonstrated that sand can move around coastal promontories, and can be transported more than 100 miles longshore from its source.

BASIC ECOLOGICAL CONSIDERATIONS

Since the sand-beach is, to a large extent, a dynamic, physically controlled environment, it presents difficult, often stress conditions for the organisms which inhabit it. The main controlling factor is exposure to wave action, which largely determines slope, and both of these are related to particle size, which in turn determines porosity and permeability. Because of the lack of stable solid surfaces, and because of the abrasive action of moving particles, a sand-beach will not support either the abundant large algae found on rocky shores or the rooted vegetation which attach in more muddy areas. Endemic primary production is restricted to micro-organisms and on a typical exposed sand-beach the plants are therefore largely diatoms which, because of the mobility of the substrata, are attached firmly to sand grains. Light for photosynthesis is confined to the top few millimetres, but because of the regular overturn of sand, living diatoms may be found as deep as 20 cm below the beach surface, where they can survive for several months and resume photosynthesis when wave action brings them to the surface again. Only in more sheltered areas where the sand is fine and begins to assume the character of a muddy shore, are unattached micro-algae common—diatoms, dinoflagellates and flagellates—which often exhibit rhythmic vertical movements and appear as visible mats on the surface of the deposit in response to optimal physical conditions.

These beaches carry, in addition to unicellular algae, large numbers of other micro-organisms, and again in exposed areas, they are mostly firmly attached to sand grains. Although such organisms have been relatively little studied, it is known that in some regions they account for the major part of the total sediment respiration. They include the important decomposer organisms which break down particulate organic matter.

An important habitat is provided by the system of spaces and channels between sand grains, the presence of which is demonstrated by the high porosity (32–40%) normally measured on sand-beaches. The fauna inhabiting this interstitial system (meiofauna) is highly specialized and adapted for such a life (Swedmark, 1964), and while nematodes, copepods, turbellarians, and gastrotrichs dominate numerically, most of the major taxa are represented [Figure 15 (a)]. These animals may, depending on the nature of the beach, be present to depths of as much as 1 m below the sand surface and occur in numbers of over one million individuals below each square metre of beach (McIntyre, 1969). They appear to feed on micro-organisms and detritus, but their exact role in the sand ecosystem is not

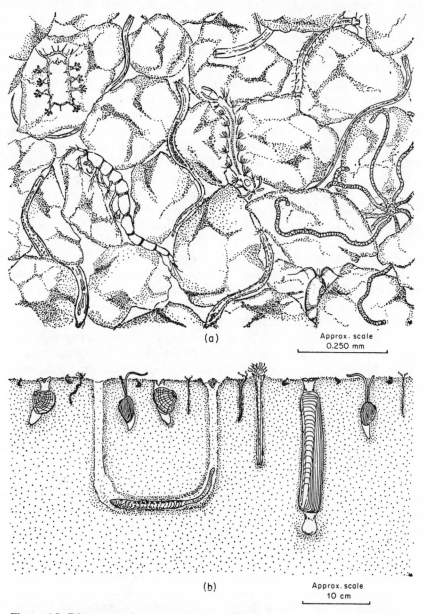

(a)

Approx. scale
0.250 mm

(b)

Approx. scale
10 cm

Figure 15. Diagrammatic representation of habitats in a sandy beach. (a) The interstitial habitat showing the meiofauna living in the spaces between sand grains (i.e. nematodes, copepods, turbellarians, gastrotrichs, polychaetes, tardigrades, and coelenterates); (b) The burrowing macrofauna, i.e. bivalve molluscs (*Ensis*, *Tellina*, and *Cardium*), polychaetes (*Arenicola*, *Lanice*, and spionids) and amphipods (drawing: A. Eleftheriou)

fully understood and is the subject of considerable research at present. Because of their short generation times and generally rapid metabolism they may be useful organisms in providing an early warning of pollution dangers.

Of the larger animals (macrofauna), a wide range of species make up the assemblage typical of the temperate sand foreshore [Figure 15 (b)]. On well sorted sand-beaches on the west coast of Britain, with a median diameter of about 200 μm, the dominant animal in terms of biomass is usually the bivalve *Tellina tenuis* which may occur in numbers of several hundred per square metre. Also common are the polychaete worms *Ophelia rathkei*, *Nerine cirratulus*, and several species of *Nephtys*, while the characteristic crusta-ceans are several amphipods, mainly of the family Haustoridae (*Urothoë*, *Haustorius*, and several species of *Bathyporeia*), and the isopod *Eurydice pulchra*. All these are typical of the fauna of the intertidal *Tellina tenuis* community, but other species may make up a significant component depending on the beach—the bivalves *Ensis*, *Cardium* and *Donax*, and the polychaetes *Arenicola* and *Lanice* and the sea-urchin *Echinocardium*. The range of species and biomass is great, and depends to a large extent on the degree of exposure to wave action. On a relatively sheltered beach more than 24 species may be common and numbers of over 8,000 individuals per square metre, with a corresponding dry weight biomass of over 20 g, may be reached. On completely unprotected beaches, open to severe wave action, the fauna may be reduced to a single species of isopod, a few amphipods and one or two polychaete species, with a total biomass of less than 1 g m^{-2} (McIntyre, 1971).

The distribution of these animals on the beach is far from uniform. Apart from irregular variations due to freshwater inflows or patchiness of the deposit, there is an important zonation correlated with tidal height, different species being adapted to varying degrees of emergence. Thus *Tellina tenuis* reaches its maximum near the low-water mark, while the several species of *Bathyporeia* are found at distinct and characteristic tidal levels.

It is significant that all the animals referred to above are burrowers, and at the time of low water an emergent sand-beach in temperature regions is often apparently barren, with most of the fauna confined beneath the sand. When the beach is submerged, however, many of those animals become active on the sand surface, and another component of the ecosystem makes its appearance—organisms which move up on to the beach when the tide is high. Probably the main element of this is the fish population and in particular juvenile flatfish such as plaice and dabs which migrate up the shore and feed in this region when the tide rises. Other groups again, the shrimps and crabs, and the sand-eel *Ammodytes* which sometimes burrows in the beach when the tide is out, mingle at high tide with immigrant individuals of their species from the subtidal zone.

A number of useful books are available for the identification of organisms of the seashore, of which Barrett and Yonge (1958) and Eales (1961) are particularly recom-mended.

SPECIAL ECOLOGICAL FEATURES

The special features of sand-beaches as ecosystems are related to the mobility and porosity of the substratum. Since sand movements restrict *in situ* primary production, this cannot support substantial production at higher levels of the food chain. However, each tide brings fresh food to the system and this material (dissolved or particulate) represents the accumulated production from an area much more extensive than the beach itself, so that the permeable sand acts as a high energy window, concentrating organic matter in the beach.

To maintain the type of community described above, an adequate supply of oxygen is required, and this too is provided by the tides. If the sand is deep and the porosity high, and if the rate of input of organic matter is not too great, aerobic conditions may exist for 1 m or more below the surface. But if internal water circulation is reduced, for example by a low profile or an impervious layer, or if large quantities of organic material (seaweed, sewage, etc.) are added to the beach, the seawater oxygen may not be sufficient to supply the decomposer organisms which normally break down organic matter, and the cycle of events may be altered. In conditions of limited oxygen, decomposition proceeds by pathways which involve sulphur, and result in blackening of the sand, unpleasant smells and alteration of the natural community. Perkins (1957) has reviewed this situation.

These features account for the type of fauna found on sand-beaches. The micro-organisms probably depend to a large extent on soluble organic matter supplied by the tides, and themselves provide the food of the meiofauna inhabiting the interstitial system. Of the larger animals, those which survive on the most exposed beaches are usually active swimmers as well as burrowers and can maintain themselves at their optimum levels on the beach, or can survive periodic transport to another zone even below the low-tide mark. On less exposed beaches, the sediment is more stable and animals more secure, but those which compose the bulk of the biomass probably use the beach sand largely for physical support and again depend to a considerable extent on the incoming tide for food.

PRESSURES ON THE HABITAT

Pressures on the sand foreshore may be from natural forces or may be the product of human populations.

Because of the inherent instability of sand-beaches they are clearly very vulnerable to excessive movement of the deposit, and this is perhaps the greatest natural hazard. When the prevailing wind is onshore, large quantities of sand may be blown inland and lost to the beach, but the main sand movements are usually caused by water, and, as indicated above, water-generated movement of sand up and down the beach is often cyclic and therefore no threat to the long term survival of the beach. But longshore movements are a different matter. Since they arise from the oblique approach of waves or from coastal currents, which may result from the configuration of the coast or from the prevailing winds, they can achieve a continuous removal of sediments which may eventually eliminate a sand-beach. Although this is a process of natural forces, it may be initiated by coastal alterations resulting from human activities. These include not only engineering works for harbour walls, but also mineral extraction (sand, limestone) which may disrupt the equilibrium of the beach, making it vulnerable to erosion.

The main man-made pressures result from the fact that the coast is an obvious site for disposal of waste materials from domestic, industrial, and commercial activities. The tipping or dumping of colliery waste is a striking example. Almost four million tonnes are dumped annually in Britain in the form of waste or pit stone. In north-east England about 10 km of coast are spoiled in this way, and beaches several kilometres from tipping sites are often affected. Even more extensive pressures are from effluents discharged directly to or in the vicinity of the beach. The most ubiquitous effluent is domestic sewage, with which trade wastes are often amalgamated. It was recently estimated that in England and Wales sewage and trade waste from six million people are drained directly into the sea or estuaries with no or only partial treatment. Sometimes all the individual sewers from coastal towns discharge directly on to the beach, but they are usually joined by an intercepting sewer parallel to the coast which terminates in one or more outfall pipes. These sometimes discharge on to the beach, but usually they terminate at or some distance below the low-water mark.

When the pipe is long and properly sited, its effects on the beach may be negligible but broken or badly placed pipes can result in untreated sewage being washed back on to the beach. The sight of crude sewage is always objectionable, and in extreme cases the use of a beach for recreation may be completely lost. Effects on human health are more difficult to determine. Raw sewage contains large numbers of organisms, some of which can be harmful to man, and when beaches are fouled, bathers could be exposed to these pathogens. The problem is whether they are present in large enough concentrations to give a bather an infective dose. If mixing and dilution are high and the sewage particles are small (the smaller they are the more readily will the bacteria be destroyed) and if the discharge is well sited, then danger may not be great, especially as the destruction of pathogens is enhanced by increased temperature, sunlight, and salinity. As well as any direct effect on human health, there is the possibility of infection through eating contaminated fish or shellfish. The ability of filter-feeding molluscs in particular to concentrate bacteria is well known, and on intertidal sand the most important species in this context is probably the cockle, though the razor shell *Ensis* is sometimes consumed.

Finally, the effects of organic material and nutrients producing a general environmental enrichment should be considered. A common result of raised nutrient levels is increased growth of algae, which may produce offensive conditions on the beach for a time when they decay. The organic input from crude or partly treated sewage leads to increased oxygen demand, but this is not likely to pose a problem in the well mixed water column over an exposed beach. Only in situations of gross sewage pollution or in the immediate vicinity of outfalls is there likely to be any long-term effect on the interstitial system and the general ecology of the benthos of an area, leading first to a reduction of species diversity and in extreme cases to azoic zones.

Because sand-beaches are found in areas of at least some exposure and water movement, they are often the preferred sites for industrial effluent, pipelines ideally leading across the intertidal area to positions and depths where the effluent will be quickly diluted and dispersed. Warm water from power stations is perhaps the most innocuous of these effluents. In temperate regions the metabolism of the beach fauna may be enhanced if the ambient water temperature is increased by a few degrees, but unless this throws some element of the food chain detrimentally out of phase, the effect is probably not harmful, although the possibility of effects on the metabolism of any pollutants present must be considered.

Where toxic substances are concerned, the situation is different. Industrial effluents containing relatively high concentrations of heavy metals or organic pollutants are numerous. They may discharge into sewers carrying normal municipal wastes, and thus complicate any attempt to estimate the effects of sewage disposal. On the other hand, pipes often run directly from works or factories to the coast. Again, a properly placed and correctly operated system will ensure that high concentrations are transient or restricted to very localized 'hot-spots'. The effluent may however result in the levels of particular substances being raised to several times the normal background concentration over a relatively wide area. This could lead to high concentrations of toxic metals or persistent organic substances in fish or shellfish, which could be dangerous to man, and regular monitoring of fishery products is increasingly being undertaken on a worldwide scale so that any threat can be quickly detected. Effects on the general ecology, however, may not be immediately obvious, especially since runoff from agricultural land and material from sewage may introduce nutrient salts, which in moderate quantities can enhance primary production and increase standing stocks at other levels of the food-chain, thus masking adverse effects due to toxic substances. However, experimental work shows that very low levels of, for example, some metals may have significant effects on metabolism or behavior. A slight reduction in the photosynthesis of phytoplankton, or in the growth rate of fish, or a small impairment of some important behavioural mechanism of a prey or predator might produce long-term effects on the whole ecosystem.

These effects may be subtle and difficult to detect, but the ever increasing threat to beaches of oil is all too obvious and has been extensively documented in recent years.

The immediate appearance of oil pollution on a sand-beach may be devastating, with unpleasant drifts up to several centimetres thick along the high-water mark, and oil mixed with and sinking into the sand. At times of sediment deposition, oil layers may be covered by fresh sand to a depth of 1 m or more, only to be brought to the surface and redistributed later. The obvious effect is thus on amenity value, and the stranding of an oil-slick could turn a pleasure beach into an area to be avoided. Further, oil can have a substantial smothering effect on organisms and, depending on its nature and degree of refinement, may be more or less toxic; so a spill of oil on a sand-beach can result in mortality of some organisms and weakening of others. However, the toxic effects seem to depend mainly on the aromatic hydrocarbon and naphthalene content of the oils, and since both these types of compound are rapidly lost by weathering, toxic effects may be significantly reduced by the time an oil-slick has reached a beach. In fact, although some components of the food-chain may be affected, threat to fishery interest from direct toxicity does not seem to be serious. However, since very low concentrations of some components of oil can be detected by the olfactory sense, taining of shellfish, either on the shell or in the flesh, may result in danger to commercial fishery interests, and these effects may take weeks or even months to disappear. Although an oil-slick may be a spectacular event, making front-page news for a time, its adverse effect on a sand-beach is not permanent. By the degrading action of sunlight and micro-organisms, and by the dispersing effect of waves and tides, oil is removed naturally at a slow but significant rate.

However, oil pollution is not necessarily caused by a single isolated event, and chronic pollution may be much more serious. This can result from small but repeated inputs of oil, as might occur near an oil port where large volumes are being handled. If the intervals between accidents are sufficiently small, there may be no time for natural elimination of the oil and recovery of organisms, so that a gradual reduction in diversity and a weakening of the population takes place. Even more serious is the long-term presence of low levels of oil in the environment, due to continuous effluents from establishments such as refineries. Here again the effects would build up with time and would be a general deterioration of the environment and recovery would not take place.

Finally, a universal blemish on exposed sand-beaches is the presence of general refuse, mostly from shipping, and often composed largely of plastic bottles or other non-biodegradable containers. These objects are stranded along the high-tide line, where they persist and accumulate. On a high amenity beach in the tourist season it may be necessary to organize their collection and disposal, but the long-term solution is to attack the problem at its source, and try to instil a more thoughtful attitude in those at sea.

METHODS OF STUDY

Because of their accessibility and inherent scientific interest, as well as their importance for recreation and coastline preservation, beaches have received much attention from the scientific community, and certain methods of study are well developed.

The initial examination of a sand-beach should include the compilation of relevant general details—over-all size, topographic setting, fetch, degree of exposure, prevailing winds, range and times of tides, and freshwater input. Much of this can be got from examination of charts and tide tables and by an initial walk across the beach. To this must be added the important physical characteristics—slope of the beach, depth of sand, level of water-table, nature, shape and size of particles, and porosity and permeability. The slope is easily measured by simple surveying techniques, and profiles should be constructed which are levelled to nearby bench marks, so that the stations examined can always be referred accurately to a specific tide level (Southward, 1965).

A description of particle size is of primary importance. If silt and clay are present and have to be separated into fractions, the analyses can be laborious, but in most sand-beaches the silt–clay content is insignificant and can be estimated as the total passing through a 0·062 mm sieve. A suitable procedure is to collect a sample (about 60 ml is a convenient size if 200 mm diameter sieves are used) with a core tube to the required depth and wash through a 0·062 mm sieve into a white basin with tapwater. This separates the sample into a sand fraction in the sieve (washed clear of salt) and a silt–clay fraction in the basin. The contents of the basin are then washed through a weighed filter paper and dried at 100 °C. A final weighing and subtraction gives the amount of the silt–clay fraction. The sand on the sieve is then dried at 100 °C and passed through a further series of sieves. A suitable series is 2·0, 1·0, 0·5, 0·250, 0·125, and 0·062 mm, but additional sieves might be inserted if a particular fraction has to be examined in more detail. The results should be expressed as cumulative curves of weight (see Holme and McIntyre, 1971). As well as the size distribution of the particles, some indication of their general shape and mineralogy is desirable.

Among other physical parameters, measurements of permeability, porosity and water content of the sand are relevant. Simple standard methods are available, and these are described in appropriate textbooks (Means and Parcher, 1964).

These physical data should be supplemented with basic chemical measurements. Notes should be made of the depth of the black layer, and the vertical profile of salinity, redox potential (E_h) and oxygen concentration should be measured. Particulate organic carbon, both free and attached, is a useful indication of the character of the beach, and chlorophyll determinations give an indication of the plant standing stock. A measure of total sediment respiration in the dark gives an estimate of the activity of micro-organisms. Detailed instructions for obtaining these data are given in Barnes (1959) and Strickland and Parsons (1968), and additional references in Hulings and Gray (1971).

Finally the animal population should be assessed. It can be conveniently divided into meiofauna, which pass through a 0·5 mm sieve, and macrofauna, which are retained on this sieve. In practice, these two components are usually sampled separately and differently.

For meiofauna a core tube of 2–4 cm diameter can be used and the core divided into lengths to give data on vertical distribution of the fauna. If samples are not to be examined immediately, they may be preserved in 5% formalin, and later repeatedly stirred in water and decanted through a fine sieve (e.g. 45 μm mesh) which retains most of the meiofauna. To obtain the animals alive, decantation is still appropriate, but an anaesthetic is first needed to prevent the organisms clinging to the sand grains—10 min in a solution of magnesium chloride isotonic with seawater. A number of more sophisticated techniques are available and these are described in *A Manual for the Study of Meiofauna* (Hulings and Gray, 1971).

For macrofauna, quadrats of about 0·1 m^2 should be dug to a depth of at least 20 cm and sieved through at 0·5 mm mesh. In areas of coarse sand this mesh may retain too much sediment, and sieves of 1 or even 2 mm may be required, but since these may lose important elements of the fauna at least some fraction of the sample should be passed through a 0·5 mm sieve.

The meiofauna and macrofauna sampling should be done at least at the positions of high water, mid-tide and low water, but it is desirable that other beach levels, particularly mean low-water neaps and mean high-water neaps, should also be sampled if possible.

These methods will produce a picture of the sandy foreshore which is essential for adequate assessments of the nature and degree of pressures on the habitat. If possible, information should be collected at several periods throughout the year so that seasonal changes can be taken into account. The regular construction of profiles at selected positions will show changes in levels of sand and the nature of the change may then be assessed. Further, it is only if such data are available that the effects on the beach of a spill of oil or some other pollutant can be assessed. Repetition of the observations at intervals indicates the long-term effects.

For specific types of pollution studies, additional methods may be relevant. If sewage is particularly in question, counts of coliform bacteria in both the water and the sediment may be required, and the additional examination of filter-feeding animals may be useful. Coliforms make up a large proportion of the microbial population of crude sewage, and while they rarely cause human illness, they are useful indicators of sewage pollution because with viable counts of 10^6–10^8 per 100 ml sewage, they are much more numerous than pathogenic organisms which appear sporadically, depending on infections in the human populations.

Unfortunately, as already indicated, some pollutants may have effects which are difficult to detect. Experimental work in the laboratory may show that levels of certain metals or persistent organics have particular effects on individual species of the sand ecosystem, but the complex environment of the natural beach makes it impossible to detect these changes in the presence of other substances which have synergistic or antagonistic effects. Comparison of the total fauna in clean and polluted areas may give some clue to such effects, but the difficulty of finding a single control area, and the magnitude of natural variation, makes this tricky. As knowledge accumulates on the natural populations and their variations on unpolluted sand-beaches, it is becoming possible to recognize unusual or unexpected features, such as a reduction in diversity or a marked dominance of particular species which may be taken as indicators of pollution.

USES

Educational and scientific importance

The coastline in general is an important region for scientific research and a useful training ground for students. Coastal sites are subject to more rapid changes in their geomorphology than most inland sites at the present time, and they display a variety of features and processes of interest to geographers and geologists. Zoologically, beaches show marked patchiness in the distribution of the fauna in response to physical factors, as well as a distinct zonation related to tidal height, and the animal species of sand-beaches illustrate a variety of adaptations for life there. These areas are therefore attractive for the study of ecological concepts and have the advantage of easy accessibility and of being amenable to relatively straightforward sampling techniques. Their importance as nursery grounds for young flatfish and as areas for commercial exploitation of some shellfish further focus research attention on them.

From the viewpoint of the ornithologist, sand-beaches are perhaps of less interest than mud-flats or salt-marshes, but sand-flats are frequented by gulls, waders, and wildfowl as feeding grounds (particularly the tide line and the wrack line) throughout the year for some species, and seasonally for others. Cockle beds in several parts of Britain are used by the oyster-catcher as winter feeding grounds during the period of low tide, and estimates as high as 30,000 to 37,000 birds have been made for flocks in Morecambe Bay. In severe winter weather, the foreshore may remain free from snow or ice longer than inland feeding grounds, and can serve a vital function as an emergency source of food.

For all these reasons, sand-beaches may be important sites for both advanced research and for training, and are used in these ways by university departments, research institutes, field course centres, and schools. Indeed, some beaches in the vicinity of marine stations, for example at Port Erin, Millport and around Plymouth are known throughout the world as the sites where fundamental concepts of marine biology have been established. This can lead to an additional pressure on the habitat not yet mentioned—that of over-collecting of specimens. However, scientists are fully aware of the danger and every effort is generally made to remove no more than necessary and to leave the habitat in good order.

Recreational importance

Recreational activities on the coast include sunbathing, picnicking, beach games, bathing, skin diving, sea angling, power boating, sailing, canoeing, surfing, water-skiing and sand yachting. The first few items on this list hardly deserve the description 'activities' and may only require sunshine and peace to enjoy them. However, although the habit of a 'day by the seaside' will remain part of the pattern of British life, the earlier pattern of coastal recreation is changing. The passive attraction of the traditional resort which provides safe bathing has little to offer the growing number of enthusiasts for the more active sports, and the expansion of private car ownership means that more beaches are within reach and more people are visiting rural seaside places outside the resorts, which were previously remote or inaccessible by public transport. For many of the active sports, however, more sheltered estuarine areas tend to be preferred, and sand-beaches, because of their exposed nature, are less favoured for such water sports as boating and skiing. The latter, because of the noise and safety problems associated with fast and powerful boats, is most unpopular with non-participants, and needs strict zoning, by either buoyed-off or restricted areas, to prevent conflict with other beach users. Divers, whether using snorkel or SCUBA, usually prefer rocky coasts, as these provide more varied and interesting bottoms. On the other hand, sand yachting, which is quite a new sport in Britain, is particularly suited to extensive sand-beaches. It is becoming increasingly popular, and on the coast of Lancashire there is a thriving club at Lytham St. Annes. This sport unfortunately requires an area clear of other beach users.

Much of the impact which most of these recreational activities might have on a sandy foreshore is quickly obliterated by the daily rise and fall of the tides. The major pressures associated with them are in fact probably located above the high-water mark and are concerned with providing for the needs of a large influx of tourists or holidaymakers—with the location of car parks; camping and caravan sites and their services and amenities; with feeding and entertaining them, and providing access to the beach. These problems are not relevant to this chapter, but they are well to the fore in the thinking of several Government departments and of such bodies as the Nature Conservancy Council, Sports Council, the British Tourist Authority, and the Countryside Commission. Several recent publications for the last named body are highly relevant to this subject (e.g. Countryside Commission, 1970a).

Economic importance

Apart from the obvious economic importance of sand-beaches to tourism referred to above, fishery interests are probably the major economic factor in such areas. The sand-beach is often used for the erection of salmon nets, of a variety of types, for the capture of fish migrating along the coast, but the main exploitation of resident stocks is probably for cockles, which are extensively exploited on many sand-beaches. These areas are also of considerable indirect importance to fisheries in that they serve as nursery grounds for juvenile flatfish—plaice, soles, dabs, and flounders—which settle inshore and spend the first months of their lives in sandy bays and estuarine areas.

Industrial uses include the extraction of silica sand and, in some areas, of shell for a variety of commercial uses. These extraction operations affect the amenity of the area and, if not controlled, can destroy important habitats of animals, alter the structure and appearance of the beach and begin the process of erosion.

Discussion of erosion leads to another feature which is certainly recognized as important by the coastal engineer. He may see the beach largely as an energy absorptive blanket in the form of an accumulation of loose material around the limit of wave action. The total depth of this material is crucial because if the cover is thin, storm waves could move it from the platform on which it rests out to deep water, leaving the platform and cliffs exposed to wave attack. The beach may thus have considerable value in terms of coastal protection.

CONSERVATION AND MANAGEMENT

If the appropriate morphology of any given sand-beach can be maintained and its natural ecosystem protected, then the value of the area for educational, scientific, and recreational activities may be assured, and the possibility of its controlled use for industrial purposes provided. This should be the aim of beach management.

Erosion

The major problem is clearly related to the dynamic nature of the sand-beach. A beach which has not been eliminated by cyclic sand movements in historical time will probably continue to exist in the foreseeable future, unless it is altered by some unpredictable natural disaster or by interference from man. Such interference could have little apparent relation to the beach in question. For example, a successful attempt to restrain erosion in one area could stop the supply of sand to a beach some distance away and thereby begin its elimination. Again, any artefact which changes the longshore currents will alter the established pattern of sand transport and change the deposition regime on a beach.

A relevant case-history is outlined by Wiegel (1959) concerning the construction of a breakwater at Santa Barbara, California. The major source of sand feeding the beaches in this area appears to be numerous mountain streams, and the supply of sand depends considerably on the rainfall and especially on the storms which produce high runoff. Local topography and sea conditions result in a wave-generated longshore drift from west to east producing, over most of the year, a corresponding movement of sand. In 1927–28 an offshore breakwater was constructed off Point Castillo (Figure 16) and when it was found that the gap between the inside and the coast caused the harbour to silt up, the breakwater was extended to join the shore (in 1930). The breakwater acted as a trap for sand moving in the littoral stream and the region west of the breakwater was quickly filled. The sand then moved along the breakwater, around its tip, and into the harbour, which again began to fill with sand. An annual sand drift of about 206,000 m³ was computed. To the east of the

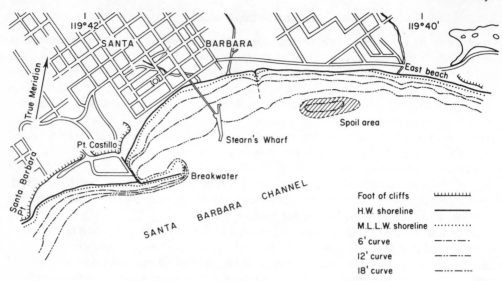

Figure 16. Map illustrating discussion of erosion in text

breakwater, the normal eastern sand movements continued, but because the breakwater acted as a barrier, this sand was not replaced, so that beaches to the east were eventually affected by erosion. It was decided to deal with the problem by dredging the harbour and then dumping the sand to the east, so that water movements would transport it to build up the downstream beaches. The first dredging took place in 1935 and material was dumped, as shown in Figure 16, some 1,600 m east of the breakwater in 6·5 m of water and about 305 m from the shore, forming a mount about 670 m long and 1·5 m high. This was not successful and the mound remained almost intact. It was therefore decided to pump the dredged material by pipeline and deposit it on a 'feeder beach' roughly at the position shown in Figure 16. This method was on the whole satisfactory and it was possible to compare the effectiveness of various dimensions of 'feeder beach', the conclusion being that the shortest and widest were least effective.

This case history is useful in illustrating the knowledge and techniques required in such a context. Extensive information was needed on topography, land runoff, rainfall, winds and storm frequency, coastal and oceanic hydrography, and on movements of sand. Models of the constructions were produced and regular monitoring of the situation was carried out. This showed where errors had been made and suggested how these might be corrected. In spite of all this, it was not possible to explain or fully understand some of the observed situations, such as the form of the sand-bank which built up at the tip of the breakwater.

The technique of sand bypassing described above, although spectacular, is not the common method of dealing with erosion. The classical approach is the building of sea-walls and groynes. While these are often effective, a sea-wall in some situations can enhance the destructive effect of waves, and a system of groynes may lead to erosion on down-wave beaches. Careful consideration must thus always be given to possible consequences of coastal constructions (including harbours and breakwaters), and to the effects of excavations such as those for mineral extraction. Even inland activities can be relevant—in areas where beaches depend on sand from streams, flood control measures could seriously increase beach erosion. In general, it should be appreciated that the coastal complex must be viewed as a unit, and that its most sensitive element is the coastal edge, which may be damaged not only by wave and wind attack, but also by interference from the landward side. The alteration or destruction of protective land just above the high-tide mark, whether by casual recreational traffic, by the creation of access roads, or by the construction of car parks, could have serious consequences for the stability of the beach, so that management of the fragile dune zone is of crucial importance and a careful balance must be struck between the conservation of an area and its proper exploitation.

Sewage pollution

Proper protection of the beach from pollution is a matter of continuous vigilance following adequate planning. The correct siting of sewage outfalls and the necessity to keep them in good repair has already been noted. It should also be remembered that any change in the coastline or in the bottom near the outfall could alter the currents, so that material previously carried offshore could be returned to the beach. In America, public health standards are set for water on bathing beaches, based on coliform bacteria counts. In Britain, such water standards are not considered acceptable, partly because the counts are so variable (there may be ten-fold differences on successive days due to such factors as sunlight and sea conditions) and partly because there is no epidemiological evidence that sea bathing would cause illness except perhaps in cases of gross pollution. The consensus of informed opinion seems to be that comminution and adequate dispersal of sewage will satisfy public health requirements for most bathing beaches. As primary treatment (removal of solids by mechanical means) becomes more widespread, involving disposal of sludge on land or on grounds offshore, aesthetic and health problems will diminish even further, especially if holding tanks are available so that the most favourable discharge conditions may be utilized. It therefore seems unlikely that the high cost of secondary sewage treatment (biological oxidation by percolating filters or activated sludge) could be justified for small towns before discharge to the open sea, but in major cities, secondary treatment plants are increasingly being installed. However, it should be noted that even if sewage is given full biological treatment at a coastal town, a system in which sewage and stormwater overflows are combined could still contribute crude sewage to the beach, and riverwater could add further significant quantities.

Considering the threat of contaminated shellfish, the temperature of normal cooking would usually eliminate any risk, but in areas of gross contamination shellfish should not be eaten, or should be subjected to approved cleansing treatment (Wood, 1969).

Finally, the presence of industrial wastes in sewage must be considered. This can not only add substantially to the oxygen demand of the sewage, but could also contribute quantities of toxic substances such as metals or persistent organics. Drainage authorities have powers to deal with such effluents, and the British Government have recently indicated that industry will be expected to carry the cost of control and disposal.

Industrial pollution

Again the vigilance of local and river authorities is essential for the maintenance of clean beaches. The first stage in management probably comes with the initial authorization of the effluent. An initial general problem is that the effluent may exert a considerable oxygen demand on the environment: this can readily be estimated by standard BOD tests (which indicate the weight of oxygen used in oxidation of a sample by biological action under set conditions—see Department of Scientific and Industrial Research, 1960), giving an indication of the degree of dilution required. A knowledge of the composition and volume of the effluent may indicate whether it is likely to be toxicologically hazardous. It may not be possible to estimate the composition with any accuracy, as when a large number of solvents and other waste products from a variety of processes are discharged from a single pipe, and the relative quantities of the components vary from day to day or even from hour to hour. In this case, experimental work on samples of the effluent may be required to test the effects of selected pollutants on selected organisms. There is a voluminous literature on such toxicity testing (e.g. Alabaster, 1970; Portmann and Connor, 1968) and much time could be spent on discussing the ideal test conditions—the time and temperature required, the species of animal to be used, its age and sex, etc. This is not the place to enlarge on such considerations, but it is perhaps useful to suggest that a 96 h LC-50 test (i.e. a determination of the concentration which kills 50% of the animals in 96 h) on some species common in the area will at least give some initial indication of the toxicity of the effluent. Armed with this information and with some knowledge of the local hydrographic conditions, it may be possible to estimate what composition of effluent may safely be discharged.

With sewage, the crucial decision may be the initial siting of the effluent pipe, and the same principles apply to industrial waste. Rapid and adequate dilution is first required, followed by rapid dispersal. The pipe should be fitted with an appropriate diffuser and the outlet should have a good depth of water at low-water springs. In some cases it is useful to restrict discharge to the most appropriate part of the tidal cycle (usually the early part of the ebb) to ensure maximum dispersal without the possibility of return on the next flood tide.

Having accepted the effluent, regular monitoring is essential to ensure that calculations are realistic. Determination of the levels of appropriate pollutants in any organism which might be eaten by man is of first importance, followed by examination of the established ecosystem in the area.

While these considerations apply to chemical effluents, the problem of oil is rather different. The appropriate treatment of oil on a sand-beach depends on the circumstances. If it originates from a spill occurring outside the tourist season, it may be best left to the cleansing forces of nature, but if a beach must be cleaned for the benefit of holidaymakers, then a number of courses of action are possible. The most satisfactory is the mechanical removal of the top layers of contaminated sand by earth-moving machinery or suction (assuming this loss is not deleterious to beach amenity or fisheries) and its dumping or burying on land. Later the beach may be ploughed to disperse any remaining oil and to expose it to sunlight and bacterial action. Another approach is to lay lines of straw or gorse at the low-water mark. This material is rolled up the beach by the tide and can be collected together with the adhering oil and sand for dispersal ashore. A method which might appear obvious, burning *in situ*, has been tried but not found successful. If mechanical methods are not possible and the oil must be removed, the last resort is the use of detergents. Unfortunately, although a detergent may disperse the oil and make it more accessible for bacterial action, the mixture of oil and detergent may be more toxic than the oil alone. This is partly because the detergent itself may be toxic, but also because, once emulsified, the oil becomes not just a surface film, but part of the aqueous environment, taken in at the gills and ingested by filter-feeders. A further point is that oil treated with detergent can penetrate much deeper into a sandy beach than untreated oil. A method used after the *Torrey Canyon* incident was to bulldoze oiled sand to low beach levels and there treat it with detergent, although due to the instability of the emulsion so formed there was a tendency for the oil to reappear and be deposited elsewhere.

However, none of these measures is entirely successful and the ideal management solution is to prevent the oil from reaching the shore by dispersing it at sea or sinking it. The final possibility is the protection of beaches by floating barriers, which may be successful for small areas. All this requires well organized and co-ordinated action and in many regions efficient administrative machinery has now been set up to determine the most effective procedure in the event of a spill.

These remarks apply to accidents; for chronic conditions the correct approach is close supervision of oil operations and a clean-up of the effluent in question.

Chapter 3

Shingle foreshores

R. E. RANDALL

INTRODUCTION

Shingle is the term applied to those sediments which are larger in diameter than sand, and have negligible capillary forces. A predominant particle size of over 2 mm (0 ϕ units) is usually held to separate sand from shingle (King, 1959). Shingle may be divided into the size fractions given in Table 6.

Table 6 Shingle size fractions

Type	Particle size (mm)
Boulders	Over 200
Cobbles	60–200
Coarse gravel	20–60
Medium gravel (pebbles)	6–20
Fine gravel (granules)	2–6

At much over 200 mm diameter, shingle becomes cliff-like in its ecology, whereas below 6 mm, it approximates to sandy foreshore or sandy marsh. Shingle is an important constituent of British coasts occurring along almost 900 km of the English and Welsh coastlines. Exact estimates are confused, but this figure does not include the many rock–shingle or sand–shingle mixtures. In the Netherlands, shingle is absent, but it is a common component of the Baltic shores (Eklund, 1924, 1931; Warming, 1906; Böcher, 1969) and the Atlantic coast of France (Géhu and Géhu, 1959; Géhu, 1960a, b, c, 1963). In the south of Britain, much of the shingle is composed of flint (Steers, 1926) derived originally from the chalk but latterly eroded from glacial cliffs by the action of the sea. This has been deposited either directly on to shingle foreshores or reworked *via* off-shore banks. In other cases, the sediments (especially cobbles and boulders) have been brought to the coast by rivers draining nearby highland areas. A third source is glacial material from the sea-bed.

Five categories of shingle structures have been recognized (Chapman, 1964; Hepburn, 1952; Tansley, 1939): (1) fringing beaches; (2) spits; (3) bars or barriers; (4) cuspate forelands or opposition beaches; and (5) offshore barrier islands. The first three of these generally have an extremely mobile substrate, and are regularly washed by spray and storm waves (Figure 17). The two latter formations are more terrestrial in nature and will

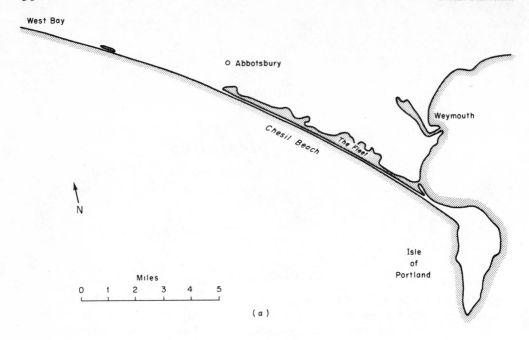

(a)

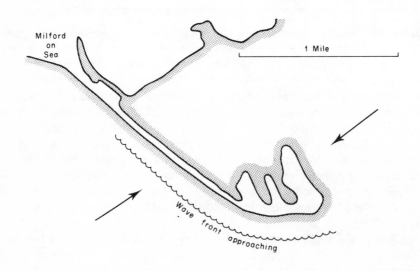

(b)

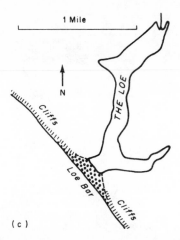

(c)

Figure 17. Shingle foreshores. (a) Chesil Beach: a fringing
beach initially hugging the coast, then forming a spit, and
ultimately tying the Isle of Portland to the mainland (the
whole unit forming a 'tombolo'). (b) Hurst Castle spit: the
distal end terminates in a number of recurved hooks. (c)
Loe Bar sealing Loe Pool. (Reproduced by permission of
Longmans from Sparks, 1972)

be covered in Chapter 10, but as much of the information contained in that Chapter is relevant to this one, the two should be read in conjunction.

Fringing beaches are the simplest and commonest type, forming a strip in contact with the land along the top of a beach. These are frequently seen on the Channel coasts of England and France.

Shingle spits form where there is an abrupt change in the direction of the coast, as for example at Hurst Castle spit in The Solent, and are commonest on coasts which have an irregular outline with frequent changes of direction. Such formations may have one or several recurved hooks and a recurved distal end: a result of the deflection of waves by refraction round the end of the spit. A spit may form across the mouth of an estuary and become a bar or barrier as at The Loe in Cornwall or across the River Exe in Devon.

Four environmental factors are responsible for the growth of a shingle beach. There must be an available supply of material at the same time as conditions (i.e. waves, winds, and tidal currents) are favourable for its movement. Since this coincidence is unpredictable, conditions without movement occurring may exist for a considerable period. Details of geomorphic growth may be found in King (1959) and Steers (1964).

An examination of the coast of Britain (Steers, 1953) shows that many of the smaller shingle features are intimately associated with neighbouring marsh or sand formations, and are thus ecologically intermediate in habitat. Others are entirely separate and display characteristic shingle ecology. Figure 18 displays the major areas of British shingle.

There appear to be three basic ecological considerations on shingle foreshores: substrate mobility, substrate composition, and water availability. These environmental factors were first recognized by Oliver (1912) and further examined by Scott (1963a).

The frequency with which shingle foreshores are disturbed by waves varies on different beaches according to fetch and prevailing wind conditions, and the resultant vegetation varies accordingly. It is important to remember that the majority of shingle foreshores are unvegetated or have an extremely sparse cover (Figure 19).

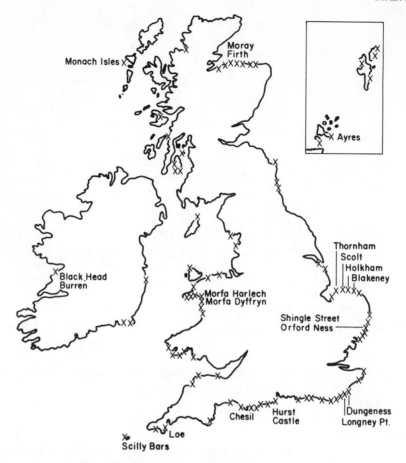

Figure 18. The location of British shingle beaches

Figure 19. The seaward face of a fringing shingle beach (Shingle Street, Suffolk) typical of Scott's (1963a) Class 1—such areas are too frequently disturbed by the physical environment to permit growth of vegetation

Scott (1963a) has recognized three foreshore stability-classes which can be observed on a vegetational basis, dependent upon the average length of time for which the shingle is left undisturbed by environmental forces. These are:

(1) No vegetation—too frequent disturbance to support plant growth: at the foot of sea-cliffs, distal points of spits, etc.
(2) Summer annual species—beach stable from spring to autumn: mainly *Atriplex* spp on drift line in Britain.
(3) Short-lived perennial species—beach stable 3–4 years: much strand and foreshore vegetation e.g. *Glaucium flavum*.

An example of the succession on the foreshore at Shingle Street is given in Table 7.

Table 7   Foreshore succession at Shingle Street, Suffolk

(1) *Atriplex glabruiscula* (muddy drift line)
↓
(2) *Lathyrus japonicus, Beta maritima* (mobile foreshore)
↓
(3) *Rumex crispus, Germanium robertanium* (semi-stable laterals)
↓
(4) *Arrhenatherum elatius, Sedum acre* (mature, semi-stabilized banks)
↓
(5) *Silene maritima, Festuca rubra* (stable banks)
↓
(6) Some specimens of *Rubus fruticosus, Rosa canina, Sarothamnus scoparius* and *Lupinus arboreus*

These habitats may, of course, occur at different levels of the same beach (Figure 20). On some shingle foreshores, mobility may be caused by accretion resulting in a specialized vegetation of *Suaeda fruticosa* and *Lathyrus japonicus*. It is because of the instability of shingle foreshores that they are frequently lacking in animal life.

Figure 20. The leeward side of Shingle Street. The lower area to the right of the photograph is a mixture of shingle and mud supporting a community of *Halimione portulacoides*; in the middle of the lee slope is a line of drift supporting *Atriplex* spp [Scott's (1963a) Class 2]; while at the crest of the ridge, where the shingle is more stable, *Rumex crispus* and *Beta vulgaris maritima* are present [Scott's (1963a) Class 3]

The predominant particle size of the substrate determines the physiographic classification, but the vegetation is controlled to a much greater extent by the proportion of the fine fraction under 2 mm in diameter. In fact, even on stable shingle beaches, the absence of a fine fraction (i.e. 'soil') gives a marked reduction in vegetation. Scott (1963a) tested this hypothesis on five British beaches and found a positive correlation between the abundance of fine fraction and plant life. This has also been suggested by Oliver (1912), Moore (1931), and Tansley (1939). Fine fraction will be more critical at the time of germination, since without it enough moisture may not be present for growth to begin. The fine fraction is usually composed of sand or silt and organic matter: sand in the west of Britain, silt and clay in the south and east. Each type has a distinctive vegetation (Table 8).

Table 8   Species commonly found on shingle with different types of fine fraction*

Fine fraction	None	Sand	Organic	Silt and clay
Species present	*Verrucaria maura*	*Festuca rubra*	*Atriplex glabriuscula*	*Halimione portulacoides*
	Placodium lobulatum	*Sedum acre*	*Rumex crispus*	*Glaux maritima*
	Xanthoria parietina	*Agropyron junceiforme*	*Potentilla anserina*	*Artemisia maritima*
	Parmelia spp	*Honkenya peploides*	*Galium aparine*	*Puccinellia maritima*
		Mertensia maritima	*Urtica dioica*	
			Iris pseudacorus	

* See Scott (1963c), Perring and Randall (1972), Nordhangen (1940), Marsh (1915) and Oliver and Salisbury (1913a).

The third basic factor is water supply, which is likely to be low because of high porosity and low water-retaining ability. Lack of capillarity in shingle usually rules out the water-table as a moisture source for most shingle foreshore plants. The principal source of supply must be pendulant water from precipitation, but this will be dependent upon fine fraction content. It has been suggested that further water supplies are obtained by internal dew formation (Olsson-Seffer, 1909; Oliver, 1912; Hill and Hanley, 1914), but Scott (1963a) questions the importance of this.

Thus it can be seen that shingle foreshores do not have a characteristic vegetation, variation everywhere being dependent upon matrix. However, there are a few species that are more characteristic of shingle than of other environments, most being those associated with extreme mobility. Of these, *Suaeda fruticosa* has been most studied (Chapman, 1947; Oliver and Salisbury, 1913b). This plant is unusual in shingle foreshore habitats in that it is woody and upstanding, reaching over 1 m in height. Usually it germinates in the driftline, rapidly sending long roots into the shingle so stabilizing the plant. Overwhelming by shingle forces the plant to a horizontal position from which it sends out new roots and new vertical shoots. After a period, the older parts of the plant decay.

Scott (1963b) has examined the role of *Glaucium flavum* as a shingle colonist. This species has a distinct preference for recently disturbed ground and may grow low enough on the beach to be inundated by winter high tides. Conversely it may grow inland on limestone and chalk quarry debris, but not in waterlogged situations. *G. flavum* is especially luxuriant on driftlines and in fact is only an important stabilizer of mobile shingle where a fine matrix and drift have collected.

Several other species have been examined from the point of view of substrate stabilization. All shingle foreshores protect valuable land behind them and any means of reducing erosion is a contribution to sea defence: vegetative stabilization of mobile shingle would be a cheap and useful advance. *Lathyrus japonicus*, *Silene maritima*, *Beta maritima*, *Sedum acre*, and *Rumex crispus* often combine to form a protective carpet on shingle, and experimental work has suggested the use of these as cover for mobile shingle. The most complete descriptions, to date, of the vegetation of British shingle beaches will be found in Tansley (1939) and Gimingham (1964).

Rare and declining species are quite characteristic of shingle foreshores. An examination of pre- and post-1930 distributions in Perring and Walters (1976) emphasizes the changing distributions of species like *Mertensia maritima*, *Lathyrus japonicus*, *Crambe maritima*, etc. Furthermore, shingle areas may often be important breeding grounds for littoral birds, e.g. *Haematopus ostralegus* and *Sterna* spp. Habitat deterioration would quickly affect population numbers and distributions of such birds.

TYPES OF PRESSURE

The major pressures on the shingle habitat directly or indirectly involve man. Birds and mammals use shingle features, but their effect is usually only marked in extremely local circumstances. Direct human pressure involves coast protection, removal of shingle for building purposes and scientific–recreational pressure; indirect pressure is associated with the grazing of man's domesticates.

The best natural defence of a coast is a high, wide beach. If this is built up artificially by means of groynes, breakwaters or other structures then the area down-wave is starved of beach material. Many of the shingle beaches of East Anglia have suffered in this way. Shingle is an important building material both for sea-defences and for terrestrial constructions. Certain shingle foreshores suffered the removal of considerable material during the Second World War for the construction of airfield runways: Shingle Street, Suffolk being a very good example. The grazing of such features is not an important economic factor but it may be ecologically important even where use is light. It is interesting to note, for example, that *Mertensia maritima* is present only on Stockay, the one ungrazed island of the Monach Isles National Nature Reserve, Scotland. Similarly, *Lathyrus japonicus* decreases markedly in frequency if grazing is allowed (Randall, 1977). Scientific–recreational pressure tends to be greatest on rare or attractive plant species. Serious depletions of rare submaritime ferns and flowering plants have been observed following the visits of university parties to the limestone boulder beaches of Burren, Co. Clare, Ireland. Likewise, the decreasing frequency of certain species on 'easy access' beaches in south-east England may have similar causes.

Natural pressures on shingle foreshores are primarily concerned with geomorphic changes and although these must be monitored they can rarely be corrected. An example of this type of pressure is the cyclic erosion and progradation of Shingle Street, Suffolk, associated with the growth of Orford Ness spit and a minimum width requirement for the estuary of the River Ore. Wildlife grazing (especially rabbits) may be of great importance in reducing vegetation cover locally but this will be dealt with in Chapter 10.

METHODS

A major method of obtaining significant ecological information about the shingle foreshore habitat is to study the ecology of individual species at different locations. Examples of such studies are Cumming (1925), Pratt (1929), Chapman (1947), Scott

(1963b,c) and Cavers and Harper (1967), but the majority of shingle foreshore species remain unstudied. An excellent programme to follow in this respect is that of the British Ecological Society's *Botanical Flora of the British Isles*. This includes the following information: geographical and altitudinal distribution, habitat, communities, response to biotic factors, gregariousness, performance, morphology, phenology, floral biology, animal feeders and parasites, and history (see Scott and Randall, 1976).

The understanding of changes in the ecology of shingle foreshores is also greatly aided by community–environment recording. One useful method is the establishment of permanent quadrats, because of the ephemeral and quickly altering nature of this habitat. Quadrats are simply sample areas for detailed examination and should be set up on the crests of the sea-ridge and also on the backslopes and laterals of spits and bars. In such a sparsely vegetated habitat, an oblong quadrat at least 4 m^2 is recommended, placed across the major environmental gradient. Siting of permanent quadrats should be carefully related to 'typical' communities, but any other form of recording should be systematic or random. Location of quadrats must never be an haphazard process. In order to relocate the quadrat, an oak peg about 4 cm^2 and 70–80 cm long should be driven into the shingle at one corner and a note made of its bearing and distance from some permanent landmarks. This will avoid human or natural damage to a physical quadrat.

Within the quadrat, accurate records can be made of changes in successive seasons through the year and also of changes from year to year. This information is usefully recorded on charts which can be visually compared [see Figures 21 (a), (b)]. The basic procedure is to map the locations of the individuals of each species included to scale on squared paper, and to outline the areas covered by spreading or predominant species. The easiest way to carry out this recording is to stretch strings across the quadrat at 20 cm intervals. It is often useful to take a vertical photograph as well as chart the quadrat. Useful ways of expressing the data from permanent quadrats are: abundance (number of individuals or shoots present), visual estimates of cover (best expressed on the standard five-point scale: (1) under 5%; (2) 5–25%; (3) 25–50%; (4) 50–75%; (5) over 75%; and visual estimates of sociability [(1) growing singly; (2) in clumps; (3) in small, well-separated colonies; (4) in large concentrated colonies; (5) in pure stands]. Note on vigour, flowering, seeding, etc. should also be made. In this way, a useful indication of speed and direction of vegetational change is obtained. Quantitative methods of recording and expressing these data will be found in Kershaw (1964) and Grieg-Smith (1957).

A second method of recording shingle vegetation is by means of line or belt transects. These are designed to give a graphical representation of the changes in composition of the vegetation along a selected line. In the case of shingle vegetation, transects should be positioned at right angles to the shore so as to include the successive zones that occur as one moves inland. Since long transects crossing several zones are more useful ecologically, detailed description is given in Chapter 10.

As have been previously noted, the nature of the soil in which a plant grows plays an extremely important part in determining the character of the vegetation. This means that no shingle vegetation study is complete without an examination of the soil. However, a soil sample as normally understood cannot be taken because of the nature of the substrate and neither can a soil profile be drawn. Thus, in the collection of soil for examination, the shingle material is ignored and analyses are based on the interstitial material alone. Although one cannot express these results by unit volume, comparisons can be made with reference to percentage weight of the fine fraction. A severe problem with shingle soil analyses is that the highly porous substrate make passage of nutrients extremely rapid. Thus if input is continuous, analysis at any one time may well reveal very low nutrient concentrations. Since plants are concerned with the total available, rather than with

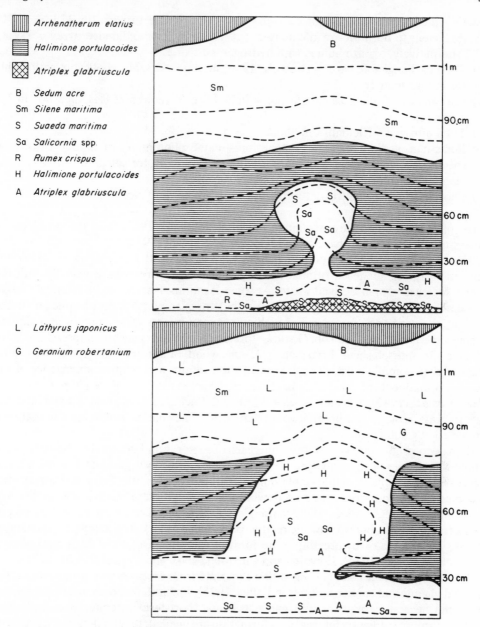

Figure 21. Quadrat charts from the edge of a lagoon at Shingle Street; (a) August 1962; (b) August 1972 (in the interval, the lagoon was breached by the sea)

concentration, any figures obtained must be 'weighted' in order to compare them with more normal soils. In fact vegetational evidence suggests that many shingle foreshores are rich in nitrogen.

The following analytical techniques are recommended:

(1) pH—1:2·5 soil:0·01 M calcium chloride suspension. Using this method, any potential error resulting from dilution or from soluble salt concentration should be overcome (Ryti, 1965).

(2) Organic matter—ignition of oven-dried soil in a muffle furnace for 12 h at 48 °C. (Sufficient to burn off organic matter, but not so high as to damage pyrex beakers.) Alternatively, oxidize humus with hydrogen peroxide.
(3) Total exchangeable bases—Brown's (1943) method. N.B. not applicable with highly calcareous matrices.
(4) Carbonates—Collins' calcimeter; or with highly calcareous matrices, digestion in 2 N acetic acid.
(5) Total nitrogen—Kjeldahl's method.
(6) Potassium, sodium, calcium, magnesium—flame photometry or atomic adsorption spectrophotometry on ammonium acetate extract. N.B. for calcareous matrices use ammonium chloride extract (see Tucker, 1955).
(7) Phosphate—colorimetric determination with ammonium molybdate.

Most of these methods are explained in detail in Black (1965), and the problems of analysing calcareous soils have been reviewed by Randall (1972).

The amount and composition of material along the drift line can be an extremely important aspect of the nutritional ecology of a shingle foreshore, especially with the recent increase in man-derived waste. At some locations, it seems to be relatively rarely present in bulk, but on the west coast of Scotland, for example, drift may be piled on to the shingle in great banks over a metre high and several metres wide, composed mainly of fucoid algae and *Laminaria* spp. On other foreshores, the drift may be composed mainly of *Zostera* spp, mussel shells, bryozoan skeletons, wood, or flotsam and jetsam. It is useful, therefore, to compile a list, at each visit, of the contents and relative importance of the various components of the drift, and also to make some estimate of weight per unit area. Pearse *et al.* (1942), Chester *et al.* (1956) and Backlund (1945) have examined the invertebrate ecology and breakdown of tidal drift. The role of *in situ* organic matter is discussed in Chapter 10.

Shingle foreshores are generally within reach of seaspray and this can be a limiting factor for certain species of shingle vegetation. The amount of spray is greatest near the sea and decreases quite rapidly, so that stable shingle features receive little. Data on the quantities reaching shingle foreshores at different wind speeds are, therefore, a valuable addition to environmental information. With a shingle substrate, measurement of soil salinity is almost meaningless, but worthwhile relative data can be obtained by analysing plant leaves during dry periods or by erecting some form of salt-trap. Details of such methods are available in Boyce (1954), Edwards and Claxton (1964), and Randall (1970, 1974).

The water supply for shingle vegetation can come from precipitation, groundwater, the sea and via internal condensation (see Fuller, 1975). Precipitation data are best obtained from the nearest permanent rain gauge because short-term evidence is of little use. Certainly, more studies on internal dew formation are needed: further details are available in Salisbury (1952) and Ranwell (1972a). Seawater will only be relevant to halophytic plants of muddy shingle, but even then it is unlikely to be important since in winter, when high seas are commonest, precipitation is usually adequate. Groundwater is more likely to be important to deep-rooted plants on shingle features and methods of examining this factor are covered in Chapter 10.

The nature and degree of human pressure on shingle foreshores is dependent upon ownership, access, and use. Maps should be compiled of multiple ownership, where relevant, and surveys made at access-points of numbers of people and vehicles entering the shingle areas. Of more importance on shingle foreshores are the natural forces and the concomitant changes of beach physiography and ecology. It is extremely important that

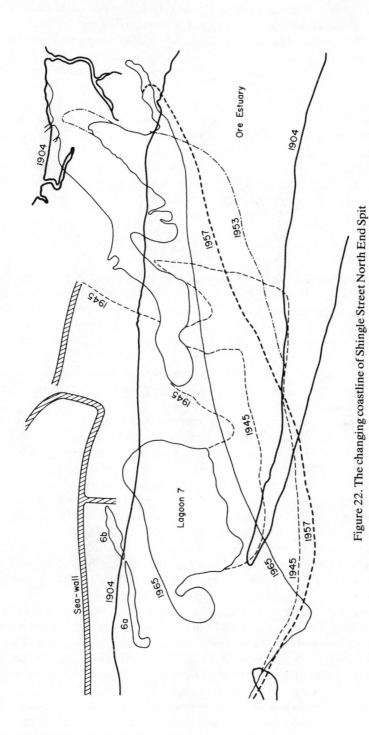

Figure 22. The changing coastline of Shingle Street North End Spit

these changes are mapped frequently: at least once each year. So long as reference is made to a permanent baseline (such as a sea-wall), plane-table surveys or levelling exercises can be carried out without great difficulty. Historical evidence of change is useful in this respect and previous editions of Ordnance Survey maps should be compared. Except for isolated instances, pre-nineteenth century maps should only be assumed to be impressionistic. Admiralty charts, River Board plans, and estate plans are all extremely useful as historical evidence of change (Randall, 1973).

By far the greatest modern aid to mapping physiographic change of mobile shingle is the aerial photograph. Many areas of the coastline are photographed frequently. In Britain, reference can be made, for example, to collections made by the Royal Air Force, the Ministry of Housing and Local Government, the Department of the Environment, the Nature Conservancy Council, various universities and aerial survey firms. Aerial photography can also be commissioned for special purposes and is usually worth the expense. An example of the sort of superimposition maps that can be obtained from these various sources is given in Figure 22 (see also chapter 10).

USES

The uses of shingle foreshores overlap considerably with those of larger shingle formations and other coastal types. However, the rapidity of physiographic change makes them ideal laboratory areas for geomorphic study. They are also good areas in which ecologists can examine pioneer seral development. It is particularly useful if local educational establishments can be brought into any programme of information gathering: much of the basic scientific detail required for understanding shingle beaches is simple but repetitive. Some shingle foreshores contain particularly good colonies of rare or declining species (e.g. *Lathyrus japonicus* (Brightmore and White, 1963) or *Mertensia maritima*); these should be reserved for scientific study.

Shingle areas have never been used for recreation in the same way as sand-dunes—they are unplesant to walk on and often dangerous to swim from because of deep water immediately offshore. However, in recent years, pressure has become so high on other forms of coastline (see Chapter 9) that more people are now visiting shingle areas. Usually they keep to the foreshore fringe and venture little on to the stable formations. Although in places this has caused reductions in rare plant species susceptible to trampling (e.g. *Lathyrus japonicus*). One particular form of recreation that has long been present on shingle foreshore is fishing. In some areas where cars are continually parked (e.g. Shingle Street, Suffolk; West Aberthaw, South Wales) this had lead to ecological change. However, it is where fishing is an economic concern that more modification occurs. On the Dungeness foreshore, for example, summer huts, miniature haulways, and boats have resulted in the removal of foreshore vegetation.

Shingle foreshores are of little economic use for domestic grazing, but this is quite common in the west of Scotland and results in considerable changes in species composition which may also affect mobility. Gravel mining is the major economic use of shingle, both on foreshores and larger formations. This is dealt with in greater detail in chapter 10, but its results on the foreshore include the development of deep pits and trackways, road building and coastal change, as seen, for example, at Shingle Street.

A final use of shingle foreshores is for protection of the area behind. Their protective use is limited because of mobility (Hoyle and King, 1955), and often a sea-wall is built. However, shingle has considerable effect in reducing the power of the waves, so that it is unwise to allow any use which is not integrated with coastal protection.

CONSERVATION AND MANAGEMENT

Conservation and management policies for shingle foreshores are difficult to separate from those of larger shingle formations. Many of the best foreshores occur at the seaward limit of features such as Dungeness and any planning policy must be integrated. Shingle foreshores have often been a neglected habitat because of their one-time lack of amenity or economic use. Most recent pressures have been created by access (e.g. the road on Shingle Street built during the Second World War) and the best form of protection is isolation. This isolation can be geographic, such as the Monach or Orkney Isles, or physical, such as the sea between the mainland and Scolt Head Island, Blakeney Point or Orford Ness.

Where shingle foreshores are near enough to areas of human habitation to be an important economic and amenity resource, management policies have to be adopted. Obviously, all shingle foreshores are not worthy of deliberate action, but certainly those with an important coast protection, scientific–educational or amenity use should be conserved and managed in an integrated fashion. The factor of chance has played its part in the ecological value of shingle foreshores as with other areas. Frequently rarities or good examples of communities 'occur' where a specialist has visited. Thus, since the ecological value of a site often depends on the length of time for which it has been studied, several of the most important sites for preservation become self-evident. British examples are Shingle Street, studied for so many years by the students of Flatford Mill Field Centre, and Chesil Beach, first studied by Oliver (1912).

Any shingle foreshore sites which are managed primarily for coastal protection will still contain significant wildlife resources and also be suitable for most amenity uses. In these locations, shingle mining must be stopped, regular aerial photography commissioned to monitor coastal changes, and experiments carried out with any species that might be useful for decreasing shingle mobility.

Shrubs such as *Tamarix gallica*, *Crataegus monogyna*, *Prunus spinosa*, *Hippophaë rhamnoides*, *Ulex europeus*, and *Suaeda fruticosa* have been suggested as stabilizers (see Kidson, 1959). *Suaeda fruticosa*, *Tamarix gallica*, and *Lathyrus japonicus* are the most hopeful species but, in Britain, all are climatically limited to the south and east. Even within this area, experiments carried out so far have not had a great success, as the survival rate on transplantation is low. Much more work is needed in this direction.

Where management and conservation for recreational–educational purposes is contemplated, convenient access and parking facilities must be provided. This avoids the severe ecological problems resulting from vehicles being driven on to the shingle and often becoming bogged down, so creating further damage. People do not penetrate far on shingle and integrated management is not difficult to practise. But without more ecological information on a national scale, so that the best areas to conserve can be judged, economic pressures for gravel from the construction industries may remove what remains.

Chapter 4

Lagoons

G. COLOMBO

INTRODUCTION

Lagoons are shallow bodies of brackish or sea water partially separated from an adjacent coastal sea by barriers of sand or shingle, which only leave narrow openings through which seawater can flow. Indeed, lagoons may become completely isolated from the sea behind such a barrier, in which case they transform into coastal ponds or lakes—of freshwater if the lagoon receives river discharge. Coastal lagoons are usually found on low-lying coasts and are normally aligned with their largest diameter parallel to the seashore. They may also originate by the transgression of seawater into previously existing freshwater lakes (these are usually deeper and extend perpendicularly to the shoreline), by the invasion of land below sea level after storms have breached coastal-defence systems, or during reclamation of salt-marshes. Lagoons in the two latter categories are normally small and ephemeral (Heerebout, 1970; Barnes and Jones, 1974).

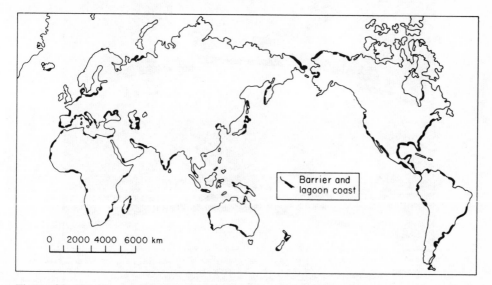

Figure 23. World map showing the distribution of coastlines with barriers and lagoons. (Reproduced by permission, from King (1972))

63

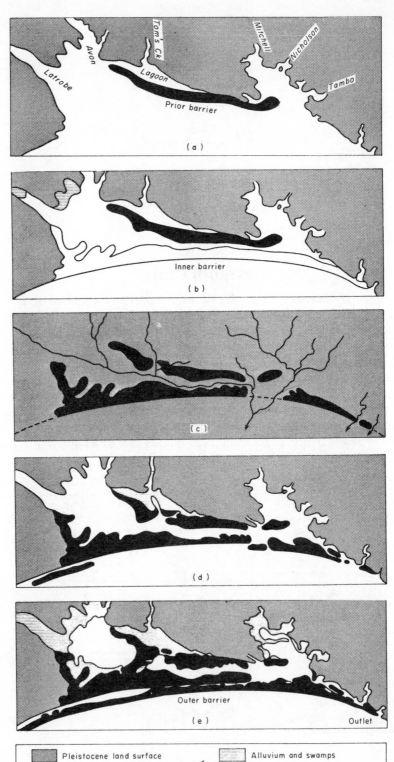

Pleistocene land surface

Barriers

Alluvium and swamps

0   10 km

0   5   10 miles

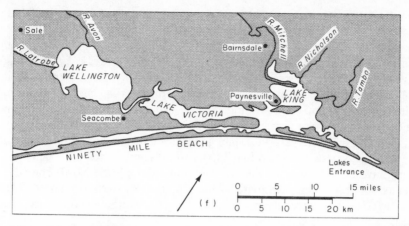

Figure 24. Post-Pleistocene evolution of a lagoon system on the East Gippsland coast of Australia. (Reproduced, by permission, from King (1972) and Bird, 1969); reprinted from *Coasts* by E. C. F. Bird by permission of the M.I.T. Press, Cambridge, Massachussetts. Copyright © Massachussetts Institute of Technology)

The primary requirement for the formation of a lagoon is the existence of material capable of forming a barrier. Such barriers, frequently enclosing lagoons, are found along some 13% of the world's coastlines (Figure 23). The largest length of barrier coastline is that along the eastern seaboard of the U.S.A. and the Gulf of Mexico, but lagoons are also common along the eastern coasts of South America and India, around the shores of the Baltic, Mediterranean, Black, and Caspian Seas, in southern Britain and western France, along the West African coast, in south-eastern Australia, northern Alaska and Siberia, etc. (Gierloff-Emden, 1961). They therefore range from equatorial to polar zones and show wide latitudinal variation in the nature of their contained and surrounding faunas and floras, the climatic regimes to which they are subjected, etc. In this chapter, only lagoons of temperate seas will be considered: tropical lagoons, particularly those enclosed by coral reefs and atolls, show a number of special features and must be excluded from consideration here.

The ecology of lagoons has been summarized and reviewed by Emery and Stevenson (1957), Hedgpeth (1967), Schachter (1969), Remane and Schlieper (1971), and by several papers in the Symposium of Estuarine Fisheries (1966) and in Lauff (1967). Lagoons, however, vary greatly from one to another and hence it is very difficult to generalize on their ecology: each is a case unto itself. They are also comparatively simple systems and are therefore easily disturbed both by natural processes and by pollution or adjacent urban and industrial development.

BASIC GEOMORPHOLOGICAL AND BIOLOGICAL CHARACTERISTICS

The geomorphological characteristics of the coast have been described by King (1959), Zenkovich (1967), and Bird (1969), to which the reader may be referred for discussion of the processes modelling the coastline and ultimately controlling the evolution of lagoons. In general, lagoons are associated with coastlines which have experienced or are experiencing change in the relative land–sea level. There is general agreement that the sea attained its present level only in the last few thousand years, and it may still be rising. Further, land subsidence can give rise to lagoons, because if subsidence is not balanced by accretion the sea may invade local depressions. They are also frequent in deltaic regions where the land–sea boundary is in a state of flux.

An example of a complex barrier and swamp system, formed in Australia during and after the Holocene marine transgression, is shown in Figure 24. Usually lagoons are present where the tide range is small, mainly because the water movements associated with large tidal ranges are strong enough to maintain sediments in a state of motion, thereby preventing deposition and even eroding existing structures. The stratigraphy and evolution of a complex pattern of barriers along the Dutch coast have been investigated by van Straaten (1965). Here a large tidal range generates strong ebb and flow currents which maintain large gaps between individual barrier islands and, instead of a lagoon, tidal flats (Waddens) have been formed (these are common all along the North Sea coasts of the Netherlands, Germany, and Denmark). One can contrast these barrier islands with the intact barriers (Nehrungen) enclosing lagoons along the southern shore of the (microtidal) Baltic Sea.

A lagoon will evolve either towards total isolation behind a complete sediment barrier, and thence from a coastal pond to swamp and marsh, or into a coastal bay again after erosion of the barrier. Such changes can occur in relatively short spaces of time, and will depend on such factors as the incidence of storms, sediment supply, wave form (including the pattern of refraction), sea level variations, etc.

Because of their small size and generally shallow nature, and because of the variable admixture of fresh and salt waters, the physical environment is often severe and subject to considerable short- and long-term fluctuations. Moreover, lagoons are also liable to occasional catastrophic events—invasion by the sea during hurricanes or storms, and flooding from the landward side by swollen rivers—and the lagoon ecosystem frequently starts its evolutionary development afresh after each such catastrophe.

These environmental features have a marked bearing on the nature of the organisms inhabiting lagoons: the flora and fauna frequently show a number of peculiarities in many ways comparable to those shown by estuarine organisms (see Barnes, 1974). First, few species are generally permanent residents of lagoons, those that are being widespread organisms with marked powers of tolerance. This does not mean that numbers of individuals and organic production are low; on the contrary, lagoons are highly productive ecosystems. Secondly, large oscillations in the populations of individual species occur. These may in part be due directly to environmental oscillations, but the low species diversities and the consequent lack of interspecific competition may mean that there are few biological controls on the numbers of some organisms and that these increase until checked by environmental change. Thirdly, many species migrate into lagoons to feed, thereby taking advantage of the considerable production of organic matter and the lack of competing species: the lagoonal ecosystem is immature in the terminology of Margalef (1963). Lastly, although phytoplankton and benthic plants are the primary producers upon which the ecosystem ultimately depends, much of their production is consumed only after its decay and decomposition by microbial organisms, i.e. the detritus food-chains are quantitatively important in lagoons. Although there have been few detailed studies on the role of detritus in lagoons, its likely importance is attested by the work of Odum and de la Cruz (1967) and by the contributors to the symposium volume edited by Melchiorri-Santolini and Hopton (1972).

SPECIAL FEATURES AND METHODS OF STUDY

Either limnological or oceanographic methods may be used for the study of lagoons, adapted according to the characteristics of the water body in question. Four basic approaches are necessary: (1) geomorphological; (2) hydrographical; (3) physical and chemical; and (4) biological. These parameters are all closely inter-related, however, and so they must always be considered together in an interdisciplinary approach. Further, the prevailing climatic factors strongly influence almost all the characteristics of lagoons and they induce periodic and aperiodic fluctuations necessitating regular measurements over long periods of time.

Excellent examples of recent investigations of lagoons are the oceanographic study of Oyster Pond, Massachusetts, investigated in a careful and well balanced manner by Emery (1969), and the research into the ecology of Swanpool, Falmouth, being conducted by Barnes, Dorey, and Little (Barnes, Dorey, and Little, 1971; Dorey, Little, and Barnes, 1973; Little, Barnes, and Dorey, 1973). Both series of investigations are recommended as models for planning an ecological survey of a lagoon, and both contain many references to techniques.

Geomorphology and hydrography

It is clear that the geomorphology of a lagoon must be investigated in order to reach judgements on its past and likely future evolution. A programme for research into this aspect requires the following types of study.

(1) Sedimentological and/or palaeontological studies to measure the thickness and nature of bottom sediments of recent origin.
(2) Physical oceanographic investigations of the adjacent coastal sea to document the tidal ranges, prevailing currents, wave action, etc. These factors are responsible for determining the evolution of the barrier; further, the entrance of a lagoon is maintained by current scour through it and this is usually determined by tidal range and current pattern. The requisite data can frequently be obtained from the published literature, but storms and other catastrophic events can cause the position of an entrance to change rapidly in an unpredictable manner.
(3) Studies on the sediment load of rivers discharging into the lagoon. The sediment discharged by rivers can, of course, considerably affect the local sediment distribution pattern, and can affect the size, shape and 'lifespan' of lagoons, particularly those in deltaic regions. Good examples of these effects can be seen in the Mississippi Delta (Morgan, 1967). One must bear in mind here that one of the natural evolutionary developments of a lagoon is to proceed by way of siltation, plus the build-up of organic matter (detritus, mollusc shells, etc.), until the lagoon is converted into marshy land.

Besides the natural evolutionary trends, human activities often affect the structure of lagoons, primarily by hydraulic works (e.g. dikes, dams, and artificial bars), but also by land-reclamation schemes and through changes in the inland watershed. Changes in the structure of a lagoon will have repercussions on its biology; indeed, hydraulic works may affect the biology directly. The evolutionary trend of the Nile Delta with its numerous lagoons, for example, would have been very different without various hydraulic works constructed by man over the last several thousand years. Instances of the use of hydraulic works as a management tool are considered later in this chapter.

The water budget of a lagoon, with its freshwater and marine components, determines the sedimentation and sediment patterns. The mechanisms involved are almost identical to those in estuaries or over tidal sand- or mud-flats (see Postma, 1961), but lagoons are generally quiet water bodies and consequently they have their own special characteristics. Sediments are usually unevenly distributed over the bottom, although perhaps the most important characteristics are the large quantities of organic matter and the high water content of the bottom deposits. These determine the mechanical and chemical properties of the sediments, which will influence any planned engineering works. It is rather difficult, for example, to build dikes in lagoons unless large quantities of stone or concrete are used, and channels must also be excavated continuously because the sediments are extremely mobile.

For an understanding of the hydrography of the lagoon itself, knowledge of its topography must first be obtained, as water movements, although primarily determined by tidal or wind-induced currents, are guided by underwater channels, bars, and by the general shape of the bottom. Topographical surveys must be repeated at intervals as the geomorphology can change rapidly. Aerial photography is very useful for the construction of a surface map, and bottom topography can be studied with the aid of an echo sounder (or similar apparatus) or, in shallow waters, by direct measurement of depth.

Variation in water level is a characteristic of lagoons as a result of their hydrographical regimes. A model illustrating the pattern of the water budget is given in Figure 25: this may

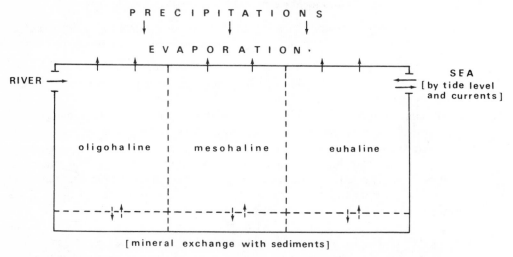

Figure 25. Diagrammatic representation of the hydrographical regime in a lagoon.

be a useful starting point for planning a hydrographical survey (see also Dorey, Little, and Barnes, 1973; Emery, 1969). One will require information on the river flow and its seasonal variation, on the rate of water exchange with the sea in relation to tidal height, currents, etc., and on the local evaporation : precipitation ratio. All these are subject to considerable variation, and therefore the hydrography of a lagoon may show fluctuations daily, seasonally, and over longer periods of time. Nevertheless, the hydrographical regime is of fundamental importance in that it will determine the distribution within the lagoon of different water types, sediments, and plant and animal communities.

Physical and chemical considerations

Because of its small depth and large surface area, water temperature in a lagoon is almost always similar to the mean air temperature. There are therefore large variations in temperature; larger than in the adjacent sea and rivers. Only where the influence of inflowing seawater is greatest, and (usually to a lesser extent) in regions directly affected by river discharge, will there be smaller oscillations. Temperature should therefore be monitored at regular intervals, preferably by continuous recording. Both low and high temperatures can be limiting to some of the organisms present, and when lagoons are used for aquaculture many devices may have to be applied to avoid the extreme conditions, sometimes by regulating the hydrographical regime.

Water chemistry plays an important role in the ecology of lagoons. Strickland and Parsons (1968) can be recommended as a manual of methods of chemical analysis, but it is often necessary to introduce modifications specifically for brackish waters (see *Colloque sur l'unification des méthodes d'analyse des eaux saumâtres méditerranéennes*, 1966). In any event, it is the experience of the author's laboratory that any method should first be checked to ascertain that it is suitable for use in brackish water before being adopted.

Because of the large quantities of organic matter on or in the sediments, and as a result of water stratification (see below), decomposition can lead to anaerobic conditions. The production of gases such as methane and hydrogen sulphide, which are either dissolved in the water or present in the form of bubbles, is therefore common: any disturbance of the bottom during the warm season can cause the undesirable rise of these bubbles. The level of dissolved oxygen in lagoon water is therefore of extreme importance.

The daily variation in oxygen content, both in the upper water layers and at different depths, is indicative of many conditions affecting organisms. This should be determined at least for each season, and more frequently during hot periods and during the plant growth period. In polluted lagoons, and also in those with large plant biomasses, measurement of dissolved oxygen is one of the easiest measures of water quantity, but it is particularly important to establish the length of time in any day during which oxygen values are low. Oxygen content can be determined by the Winkler methods and by polarography (provided that the sensor is carefully checked in proper conditions!). Surveys should be planned carefully because of the considerable variation in dissolved oxygen in relation to the activity of plants and other factors; and, when comparing values from different stations, different lagoons or different seasons of the year, one must allow for differences in temperature and salinity and make the appropriate corrections.

Lagoons show considerable variation in salinity both in time and in space, and several attempts have been made to erect classification systems based on this parameter. D'Ancona (1959), with reference to north Adriatic lagoons, proposed a system which incorporated fluctuations in salinity, and more recently Heerebout (1970) incorporated both median chlorinity and variation in chlorinity into a scheme (derived from his studies on inland lagoons in the Netherlands). Both systems have a practical value in facilitating comparison of different lagoons, but in any given lagoon it is more important to know the spatial and temporal distribution of salinity and its range, than to assign it a place in a classification scheme.

In shallow lagoons (*c.* 1·0–1·5 m depth) only localized stratification occurs, but in deeper lagoons (>2 m depth) density stratification on the basis of salinity and temperature difference between various water masses is the rule, although this will be subject to seasonal variation. Salinity differences can also be found along a horizontal gradient, varying from oligohaline near a river mouth (using the Venice System, 1959, terminology), through mesohaline, to polyhaline near the lagoon entrance (Figure 26). In the tropics and subtropics, hyperhaline lagoons occur, and completely oligohaline lagoons can be found in the Baltic. Changes in the size of a lagoon entrance can have a marked effect on the salinity of a lagoon.

Salinity is most easily determined by the conductivity method, although it should also be checked by measurement of the chlorinity. Again continuous recording is preferable. Information on salinity fluctuation in relation to the ebb and flow of the tide and its

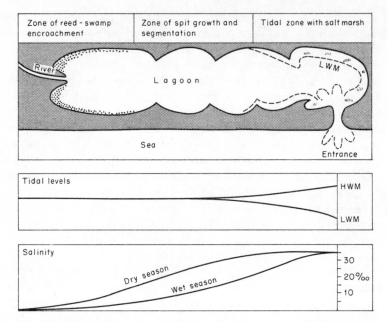

Figure 26. The essential geomorphological and ecological gradients of a lagoon in relation to the sources of fresh and saltwater input (reprinted from *Coasts* by E. C. F. Bird by permission of the M.I.T. Press, Cambridge, Massachussetts. Copyright © 1969 Massachusetts Institute of Technology)

seasonal variation can also be used to estimate the water budget. The combined annual variation of salinity and temperature can be used to characterize a lagoon, and this can be quickly and easily shown graphically: Figure 27 shows the halothermograms of lagoons compared with those of the open sea.

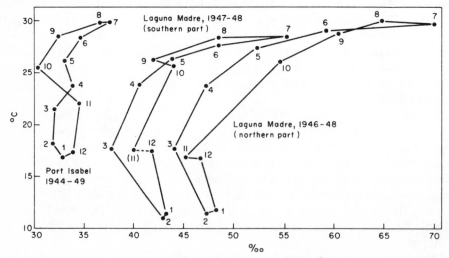

Figure 27. Variations in temperature and salinity plotted as halothermograms in the Laguna Madre, a hyperhaline lagoon in the Gulf of Mexico (Reproduced, by permission, from Hedgpeth, 1957, after Hedgpeth, 1953)

The chemical composition of lagoon water shows deviation from that of seawater as a result of the mixture of fresh and salt waters with their different ionic ratios (see Phillips, 1972) and the quantities of suspended solids: the same phenomenon occurs in estuaries. Other deviations related to the increase in salt concentration by evaporation are found in hyperhaline lagoons. The borate and sulphate concentrations, for example, are frequently different from those in seawater, and this also involves differences in the concentration of calcium and magnesium ions. Moreover, the presence of organic compounds with chelating properties modifies the equilibria attained. Another chemical parameter of some importance in lagoonal water is the boron:silicon ratio. However, in the author's experience, dissolved silicon shows considerable seasonal variation which seems to be more related to the diatom cycle than to other chemical equilibria.

One of the most important differences between brackish and fresh or salt water concerns the variation in the buffer system with respect to salinity and temperature. Both affect the solubility of carbon dioxide in water and the dissociation of carbonic and boric acids; further, different equilibria are reached in relation to the solubility of calcium carbonate. This will affect the pH, total alkalinity and buffer balance. Because of their importance with respect to the pCO_2, daily and seasonal variations in pH and total alkalinity should be recorded wherever possible.

Nutrient concentrations also show wide variation during the year and should be investigated in relation to both the hydrographical regime and the seasonal cycle of biological production (Figure 28). The analytical methods are generally those used in oceanographic research (see Strickland and Parsons, 1968), although in oligohaline water, the modifications suggested for freshwater investigation (Golterman, 1969) should also be

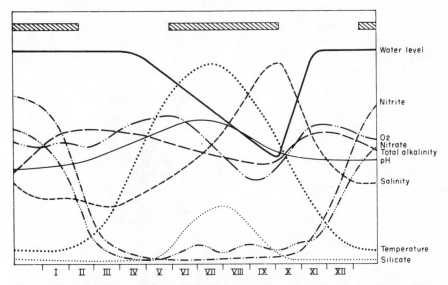

Figure 28. Seasonal variation of physicochemical conditions in the Comacchio lagoon system, Italy (after Colombo, 1972; reproduced by permission of *Bolletino di Zoologia*). The ranges of the various measured parameters in this system, which is connected only to the sea (the north Adriatic), are:

water level 0·6–1·1 m   total alkalinity 1·7–3·1 ml HCl
temperature 8·5–29·0 °C   nitrite–N 0·01–5·1 μg-at 1^{-1}
salinity 26–48‰   nitrate–N 0·05–9·3 μg-at 1^{-1}
O_2 saturation 50–150%   silicate–Si 0·1–4·1 μg-at 1^{-1}
pH 7·7–9·5

followed. Usually lagoons are shallow and thus there are no technical problems associated with sampling at different depths—a depth profile of chemical quality can easily be obtained and should be undertaken in all but the shallowest lagoons.

The movement of the water masses and the nutrient cycle are of great importance with respect to a lagoon's productivity. However only in rare cases do nutrient concentrations limit the production; other more stressful conditions usually act as a brake and nutrients are generally plentiful.

Biology

Although one is continually aware that lagoons are greatly influenced by the adjacent marine, terrestrial and freshwater ecosystems, one can also justifiably regard a lagoon as an entity in its own right—in fact, one of the most helpful approaches to the biology of lagoons is the formation of a conceptual model for each system based on the 'semi-isolated unit'. To create a model of a given lagoon, the basic requirements are information on its productivity and on the nature of its food-web, both of which require information on the types of organism present. We have noted above that productivities are frequently high and that the faunas and floras are not rich in species, both aspects being mainly controlled by the nature of the physical environment.

The most satisfactory method for measuring primary production is that developed by Steeman-Nielson on the basis of the uptake of carbon-14, and for the measurement of the primary production of phytoplankton, this technique can be recommended. The procedures of this method are now standardized, and the International Agency for ^{14}C Determination (Charlottenlund Slot, Denmark) established in 1958 by UNESCO helps and provides facilities for such investigation. In most lagoons, however, benthic plants are by far the most important producers, and although the modification of Steeman-Nielson's method published by Grøntved (1960, 1962) will enable the productivity of benthic diatoms and flagellates to be measured, the determination of that of the larger plants (both algae and angiosperms) by the ^{14}C-method is more difficult.

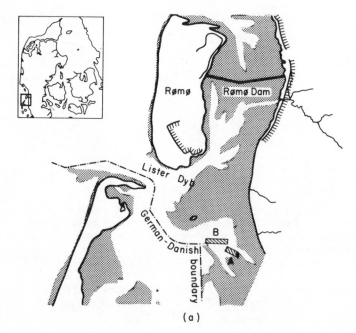

(a)

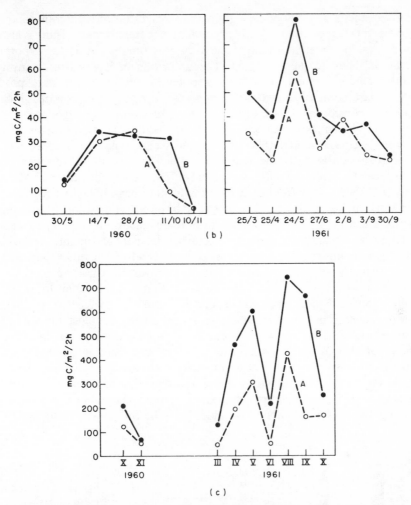

Figure 29. Productivity of the plankton (b) and microbenthos (c) of the two areas
in the Højer Dyb, Denmark, indicated in (a) (after Grøntved, 1962). The ratio of
planktonic productivity to that of the benthos in such areas is in the order of 0.1

Indirect methods are still useful, although they lack the precision of the ^{14}C-method, e.g.
the determination of dissolved oxygen in light and dark bottles (see Strickland and
Parsons, 1968). In particular, the 'diurnal cycle of dissolved oxygen method' developed by
Odum (1956) is useful because it can be used for the underwater macrovegetation. Of
course, one must remember that the ^{14}C-method estimates the net production, whilst the
oxygen methods estimate the mean gross production. For recent work on the measurement of primary production, Vollenweider (1969, 1973) should be consulted. Postma and
Rommets (1970) have investigated primary production in the Dutch Wadden areas, which
are in some ways similar to lagoons, and Vatova (1961) has worked on lagoons in the north
Adriatic. In Vatova's papers one can find a discussion of the conditions which lead to high
productivity. Grøntved's (1960, 1962) papers are also good models for the type of research
to be undertaken in lagoons (Figure 29).

When we come to secondary production, we experience difficulties, as except in a few
clear-cut cases it is difficult to assign many of the consuming species to distinct trophic

levels. Further, the large number of migratory species, often with different food requirements at different stages of their life history, complicate the situation. There is also a large
import of detrital food materials both from rivers and from the sea, and an export of some
of this material back to the sea again. The utilization of lagoons for many types of
aquaculture is, however, evidence of their potential for high secondary production.
Although, in part, some of the examples of so-called high production of fish can more
realistically be considered as examples of the effectiveness of lagoons as traps for
migratory fish, which can be caught with various techniques inapplicable in the open sea!

In general, the complicated pattern through which energy flows in a lagoon is similar to
that outlined for estuaries by Barnes (see Figure 3.8 in Barnes, 1974), although, of course,
the actual species involved will differ (the faunal differences are illustrated by Muus, 1967;
Schachter, 1969; etc.—they frequently take the form of pairs of sibling species).

Detailed links in the food-web can be identified by direct observation, by examination of
gut contents, etc. and by so doing the nature of the whole food-web may be discovered.
One must here emphasize that although a lagoon may be managed mainly for the benefit of
a limited range of commercial species, the place in the food-web of other species should be
ascertained, as many will play important roles in the over-all equilibria.

The ease with which such a model can be assembled and the extent to which a given
model will be applicable over an extended period of time will obviously vary with the types
of organism present. Planktonic communities are frequently the most difficult to analyse,
because not only do they show temporal and spatial variations in distribution and
abundance, but also in species composition—many zooplanktonic species have short
lifespans and other species are only temporarily planktonic, e.g. in their larval stages
(Figure 30). Populations of planktonic algae (diatoms, dinoflagellates, microflagellates,
etc.) are also strongly seasonal, and they show marked fluctuations from year to year
dependent on various environmental variables (Marchesoni, 1954; Voltolina, 1973a, b).

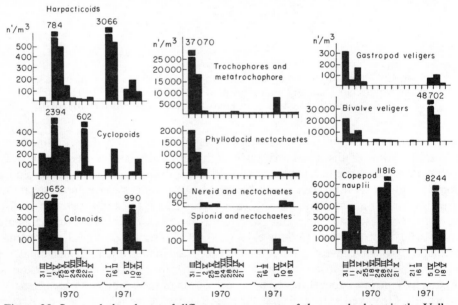

Figure 30. Seasonal abundance of different components of the zooplankton in the Valle
Campo, one of the Comacchio lagoons (after Colombo, 1972; reproduced by permission of
Bolletino di Zoologia). Note that the larvae of benthic animals (polychaetes and molluscs)
are the most numerous component

Therefore, although marine planktonic communities have been successfully modelled, such has not yet proved possible for lagoonal ones. Nevertheless, knowledge of the dominant species and the hydrography of a lagoon does permit prediction of planktonic abundance in space and time with a fair degree of accuracy.

A picture of the benthic situation is more easily obtained, however; in part at least because the organisms in this zone have longer lifespans. The community approach to benthic animals developed by Petersen over 50 years ago (see Thorson, 1957), and the parallel phytosociological approach to larger plants can both usefully be applied to lagoonal situations (see, e.g. Vatova, 1940). The works of Pérès (e.g. Pérès and Piccard, 1958; Pérès, 1961) may be consulted for the appropriate methodologies, etc.

The faunas and floras of various lagoons have been well documented; for example, Hedgpeth's (1957, 1967) papers on subtropical lagoons in the U.S.A. and Muus's (1967) study of Danish lagoons have extensive lists of macrofauna and flora. Planktonic organisms have received less attention, although Marchesoni's (1954) work on the phytoplankton of the Venetian Lagoon is of interest in this respect.

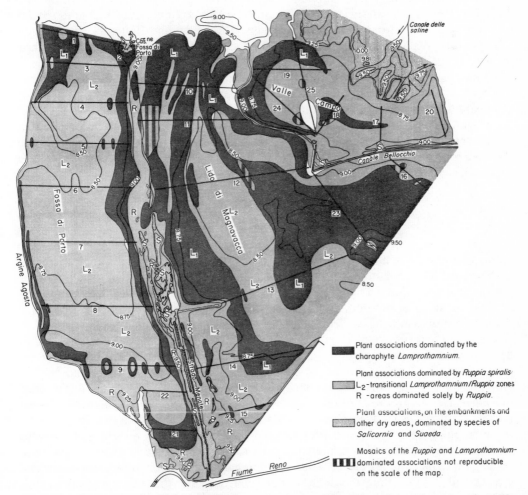

Figure 31. Map of the submerged vegetation in the Comacchio lagoons (after Ferrari *et al.*, 1972; reproduced by permission of the University of Ferrara)

Essentially, the methods for sampling lagoonal fauna and flora are the same as those generally used in limnological or oceanographic research, with allowance made for the fact that lagoons are generally shallow, soft-bottomed bodies of water. It is difficult to recommend a general procedure to be followed, as the species present vary considerably from lagoon to lagoon. In recent years, many textbooks on methods for ecological investigations have been published (e.g. Schlieper, 1968; and the I.B.P. handbooks published by Blackwell—that of Holme and McIntyre, 1971, is particularly relevant here) and these contain a wealth of different techniques. These should provide a pool from which techniques suitable for the lagoon under study can be drawn. However, it is worthwhile emphasizing a few basic points:

(1) Sampling should be repeated regularly (at least seasonally), as many organisms enter and leave lagoons at different times of the year;
(2) It is necessary to select sampling stations carefully, making due allowance for the nature of the different sediments present, the hydrographical regime, and the other environmental factors to which organisms respond, sampling points should therefore cover all available types of habitat;
(3) Many lagoonal organisms are widely distributed and common species, but as speciation may be in progress in several groups (D'Ancona and Battaglia, 1962), their systematics may often be controversial and their identification a matter for specialists (see Muus, 1967);
(4) More than a species list is required to evaluate the nature of the communities present—information on density and biomass per unit area should always be obtained;
(5) Maps, such as that for the vegetation of the Comacchio Lagoon (Ferrari *et al.*, 1972) (Figure 31), are useful in that they serve both as indicators of the different communities present, and preserve a record of those present at a given point in time. Changes can then be assessed against a fixed point; marked change from year to year is, as we have seen, a characteristic of lagoonal environments.

USES

Human use of lagoons has had a very long history, mainly because they are natural sites for harbours and for aquaculture. In more modern times, lagoons have also been put to use for recreational and industrial purposes (cf. estuaries). The pattern of usage differs markedly between the so-called developing and developed countries, but in each case it is the form of the local economy which has dictated the way in which lagoons are utilized.

The oldest form of exploitation is the culture and removal of mussels, shrimps, crabs, fish, etc. for human consumption, and this still forms a valuable source of protein for the populations of several countries. As stated earlier, lagoons form nursery and feeding ground for many migratory marine animals—several species entering lagoons during their juvenile stages and returning to the sea to breed as young adults. The period of fastest growth of such species therefore takes place in lagoons (and estuaries). In principle, exploitation of these food stocks is carried out by (1) encouraging the entry of juveniles (crustacean larvae, fingerling fish, etc.), if necessary by manipulating the hydrographical system; (2) by maintaining the lagoonal environment in a condition suitable for their rapid growth, and (3) by erecting traps in the outlet channel to catch the outgoing adults on their migration back to the sea, or by catching various species within the lagoon itself. Such systems have been evolved in many parts of the world, each system being designed to suit the behaviour of the exploited species.

In many cases, lagoons are also invaded by adult fish (e.g. flounder, mullet, bass) and once they are in the lagoon they can be caught comparatively easily. In the Orbetello Lagoon in central Italy, for example, mature mullet are fished for their large ripe ovaries (*bottarda*).

Two environmental conditions are required for successful aquaculture in lagoons: the hydrographical regimes must be suitable for the entry of the species in question, and the environment within the lagoon must be suitable for their growth. Technological developments have made possible modification of the hydrographical regime: ponds, channels, dikes, etc. can comparatively easily be constructed to improve a lagoon for aquaculture. However, hydrographical change may also alter the environment in an adverse manner, including its productivity. Some types of aquaculture rely on the natural productivity, e.g. that of oysters and mussels feeding on phytoplankton, and hence one can interfere in such situations only with extreme caution. Other lagoons are managed for 'intensive' aquaculture, food materials being added in a controlled manner. Such systems do not rely on the natural ecosystem and hence hydrographical changes may be implemented more successfully.

In any event, in order to culture potential food organisms, water quality must be maintained at a high level. Therefore, aquaculture particularly that based on natural productivity, is in many ways an ideal use to which lagoons may be put, as it yields considerable revenue for little expenditure, and conserves the lagoonal ecosystem. However, in developing countries, such a 'low intensity' use is probably only possible in regions of low population density and in those without severe food shortage.

Recreational uses of lagoons are frequent, particularly in areas of high population density in developed countries. In coastal resort towns, for example, small lagoons are used as boating lakes and natural swimming pools, mainly because of their shelter and consequent lack of strong wave action. Although there would not seem to be any particularly good reason why it should be so, recreational use of these lagoons is the occasion for modifying both the hydrography and the surrounding land, so that the lagoon is frequently converted into a totally artificial, concrete-walled pool of little or no ecological interest. Larger lagoons, however, can accommodate this recreational pressure (although propeller-driven craft can damage submerged vegetation) and are too large to modify in such a manner.

Lagoons situated away from centres of human population and resort areas may still be in a comparatively natural state. These are in many senses ideal centres for education and scientific research. Lagoons form natural aquaria in which scientists, taking advantage of the restricted fauna and flora in terms of numbers of species, can endeavour to elucidate the basic ecological interactions and relationships between different parts of a complete ecosystem. Lagoons can be considered as models of the larger marine ecosystem and have all the advantages of models with respect to scale and cost of investigation. Being immature systems, they also respond quickly to changes taking place in their watersheds and may serve as useful 'indicator communities'. Lagoonal organisms have great physiological flexibility and are therefore of considerable interest to physiologists studying the mechanism of osmotic, ionic, and thermal regulation, etc. Further, as mentioned earlier, they are also of great interest to geneticists and offer valuable material for the study of the process of speciation (see, e.g., Battaglia, 1965; Schachter, 1969).

Finally, lagoons are also utilized by industry and commerce, in part because they provide a safe natural harbour. Developments in lagoons are here paralleled by those in estuaries (see Chapter 7), and under conditions of excessive discharge of waste products, the only use to which some lagoons can be put is as an open sewer. In such lagoons, ecological interest reduces to the level of monitoring near lethal conditions for the life of the system.

RECOMMENDED CONSERVATION AND MANAGEMENT POLICIES

The first decision that one must take with respect to any given lagoon is whether or not it should be preserved in anything approaching its natural state. This question is by no means purely academic: in several parts of the world, there has been pressure for their destruction. Along Mediterranean coasts, for example, lagoons (especially oligohaline ones) were and are the breeding grounds of the vectors of malaria. Consequently, such water bodies were systematically eliminated at the beginning of this century, and, as an additional benefit, the land so gained was given over to agriculture. This anti-mosquito campaign almost completely eradicated malaria, but mosquito populations subsequently increased again and they remain a problem which is now tackled largely by the use of insecticides. In other areas, e.g. Denmark and the Netherlands, lagoons were reclaimed for agricultural use because of the shortage of land.

Today, there is less pressure for the destruction of lagoons, partly because the increased efficiency of agricultural practice has reduced the need for such large-scale land-reclamation and also because of the decrease of malaria. The preservation of lagoons is currently said to be motivated by conservationist policies, but in fact it is more a result of the decreasing need for their deliberate destruction and of their economic potential for recreation, etc. The decision to preserve or to destroy must therefore take into account the possible uses of lagoons (see above).

If a lagoon is to be conserved, one must further decide which of two situations apply. Either one may leave the lagoon to evolve naturally in conjunction with its surrounding area, or else one may adopt policies designed to prevent the natural evolutionary course of the water body. As has been stressed above, every lagoon is a separate system and each will have its own potential evolutionary development in relation to the local geomorphology, coastal physiography, etc. Examples described above (see Figures 24 and 25 and associated text) outline such evolutionary changes; a feature of which is the short time-scale involved—marked changes may occur within a few human generations. In part, one's decision will be based on the nature of the surrounding land.

Large lagoons and complexes of lagoons within larger marshy areas often occur within sparsely-populated 'wilderness' areas. These have frequently been designated national parks or nature reserves, with the intention of either using them as areas for education in the broad sense of that word or of leaving them as enclaves of 'wild nature'. Where the surrounding land is both marshy and of considerable extent, such lagoons can be preserved simply by leaving them alone. Lagoons will fill with vegetation and others will form, in a complex but natural sequence. Examples of such wilderness policies can be seen in the Everglades of Florida, in the Camargue and Danube delta in Europe, etc. However, even when large areas are left in a wild condition, they will remain susceptible to deterioration caused by remote changes in their watersheds, e.g. by pollution or hydraulic works. The projected diversion of water resources in Florida, for example, will modify the water balance of the whole Everglades region including the preserved areas. In other words, a lagoon must be considered in relation to its surrounding area: it is impossible to preserve a lagoon in its natural state if the surrounding land is undergoing development or deterioration—the lagoon will inevitably show signs of degradation.

An example of such a situation is given by Bellan (1972), namely the effects of an artificial stream on the Étang de Berre (area 150 km^2) near Marseilles (Figure 32). Since 1966, a canal from the River Durance has flowed into the northern part of this Étang (which is well known from the investigations conducted by the staff of the Marine Station at Endoume). A survey in 1970 showed that the original marine communities were being replaced by brackishwater ones as a result of this freshwater inflow, but continuous

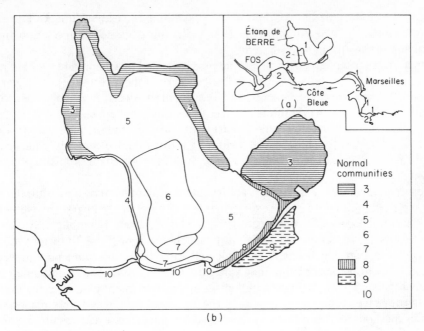

(b)

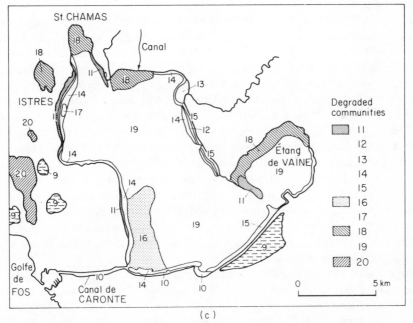

(c)

Figure 32. Changes in the benthic communities of the Étang de Berre, on the French Mediterranean coastline, between 1964 (b) and 1970 (c) consequent on hydrographic changes (Reproduced by permission, from Bellan, 1972). 1. degraded zone. 2. Disturbed balance zone. 3. *Zostera* beds. 4. Photophilous algae. 5. Muddy sands in sheltered areas. 6. Impoverished muddy sands in sheltered area. 7. *Mytilus* beds. 8. Fine sands in shallow waters. 9. Euryhaline and eurythermal community. 10. Polluted zones. 11. *Zostera* beds on muddy sand in sheltered areas. 12. *Zostera* beds on fine sands in shallow waters. 13. Photophilous algae. 14. *Mytilus* beds. 15. Fine sands in shallow waters. 16. Fine sands and laguna communities. 17. *Capitella capitata* settlement. 18. Heavily polluted zone. 19. Heavily degraded muddy sands. 20. Salinas.

change in the environmental conditions was preventing the establishment of an equilibrium. Moreover, the ecosystem of the Étang is being degraded by pollution from the surrounding urban and industrial areas.

Further discussion of comparable problems is provided by Chapman (1966) in relation to the consequences of the Texas Basins Project on the fishery resources of the very large mixohaline lagoons along the Gulf coast of Texas. This project, involving the construction of canals and reservoirs, will result in changes to the physicochemical characteristics of the water in the lagoons (particularly the salinity), with an expected reduction in the commercial fisheries (and sport fishing) harvest running into millions of pounds. Changes in the yield of the fisheries can here be taken to be indicative of changes in the productivity generally.

In heavily populated areas, lagoons are frequently surrounded by agricultural, residential, or industrialized land (as, for example, in many parts of Denmark, the Netherlands, France, and Italy), and here the accent is on the lagoon itself, being the only part of the system to survive. In many areas, this has taken the form of trying to restore to healthy conditions a lagoon which has been subjected to considerable (and now unwelcome) pollution. The problem of maintaining water quality in such lagoons is not a new one: in the sixteenth century, hydrologists, biologists, and Government officials called *Savi delle acque* ('water sages') surveyed the state of the Venetian lagoon and were responsible for proposing methods of maintaining it in a satisfactory state. These activities have continued to the present day.

In the region of Venice, for example, a river diversion project extending over four centuries has been accomplished. Rivers draining into the lagoon have been diverted to discharge directly into the sea, so that their sediment loads are now delivered to the north and to the south of the lagoon. This is therefore now artificial, being maintained solely by marine influences (tides, waves and currents) and by anthropogenic hydraulic works.

It is instructive to compare this lagoon with those on the River Po delta, although human activities have also modified, and continue to modify the geomorphological balance of the latter. Besides land-reclamation projects started several centuries ago, human intervention was directed towards the hydraulic system of the River Po. The adjacent shoreline runs north–south, and both the Po delta and the mouths of the river tend naturally to move towards the north under the influence of the prevailing wind and water movements. In 1600–1604, the Venetian Government constructed a large channel (*taglio di Porto Viro*) about 50 km from the river mouth and diverted the main branch of the river to the south. Thereafter the southern part of the delta grew by accretion of sediment, and in this way it was possible to prevent silting of the lagoons on the north side of the delta and also of the Venetian lagoon even farther to the north.

Even at the time of the Venetian Republic, pollution was a problem and, in an effort to keep the water clean, strict laws forbade the discharge of wastes into the canals. Other severe laws regulated the maintenance of the lagoon mouths, strong rocky dams being constructed in various gaps in the sand-dune barrier. In fact, even in those days, a considerably body of hydraulic and conservation experience existed, on which the Government could draw.

Today, both our knowledge and our technology are sufficient to accomplish projects of the magnitude of the Zuider Zee and Delta Plans in the Netherlands. These projects, described elsewhere in this volume (Chapter 1), have created large basins in many ways equivalent to lagoons. The physical, chemical, and biological changes consequent on isolating the IJsselmeer have been carefully documented by the Dutch research workers (e.g. Beaufort, 1954), and the results of similar studies on the Rhine–Maas–Scheldt estuary are appearing in the *Netherlands Journal of Sea Research*. These studies form a model for the investigation of any lagoon likely to be subjected to the effects of major hydraulic works.

Many problems arise when one is forced to manage a lagoon situated within a built-up or agricultural area, and in view of the variety displayed by lagoonal environments, generalization is difficult. Some fairly general points can, however, be made.

As water currents inside a lagoon are invariably weak in comparison with those of the adjacent sea, and as the flushing time of lagoons is generally relatively long, dilution and dispersion of materials discharged into lagoons occurs at a slow rate. If the discharged material is organic in nature, lagoons will therefore easily become eutrophic. The discharge of sewage, for example, should therefore be avoided, unless a careful hydrographic survey indicates that, in relation to the lagoonal circulation system, etc., the amount of material discharged will not lead to eutrophic conditions. Even more caution should be displayed in relation to the discharge of industrial waste. Although this point is an obvious one, unfortunately it has been overlooked in several areas with the result that water quality and biological productivity have declined markedly. Accumulation of toxic products is another obvious danger, especially in lagoons used both for aquaculture and for the discharge of waste products.

Even when industrial pollution is avoided and a lagoon is used solely for recreation, problems sooner or later arise, especially when the 'recreation' is in the modern tourist form. The development of lagoons for boating in all its guises frequently involves extensive modification of the shoreline, damage to submerged vegetation, and pollution—especially by oil. Many lagoons can withstand light recreational pressure, but if, for example, boating pressure is high, one may have to ask the question 'does one wish to preserve the lagoonal system or to manage the lake purely for boating?'—and act accordingly.

A lagoon is a comparatively simple system and has less homeostatic controls than most other systems; it is therefore comparatively easy to disturb, although it is not too difficult to preserve it in a moderately good condition. In many cases, ecological and hydrographical investigation of a lagoon before specific projects are initiated will allow estimates to be made of the likely future evolution of the system and of its susceptibility to deteriorative changes. This will provide a framework against which prospective changes can be viewed, and, once a decision has been taken, monitoring of the environment should provide an index of the effect of the policies adopted.

ACKNOWLEDGEMENT

The author is deeply obliged to Dr R. S. K. Barnes for his careful revision of the text and for his help in formulating the chapter in English.

Chapter 5

Muddy foreshores

C. R. TUBBS

Mudland has received rather scant attention in the past, presumably because of the discomforts and difficulties involved in working in mud and also, perhaps, because most mud organisms, once unearthed from their clogging environment, tend to be visually unimpressive. This environment, however, is no less rich in invertebrate life than other littoral habitats. Moreover, mudlands possess special ecological significance because they are the main feeding medium for many wading birds and wildfowl, which congregate on the coast in great numbers during migration periods and in the winter months.

ORIGIN AND NATURE

Mud-flats are formed by the deposition of finely inorganic material and organic debris in particulate form, which has been held in suspension in the sea or in estuaries. Because of the fine texture of this material, deposition tends to occur where the turbulence of the sea is most abated and the gradient of the underlying land slight. Mud-flats are thus formed in the sheltered parts of embayments, inlets, and estuaries or behind the protection of shingle spits or dune systems. Mud has an adhesive quality which is enhanced by the abundant organic matter which causes cohesion of the particles. Once deposited, it requires a strong tidal scour to remove it. In sheltered situations the deposits accumulate to form extensive, level or gently sloping expanses drained by networks of deeply incised channels, some formed by the flood tide and some by the ebb.

In great embayments exposed to the sea such as The Wash or Morecambe Bay, where there are vast areas of intertidal deposits, there is a progressive increase in particle size from the landward margin to the sea. In The Wash, the mudlands occur around the sheltered estuaries of the Withan, Welland, Nene, and Ouse, yielding to sand-flats to seaward. In a recent study of Morecambe Bay, Anderson (1972) showed, similarly, that the finer sediments occurred in the sheltered, innermost areas of the Bay on the higher level flats, whilst the deposits became increasingly coarser towards the open sea and lower down the shore. Similarly, in many smaller estuaries and inlets, sand-flats and bars tend to occur at the exposed mouth, the finer silts and muds occurring behind them. The most finely divided (and most treacherous) sediments have accumulated in the most sheltered sites, such as the estuaries and inlets of Essex and of the south coast of England between Pagham Harbour in Sussex and Poole Harbour in Dorset. On exposed coasts, mud accumulates only where shelter is afforded by offshore dune systems such as the often quoted examples of Scolt Head and Blakeney Point in Norfolk. Similar developments have taken place on a massively greater scale behind protective dune systems on the Biscay coast of France and between the Friesian Islands and the mainland of the Netherlands, north-west Germany, and Denmark.

The process of accretion over level expanses of mud leads naturally to the development of salt-marsh, notably where the silt content of the deposit is high and the flats are formed against a lee shore or elsewhere where there is minimal tidal scour on the ebb. The development of salt-marshes lies beyond the scope of this chapter, but it is relevant to dwell briefly on the formation of marshes dominated by the cord-grass, *Spartina anglica* (*S. townsendii sensu lato*) from the mud-flats because *Spartina* exhibits exceptional powers of colonizing muds, including those at a much lower level in the tidal range than can be colonized by the typical pioneer plants of true salt-marsh. The rapid spread of *Spartina* marsh since the 1870s has involved a substantial reduction in the total mud-flat resource of north-west Europe and this in turn has greatly reduced the areas available to waders and wildfowl as feeding grounds.

Spartina townsendii (*sensu stricto*) Was first discovered at Hythe on Southampton Water, Hampshire in 1870. It appears to have arisen as a natural hybrid between the indigenous salt-marsh grass *S. maritima* (now comparatively rare) and an American species *S. alterniflora*, which it is generally assumed was accidentally introduced by transatlantic vessels docking at Southampton. After hybridization, a doubling of the chromosomes seems to have occurred to produce the fertile tetraploid *Spartina anglica* which proved to have a remarkable capacity for colonizing mudland compared with its parent species. It is a sturdy, rhizomatous grass sometimes growing a metre in height and sending down roots a metre below the mud surface. The initial clumps spread rapidly to form a continuous sward drained by steep-sided channel systems of tortuous complexity. Besides consolidating the mud in which it is growing, *Spartina* marsh slows the movement of tidal water and thus increases the rate of silt and mud deposition so that the marsh grows rapidly upwards. On the south coast of England, the area of initial colonization, the point appears to have been reached where upward growth has ceased and for reasons which are imperfectly understood the *Spartina* is dying-back, other salt-marsh plants are failing to colonize, and the platforms of accreted material are slumping and eroding.

During the last two decades of the nineteenth century, the hybrid spread rapidly across the mud-flats of Southampton Water and appeared in many localities elsewhere on the central south coast. Subsequently it spread to most suitable areas in north-west Europe, aided by deliberate introductions once its potential as an agent for mud-flat reclamation became recognized. It has been estimated that there are today between 21,000 and 28,000 ha of *Spartina* marsh on the coast of Britain, France, the Netherlands, Germany, and Denmark (Ranwell, 1964c; Hubbard, 1965; Ranwell, 1967; Hubbard and Stebbings, 1967).

THE INVERTEBRATES OF MUDLAND

It is a fair generalization that the organic content and nutrient status of intertidal sediments increases with decreasing particle size. Anderson (1972), for example, showed that in Morecambe Bay the percentages of organic carbon, organic nitrogen, and phosphorus increased in inverse proportion to the particle size of the sediments. She quoted Newell (1965) in support of the view that the high carbon values were due mainly to the presence of organic debris, and that the high nitrogen values resulted from the presence of micro-organisms on the surfaces of the particles. Clearly, in the finer sediments the surface area per unit volume available to micro-organisms is greater.

Since the intertidal invertebrate population is dependent on the presence of organic matter, either because they exploit it directly or because they prey on the detritus feeders, the invertebrate populations will naturally tend to increase across the spectrum of particle size between coarse sand and fine mud. At the same time, the clogging, water-retentive environment of the finest muds places considerable constraints on the variety of organisms which have adapted to them. There is thus a tendency for the greatest diversity of organisms to occur in materials intermediate in texture. Anderson concluded that sediments with a silt content of 30% or more seemed to provide the most favoured conditions for an abundant mud fauna—conditions which were fulfilled in Morecambe Bay on the higher, inner flats, though none of the deposits could be described as 'true' mud. Longbottom (1970), who investigated the status of the burrow-dwelling lugworm *Arenicola marina* in a range of sediments on the north Kent coast, also found that the amount of organic matter in the sediments varied inversely with median particle size and that the biomass of the lugworm exhibited a similar relationship with particle size, except that it was absent from the finest muds. He considered the possibility that low oxygen levels in fine mud might be a restricting factor, but thought it more likely that this particular species, which is relatively large and needs to maintain a substantial burrow, probably found this difficult to achieve in fine, sometimes almost 'liquid' mud, which in turn prevented it from maintaining contact with the oxygenated sea water above the mud surface.

A further generalization which can be made is that the coarser the texture of the sediment, the larger and thicker shelled the macro-organisms have to be to burrow successfully. Conversely, in mud the animals tend to be relatively small and thin-walled. Furthermore, the coarser the material, the deeper its inhabitants need to burrow at low tide to maintain contact with the water-table. In fine muds, most of the invertebrates occur in the top 5 cm. All these general factors have implications for many of the birds which feed in the intertidal zone, because in general they mean that the finer the sediments are the more readily available the invertebrates will be as food. Exposed flats of coarse sand are seldom used as feeding grounds by wading birds, ducks, or geese. At low water the flocks are concentrated on the muds where the greatest biomass of prey of the appropriate, that is relatively small, size is most easily found.

Despite the physical constraints on the invertebrate life of fine muds, the organisms which have adapted to feeding and respiring in this medium can occur in very high densities. The macro-invertebrates of the muds comprise three broad groups: first, animals which occupy burrows and are completely concealed at low water; second, those which are exposed on the mud surface at low water, or which retreat into the subsurface of the mud; and third, surface dwellers which are associated with cover such as shell or shingle deposits on the mud surface, algal mats, or beds of eelgrass (*Zostera* spp). Among the first group, ragworm (*Nereis* spp) and some other polychaete worms, the amphipod *Corophium volutator*, and bivalve molluscs such as *Macoma balthica*, *Scrobicularia plana*, *Mya* spp and the common cockle *Cardium edule*, are among the most abundant species. The second group mainly comprises gastropods such as *Hydrobia ulvae*, and the winkles *Littorina littorea* and *L. Saxatilis*. The third group is large and embraces many animals not strictly associated with mud. It includes many crustaceans—the shore crab *Carcinus maenas*, *Gammarus* spp, *Idotea viridis*, *Melita palmata* and others—and a variety of molluscs including mussels *Mytilis edulis*, dogwhelks *Nucella lapillus*, besides other species perhaps more typically associated with rocky shores, such as limpets (*Patella* spp) and barnacles.

The abundance of invertebrate life in the muds can best be emphasized with examples of population densities recorded for three of the most widespread and characteristic species

(Table 9); all of which are of great importance as food sources for estuarine birds—the polychaete worm *Nereis diversicolor*, the crustacean *Corophium volutator*, and the snail *Hydrobia ulvae*.

Table 9   Densities achieved by three mudflat invertebrates

	Numbers m^{-2}	
	Maximum	'Typical'
Nereis diversicolor	100,000	100–500
Hydrobia ulvae	100,000	5,000–9,000
Corophium volutator	24,000	5,000–9,000

All three species are detritus feeders and depend on the organic matter incorporated into or present on the surface of the mud. *Hydrobia* is also said to feed during its peregrinations about the estuary. Adult *Nereis* are large enough to be dug for bait. *Hydrobia* can achieve a length of about 0·5 cm and *Corophium* a length of about 1 cm. That these animals not only occur in vast numbers in the muds but are far from being microscopic organisms, emphasizes the enormous productive capacity of the mudland ecosystem.

THE BIRD POPULATIONS OF MUDLANDS

In north-west Europe, including the British Isles, at least 18 species of wading birds (Charadrii) occur in substantial numbers in the intertidal zone, either over-wintering congregations or as migrants between northern breeding areas, such as the Arctic and subarctic tundras, and more southerly wintering grounds such as north Africa. At least 13 of these species depend heavily on mud-flats as feeding grounds. Two others, the knot *Calidris canutus* and the bar-tailed godwit *Limosa lapponica*, tend to feed where the sediment is intermediate in texture between mud and sand. The turnstone, *Arenaria interpres*, although particularly associated with rocky shores, also feeds extensively on mud-flats and sands, mainly where these are intermingled with shingle and seaweed. To some of these species, the intertidal muds and silts of the north-west European costline are vital because they sustain large proportions of the world population during the winter months. For others, the estuaries and other intertidal areas of Europe are equally vital as migratory staging posts where they can accumulate fat reserves to complete their journeys northward (in spring) and southward (in autumn).

From the evidence available it seems that polychaete worms, small bivalve molluscs, small gastropods (mainly *Hydrobia*) and small crustaceans (mainly *Corophium*) are the main prey sources for wading birds whilst they are on the coast. To some extent, the prey taken by the different species of waders depends on the lengths of their bills. Plovers are best equipped in this respect to obtain their food from the mud surface, and *Hydrobia* probably feature highly in their diet on the shore. Dunlin and other waders with medium length bills are able to probe 3–4 cm beneath the surface and probably mainly exploit small worms, small bivalves and *Corophium*. The longer billed waders such as curlew, *Numerius arquata*, and godwits possess the ability to exploit an even wider range of prey species by probing deeply in the mud. Indeed, the curlew with its long, curved bill, would seem to be especially well adapted to seek deep-burrowing worms, which generally occupy a curved or U-shaped burrow; and one can often watch a curlew on the mud manoeuvring itself so that the curvature of the bill can follow the burrow of the worm.

Besides waders, the intertidal muds harbour large populations of several species of ducks. Most surface-feeding (as opposed to diving) ducks depend to some extent on the muds for feeding. Shelduck feed mainly on marine molluscs—especially *Hydrobia*—and the intertidal muds and muddy sands of the north-west European coastline supports something in the order of 130,000 birds. Mallard (*Anas platyrhynchos*), teal (*A. crecca*), wigeon (*A. penelope*), pintail (*A. acuta*), and shoveller (*A. clypeata*) all feed partly on intertidal invertebrates, but mainly depend on vegetable matter obtained from freshwater marshes, woodlands and agricultural land, as well as from the mud-flats. Indeed, for the large flocks of mallard and teal which congregate on the coast during the winter months, the intertidal zone is often less important as a feeding ground than as a relatively safe roost from which they can venture inland to feed at last light. Wigeon and pintail, however, are more essentially coastal ducks and exploit two main food sources on the coast; *Puccinellia maritima* and *Agrostis stolonifera* swards on salt-marshes, and *Zostera* and the green algae, *Enteromorpha*, on mud-flats. In Britain alone, there are probably upwards of a quarter of a million wigeon present in midwinter and most of these are concentrated on the muddy coasts.

Although geese tend to be popularly associated with the coast, the only species dependent on the intertidal zone is the brent, *Branta bernicla*. Once it has left its breeding grounds in the high Arctic, the brent is mainly a littoral organism, feeding largely on the eelgrasses, *Zostera angustifolia* and *Z. noltii*, both of which are confined to intertidal muds; and on green algae, mainly of the *Enteromorpha* genus, which also occurs abundantly in similar conditions. The brent-goose is of special interest because the nominate race, the dark-bellied brent, *B, bernicla. bernicla*, which winters exclusively in north-west Europe, was until recently at a low population ebb and there were fears for its ultimate survival should more of its winter feeding grounds be lost to reclamation and development.

The brent goose populations on both sides of the North Atlantic declined drastically earlier this century, a decline exacerbated by the widespread disappearance of the *Zostera* beds which were attacked by a 'wasting disease', the causes of which remain partly obscure. Populations declined alarmingly. For example, that of the American pale-bellied brent-goose, *B. bernicla hrota*, may have been in excess of 250,000 birds in the late 1920s. By 1935 it had slumped to about 22,000 (Cottam, 1935).

The brent-goose population wintering on the north Atlantic seaboard of America seems to have recovered to somewhere near its former numbers by the 1950s; though the small population of this race which winters in north-west Europe remained at a low ebb. The population of the dark-bellied race, wintering entirely in north-west Europe, climbed slowly from 16,000 in the mid-1950s to about 35,000 in 1972. At this point it was widely predicted that limited food supplies would prevent any further increases; yet in the succeeding four years the population 'took off' to reach no less than 110,000 in winter 1975/76 (Ogilvie and Matthews, 1969; Ogilvie and St. Joseph, 1976). Simultaneously, the species proved that it was able to adapt to feeding on farmland bordering the mudflats when the food resources of the latter were depleted. The main dark-bellied brent-goose centres in Britain today are the Essex coast, the north Norfolk coast, The Wash and Lindisfarne on the east coast; and Langstone and Chichester Harbours on the south coast. The vast *Zostera* beds of Foulness, in Essex, have been described as the most important early winter refuge for the dark-bellied brent in the world, and at times they support as much as 25% of the world population (Rudge, 1970). The reprieve of Foulness from obliteration beneath a third London airport is thus fortunate, but the long-term future of many other *Zostera*-bearing mud-flats, not only in Britain but elsewhere on the north-west European coast, remains overshadowed by the possibility of reclamation. It is a sad reflection that interest in bird populations and their food resources is only now largely being engendered by the threat of destruction which hangs over so many muddy estuaries and inlets and by the consequent need to construct the scientific arguments for their conservation.

THE EXPLOITATION OF MUDLANDS

The mud-flat ecosystem, based on organic detritus, can be likened to an enormous storage battery for energy derived both from the sea and the land. An important feature of the system, particularly as regards its capacity to support migrating and over-wintering populations of wildfowl and wading birds, is that the heat storage capacity of the sea reaches a maximum in late summer, so that whilst the productivity of terrestrial ecosystems declines during the autumn, that of the mud-flats actually reaches its peak then. Thus the maximum biomass of prey species is available near the time when the maximum numbers of predators are present. The immense nutrient resources of mud-flats have been utilized by man probably from his earliest beginnings and in recent—post medieval—times in Britain, sophisticated shellfish industries grew up in many estuaries, whilst wildfowl and wader populations were widely exploited by professional wildfowlers. At high water, the flats are important feeding grounds for fish such as flounders and the channels and creeks which drain them supported substantial local seine and stake-netting fisheries.

One hundred years ago or less, the mud-flats represented a rich natural resource of considerable importance to the human communities living beside or near them. In this century, exploitation has tended to polarize towards either the casual or the highly commercial. Professional wildfowling was a way of life in the past, but many enjoy wildfowling as a sport without making significant inroads on wildfowl populations, which the older fowlers undoutedly did. Shellfish exploitation, once part of the local economy wherever there were mud-flats, has tended to concentrate mainly on sites where a combination of minimal pollution and specialization in harvesting and marketing methods have permitted survival—but there are still many people who collect a pail of cockles from the flats as a weekend pastime.

Other recreational uses dependent directly on mud-flats are few today. The muds in sheltered creeks and bays provide safe moorings for small craft and thus may be said to contribute to the over-all capacity of a given inlet to support sailing and other forms of marine recreation. The bird populations dependent on the mud-flats are the object of another recreational 'growth point'—birdwatching—which besides giving pleasure to a large and rapidly increasing number of people also supplies scientific feedback in that it provides much of the basic data about the birds which depend on the muds: for example, our information about the populations of waders and wildfowl is derived from organized counts carried out mainly by volunteers. In the formal education field, mud-flats offer only limited opportunities as teaching arenas: sandy or rocky shores are less difficult of access and offer more diverting organisms than the muds. The planner surveying the uses of muds may perhaps be forgiven for regarding them as a kind of land-use vacuum. It is perhaps less forgivable if the assumption automatically follows that the vacuum must somehow be filled.

The modern industrial society of Britain tends to see intertidal mudlands not as a rich natural resource, but as potential new land for development. The reclamation of mud-flats for grazing and even arable land has, of course, occurred widely in the past and the methods used have included both the deliberate acceleration of natural processes of silt accretion and direct embanking against the tide. Such reclaimed land has an inherently high fertility because of the organic content of the muds. Along the seaboards of the

Netherlands, West Germany, and Denmark, reclamation has taken place more or less systematically since medieval times. In Britain, it has occurred in a more piecemeal fashion, but nevertheless a large part of the fringe of land along the east coast from Lincolnshire to Kent was once mudland covered twice daily by the tides. Today there is some likelihood of a significant proportion of the remaining national mudland resource becoming lost, not to farming as in the past, but to industry and water storage.

Large industrial complexes such as steelworks and oil refineries, which require seaborne bulk supplies of raw materials, demand sites on the coast, and current national policy is to encourage the siting of these new industrial plants not only on existing industrialized coastline, but on relatively undeveloped sites designated 'Maritime Industrial Development Areas'. The outer Thames estuary is a good example. Further north, the Humber, Tees, Forth, and the estuaries of north-east Scotland are all threatened with industrial development which is likely or certain to involve reclamation. On the south coast, most of the remaining intertidal muds of Southampton Water may be lost to industrial and dock development tentatively planned for the next decade or two. Even more consumptive of intertidal land are estuarine barrages, and impoundments for water storage, such as those which have been proposed for the Wash, the Dee, Morecambe Bay, and the Solway Firth. Studies of the feasibility of impounding huge areas of The Wash, Dee, and Morecambe Bay are either in progress or have been completed. Because of the location of the finer sediments in the inner parts of these embayments, barrages would inevitably involve the loss of most or all of the rich muds, leaving only the more infertile sands to seaward. At times, the muds of Morecambe Bay, The Wash, Dee, and Solway between them support as much as 30–35% of the waders present on the coastline of Britain—which have been estimated to be $1-1\frac{1}{2}$ million birds in midwinter by the British Trust for Ornithology–Royal Society for the Protection of Birds, *Birds of Estuaries* survey. It is doubtful if the displaced wader populations could be absorbed by mudlands remaining elsewhere in north-west Europe. The inevitable consequence would therefore be a serious diminution in the world population of at least some of the species involved. The knot is an example. In the early 1970s Morecambe Bay, the Solway, Dee, Ribble, and The Wash, together supported over 30% of the 435,000 knot which wintered in Europe and north Africa. Morecambe Bay alone supported about 16% or 70,000 birds (Prater, 1972). These five areas are vital, not only as wintering grounds, but as migratory staging posts. In spring the birds need to arrive in the Arctic in breeding condition, so that the loss of migratory feeding grounds would be likely to have direct repercussions on breeding success in the same year.

On the south and south-east coasts of Britain, the most important single factor in reducing the area of intertidal mudland during this century has probably been the spread of *Spartina*. In addition to losses of mud-flats to *Spartina*, piecemeal reclamation has been widespread in the south and south-east. Chichester, Langstone, and Portsmouth Harbours, on the Hampshire–Sussex border, which at times support, for example, more than 20% of the world population of dark-bellied brent-geese and more than 5% of the dunlin wintering in Europe and north Africa, have recently been under constant pressure from proposals for refuse tipping, and reclamation for residential development, roads, commerce, and recreational facilities such as yacht harbours. Some 5% of the Langstone Harbour muds have been lost since 1963. About 25% of the muds of Portsmouth Harbour have recently been reclaimed for motorway construction, industry, and other development.

A third factor which would appear to be influencing the mud-flat communities of south-east England is the discharge of sewage effluent into estuaries and inlets. A phenomenon associated with the resulting elevation of nutrient levels is the spread of green algae, mainly *Enteromorpha* and *Ulva*, across the mud-flats. This sitmulation of algal growth following the discharge of effluent, treated or untreated, into semi-enclosed tidal waters, is well known, and was first recorded from Belfast Lough early this century. Within limits it may have beneficial effects; the algae are eaten, for example, by brent-geese and wigeon. Beyond a certain threshold, however, the physical blanketing of the muds by the algae, especially in the autumn and early winter, reduces the diversity and abundance of mud-living organisms by depleting the upper horizons of oxygen, reducing in turn the capacity of the muds for supporting waders and some other mud-flat birds. In Langstone Harbour, for example, an increase in the quantity of *Enteromorpha* on the mud-flats since the 1950s has coincided with an increase in the volume of treated sewage effluent entering the harbour. During the same period, the winter brent-goose population has increased from about 100 to 4,000, and although external factors are almost certainly involved it is highly unlikely that the combined *Enteromorpha* and *Zostera* resources of the harbour 20 years ago could have supported 4,000 geese. The autumn and winter populations of wigeon and teal have also increased over the same 20-year period, whilst the numbers of some waders have fallen somewhat. It is probable that the amount of effluent entering Langstone Harbour today has passed a critical point and is now having ecologically undesirable effects on the plant and animal communities of the mud-flats (Dunn, 1972; Southgate, 1972; Tubbs, in press).

THE CONSERVATION OF MUDLANDS

There are usually strong economic or social reasons why any individual development involving mud-flat reclamation should go ahead. As the leading article in the *Marine Pollution Bulletin* for November 1972 pointed out, the Tees estuary may be the only estuarine feeding ground between the Humber and the Forth but it is difficult to argue with the people there that the conservation of the flats and their bird life should take precedence over development intended to solve acute unemployment. There are, however, sound reasons for conserving intertidal mud-flats. First, it is foolish policy to destroy a biologically rich natural resource whose productivity may one day be of even greater benefit to man than at present, without first determining how and where similar ecosystems can be created. Second, if the conservation of populations of other vertebrates with which we share the earth is a valid concept (and there seems to be a general assumption that it is) then we have to conserve sufficient of their habitat. Outside their brief breeding season, the Arctic geese, ducks, and waders are largely dependent on mud-flats in temperate regions. Without great care we are likely to lose them by destroying and polluting their wintering grounds.

There is clearly a case for pausing and taking stock of the total mud-flat resource, the birds dependent on it, its vulnerability to pollution and disturbance, and the means of recreating mud-flats to replace those lost to barrages or other manifestations of modern human culture. We need to devise means of recreating elsewhere, for example, the vast *Zostera* beds which will be lost in the event of the third London airport proposal being resurrected. We need to know much more about the ecology of mud-flats and the capacity of different sediment types for supporting invertebrates and their bird and fish predators. It is far from impossible that, given a little more knowledge, new mud-flats of a given sediment and organic composition could be created by the construction of carefully aligned artificial arms of the land.

In the past, ecological research into mud-flat ecology has been more or less fragmentary, but the major barrage proposals, together with the third London airport and other developments proposed on the coast, have sparked off fresh interest, both in university departments and in Government sponsored research laboratories. Importantly in 1972, the Nature Conservancy was given the resources to carry out comprehensive studies of The Wash and the Essex coast. The objects of this research included determining means of creating alternative habitats to those which would be lost to impoundment for water storage on The Wash and the construction of an airport at Foulness. It is important that this momentum, so recently gained, should not be lost. It is equally important that research should not be regarded politically as an alternative to conservation: further destruction of the rich intertidal muds of Britain and the rest of Europe must not be countenanced in the absence of clear solutions to the biological problems posed by their loss.

Research takes time. What immediate recommendations can the ecologist make to the planner? First, the conservation of mud-flats and their fauna, as a limited natural resource, demands explicit recognition as a valid planning objective. Given this, then the destruction of the resource by reclamation, on however limited a scale, should not be countenanced unless it can be clearly shown that no alternative course of action is open, or that by very limited sacrifice it will be possible to more adequately safeguard the larger area.

A more positive step towards conservation which can be taken is the allotment of mudland areas to nature reserves. In Britain, many important mud-flats are already managed as reserves by the Nature Conservancy Council, Royal Society for the Protection of Birds, County Naturalist's Trusts or local authorities, but the last named still make too little use of their powers, under the National Parks and Access to the Countryside Act, 1949, to acquire and manage areas as Local Nature Reserves—partly because there is a feeling that declaration as a nature reserve will sterilize the land for other uses and, in particular, for public access. This need not be so and, indeed, it should not be so. With care, it is normally possible to devise means of allowing controlled public access without damaging the habitat or, especially in the case of mud-flats, disturbing its birds. Management devices such as hides well placed to observe spectacular numbers of waders or wildfowl and approached by screened walks, are sometimes appropriate in a wetland context and provide both entertainment and the opportunity to educate the public to a wider appreciation of their varied surroundings.

In practice, most mudlands are components of larger intertidal systems which may include sand-flats (or dunes), shingle and salt-marsh, together with the channels and creeks which will drain them. Such systems will often have to meet the demands of many different activities, both commercial and recreational. The problems arising from these demands are more properly the province of other chapters, but it is relevant here to make the point that it will often prove possible for the mud-flat element of the system to remain relatively undisturbed within the context of a multiple use plan.. This, for example, would appear to be the general intention of the Chichester Harbour Conservancy, a body which has received generous powers from the Chichester Harbour Conservancy Act, 1970 for the over-all control of recreation and the conservation of Chichester Harbour in West Sussex. In neighbouring Langstone Harbour, just over the county border in Hampshire, the Langstone Harbour Board accepted recommendations made as a result of a study of harbour uses and conservation by Tubbs (1966) on behalf of the Nature Conservancy and have adopted a policy of maintaining the main area of mudlands free of access points or other development and concentrating recreational (and the limited commercial) shore facilities and moorings in locations where they have the least adverse effect on the ecology of the harbour as a whole. Much can be achieved in this and similar situations by time and space zoning of activities.

Eutrophication arising from the discharge of sewage effluent into the intertidal zone is, similarly, a problem which demands attention from the standpoint of the system as a whole rather than from that of one of its component habitats. Nonetheless, it is relevant here to stress again the damage likely to occur to the invertebrate fauna of the muds from the blanketing effect of green algae—besides other pollution hazards. Ecologically there is a strong case for discharging some distance at sea rather than in an enclosed area and it is gratifying to observe that this has been recognized, for example, in the *South Hampshire Structure Plan* (1973), which considers the problems of population growth in the Southampton–Portsmouth area of Hampshire and stresses both the limits inevitably set on growth by existing methods of sewage disposal and the environmental advantages of exchanging existing methods for discharging some miles out in the sea at Spithead.

However convincing the exhortations of the conservationist, mud-flats are often likely to seem to the planner to be worrying under-used areas, and especially where they occur in proximity to large conurbations he is likely to seek some further rationale for their continued existence. There is a temptation to resort in reports to their 'natural history value' or their 'ecological interest', without clearly defining either term, and perhaps to stress their importance for 'birdwatching', without incidentally acknowledging the scientific contribution of the activity. I do not suggest that these are irrelevant matters, but I would like to conclude this chapter by asserting again that the real issue is that the mudlands demand conservation because they are a limited and, at present, irreplaceable natural resource of great biological richness and productivity.

Chapter 6

Salt-marshes

W. G. BEEFTINK

DEFINITION, DISTRIBUTION AND GEOMORPHOLOGICAL HISTORY

Definition and range

Salt-marsh can be defined as natural or semi-natural halophytic grassland and dwarf brushwood on the alluvial sediments bordering saline water bodies whose water level fluctuates either tidally or non-tidally. The salinity of the adjacent water bodies may vary from >38‰ to 5‰: if the salinity is below 5‰, however, salt-marsh is replaced by reed and rush marshes, willow coppice and tall herbage, etc. On coasts with a high rainfall, landward penetration of salt-marsh may be curtailed. Where coastal rainfall is low, salt-marshes extend further landwards owing to the relatively high evaporation: on these coasts, they easily pass into salt-steppe and salt-desert conditions.

Developing on alluvial substrates, salt-marshes require conditions of shelter sufficient to ensure that sedimentation occurs and also to prevent too much erosion. The texture of the soils varies from heavy to sandy clay, and even to silty sand or silty peat. Salt-marsh vegetation on sandy sediments is found on coastal dune-slacks or beach plains partly enclosed by sand-bars, dunes or shingle ridges and thus protected against surf exposure. Salt-marshes on peat formations are found along the New England coasts of North America and in south-west Ireland (Chapman, 1960b).

In tidal areas, salt-marshes develop on mud-flats which are covered by seaweeds and eelgrasses. Here terrestrial halophytic vegetation begins to colonize between the tidal levels of mean high-water neap (MHWN) and mean high water (MHW). *Zostera* beds are excluded from the salt-marsh vegetation proper as their vegetation must be considered as truly aquatic communities. Salt-marshes find their upper limit where saline influences are so far reduced that halophytes are in the minority or are totally lacking. Under the sheltered conditions appropriate to salt-marsh development, this upper limit is found normally between the levels of mean high water and extreme high water of spring tides (MHWS–EHWS).

For rocky coasts, the marine-influenced zone above MHW is subdivided by various authors (e.g. Sjöstedt, 1928) into a lower supralittoral ('swell' or 'wash' zone) and an upper supralittoral ('storm' or 'splash' zone). This subdivision is also useful for the higher marshes: the lower supralittoral zone is reached by the higher tides, more or less regularly throughout the whole year owing to the lunar tidal variations. The upper supralittoral zone, however, is only exposed to saltwater during seasonal storms. How far upwards the halophytes will reach in these zones and how far upwards the ecosystem can be considered as a salt-marsh will therefore depend on the coastal climate (e.g. amount of precipitation and evaporation).

Distribution and types

Maritime salt-marshes are distributed all over the mid- and high-latitude regions. Even in the Tropics, where their place is largely taken by mangrove swamps, there are areas with a tropical version of salt-marsh (Macnae, 1963, 1966). In temperate regions, salt-marsh is found in and around river-mouths (estuaries, deltas), in bays, Wadden areas, lagoons, and on coastal plains protected by sand- and shingle-spits. Dependent on conditions of salinity and tidal range, we can distinguish six types of salt-marsh: (1) estuarine; (2) Wadden; (3) lagoonal; (4) beach plain; (5) bog; and (6) polderland (see Chapter 13).

Salt-marshes of the estuarine type are represented in every western European river-mouth. They are characterized by widely fluctuating salinities and mostly by strong tidal currents. The Wadden type is protected from the sea by a chain of islands acting as a barrier; in western Europe such a barrier is found in front of the Dutch–German–Danish Wadden Zee. The lagoonal type is protected by split-like sand or shingle barriers, which usually leave only a narrow entrance to the tides: good examples are the salt-marshes on the inner side of the 'Nehrungen' on the south Baltic coast and of the Fleet landwards of the Chesil on the Dorset coast. This type is characterized by subdued tides owing to the narrowness of the inlets. The salinity here depends on the discharge of river-water into the lagoon and may vary greatly. Beach plains are also protected by sand- or shingle-spits, but only against wave action, and they are characterized by their sandy soil. Accretion is here effected by the tidal floodwater as well as by the wind, and so their geomorphological pattern has elements of both hydraulic (creeks) and aeolian (dune ridges, recurved spits) origin. Examples are the Boschplaat, the eastern part of Terschelling, a Dutch Wadden island, and the coastal formations of Blakeney Point and Scolt Head Island. The bog type is known from south England, south-west Ireland, and the Baltic. Marine peat-marshes can be built from subsiding freshwater or brackish bogs under a heavy deposition of autochthonous plant debris and a limited supply of silt (Chapman, 1960b). Other peat formations develop in estuaries and on the inner side of coastal barriers but only under peripheral marine influences. Finally, the polderland type is found along former creeks, in bottom land and other depressions remaining saline in embanked former marshland now under cultivation. They are common in the Dutch estuarine polder areas and undoubtedly also occur in other embanked marshlands (e.g. the north-west German and Danish polderland and the fens of East Anglia). Such areas are considered in detail in Chapter 13.

Geomorphological history

Excepting peaty formations, tidal salt-marshes originate by accretion of the foreshore through sedimentation of mineral materials of either fluvial or marine origin. Subsequent erosion processes may provide secondary material which will be transported and after-wards deposited elsewhere. The lower tidal flats exhibit a convex pattern showing shelving slopes between the channels (maxima of erosion) and the centres of the shallows (maxima of sedimentation). This pattern originates essentially by the lateral transport of sand and silt resulting from differences in velocity of the tide-runs from the channel to the top of the shallows (Figure 33c (I → II → III)]. At this initial stage of development, the geomorphology is largely defined by the levelling effects of the tide-runs forming the horizontal

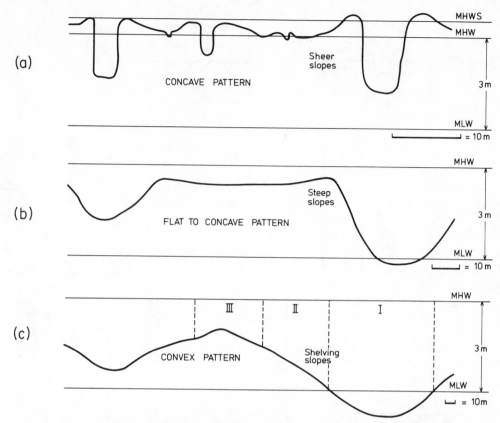

Figure 33. Genesis of salt-marsh from sand and mud flats. (a) Salt-marsh. (b) High tidal flats. (c) Low tidal flats. Explanation in text (after Beeftink, 1966)

component of the tides. This unstable system is differentiated by accretion into a gradually more concave pattern diversifying the maxima of sedimentation and erosion more and more, and dividing the marshes into partial entities. Parallel to these processes of increase of diversity, the spatial decrease of the lateral transport of material creates a second type of transport resulting in a new process of sedimentation: while lateral transport leads to sedimentation of mostly sandy material on the creek banks, the apical type chiefly brings fine particles into the basins via the smallest ramifications of the creeks. The result of these processes is a fine-grained pattern of high-levelled and aerated creek banks enclosing mostly waterlogged and only superficially aerated basins. Ultimately, sedimentation via apical transport will dominate over lateral flooding, resulting in a levelling tendency of the surface-relief (Beeftink, 1965, 1966).

At the back of bays, lagoons, and Wadden areas a terraced type of salt-marsh can occur (Figure 34). Owing to the feeble tide-runs in relation to the tidal range in semi-enclosed areas, wave action takes a proportionally greater part in modelling such salt-marshes. Wave action attacks the marshes mainly at the mean high-water level. This process results in the formation of a cliff-like scour, mostly intensified by the development of a parallel-running creek. At the opposite side, this creek may contribute to the formation of a sand-bar. Later on, a new terrace may form upon the bar. This process is assisted by the developing vegetation (Jakobsen, 1954, 1964). A similar cyclical balance between erosion and deposition is found in drowned river valleys e.g. at Milford Haven (Dalby, 1970).

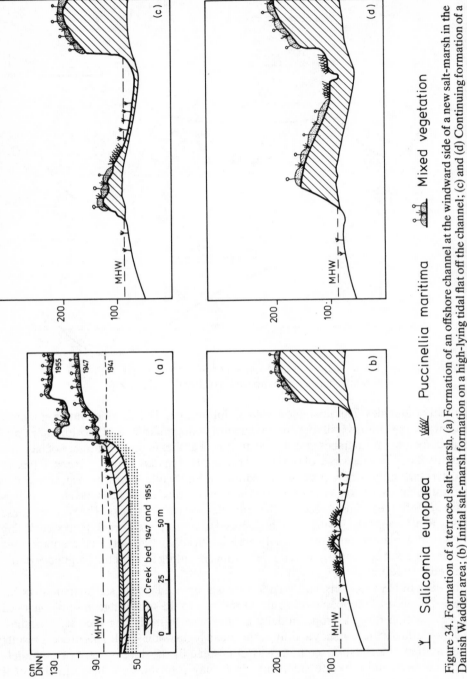

Figure 34. Formation of a terraced salt-marsh. (a) Formation of an offshore channel at the windward side of a new salt-marsh in the Danish Wadden area; (b) Initial salt-marsh formation on a high-lying tidal flat off the channel; (c) and (d) Continuing formation of a new terrace, the channel filling with mud [(a) after Jakobsen, 1964, reproduced by permission of the Editor of *Folia Geographica Danica*; (b)–(d) after Jakobsen, 1954, reproduced by permission of the Editor *Geografisk Tidsskrift*

TROPHIC STRUCTURE, MATERIAL CYCLES, AND ENERGY FLOW

All authors agree that the trophic structure and energy pathways in the salt-marsh ecosystem are relatively simple. Keefe (1972) distinguished two food-chains. In the first one, comparatively large herbivores feed on the standing marsh plants and transform energy seasonally. Only a limited number of species are involved, and human interference has replaced most of these by domestic cattle or sheep. Of the remaining macroherbivores the following can be mentioned: root vole, brown rat, musk-rat, rabbit and brown hare, geese (Ranwell and Downing, 1959), and duck (Olney, 1970). It is significant that there are relatively few herbivores which browse directly on living *Zostera*, and almost none that can feed on *Spartina* (Russel-Hunter, 1970). The present author observed that grazing pressure by rabbits, hares, and root voles increased immediately after the tidal influence had been suppressed by damming. Grazing by mammals thus seems to be limited by the tides.

The second food-web distinguished by Keefe (1972) consists of smaller animals feeding on algae and plant detritus on and in the sediment. The energy flow through these populations is more constant during the year, since they are more adapted to tidal influences and their food is produced more continually. The microconsumers are spread over many taxonomic groups; e.g. bacteria and other microbes, molluscs, and arthropods (for instance, Newell, 1965, and Odum and De la Cruz, 1967).

In his excellent book on the ecology of salt-marshes and sand-dunes, Ranwell (1972a) stressed the point made by Odum and Smalley (1959) that numbers tend to overemphasize, and biomass to underemphasize the importance of small organisms, while the reverse tends to be true with large organisms. This tendency is also observed in the food-web outlined by Paviour-Smith (1956) showing the trophic levels of a New Zealand salt-marsh (cf. Ranwell, 1972a). Numbers and biomass can be integrated by consideration of productivity or energy flow. Odum (1962) and Teal (1962) found that the major energy flow between the autotrophic and heterotrophic levels on a marsh is by way of the detritus rather than the grazing food-chain.

Although nearly all studies on trophic relations and energy flow in salt-marshes have been carried out outside Europe, there are no reasons to suppose that European relationships are of a different nature. The most important vegetational difference between North American and European salt-marsh is the great importance of *Spartina* species in the former, while *Puccinellia* species and dwarf brushwood (especially Chenopodiaceae, *Artemisia*) are in a minority.

In Figure 35, a scheme is given which may characterize the trophic levels of the European salt-marsh ecosystem, and their inter-relationships. Similar schemes have been produced by Paviour-Smith (1956) in her food-web diagram and by Teal (1962) depicting the energy flow through a Georgia salt-marsh. Studies on trophic structure in relation to management for nature conservation are still lacking, but some trends may be outlined here:

(1) The importance of the detritus food-chain emphasizes the importance of contamination through organic materials, both those produced locally and those washed ashore.
(2) Grazing by large herbivores seems to exert a more indirect influence on the ecosystem, keeping biomass within bounds and maintaining diversity among other consumers and their predators, but not directly enhancing the rate of energy flow through the ecosystem to any marked extent.

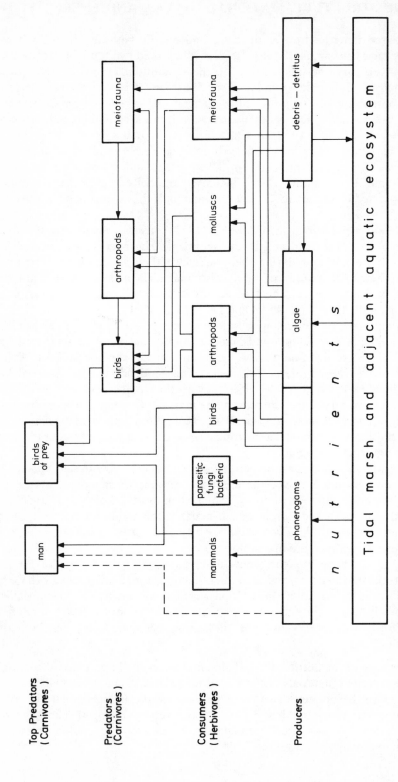

Figure 35. General scheme of the trophic structure of a salt-marsh. (The complicated role of microbial organisms as decomposers and as a source of food has not been included as their role, although probably of great importance, is imperfectly understood)

ZONATION AND DYNAMIC RELATIONSHIPS IN VEGETATION

The zonation pattern in vegetation from bare mud to the upper limit of salt-marsh has been presented by various authors in the form of diagrams (Chapman, 1960b). Although zonation can be the spatial expression of temporal succession, comparison of the situation in subsequent stages of the process without conclusive research over a period of time neglects the different ways in which diversity in the ecosystems actually increases. To give an example (Beeftink, 1965): the Halimionetum (an '-etum' is a community of plants characterized by the genus stated to which has been added the suffix '-etum', e.g. a Salicornietum is characterized by Salicornia) appeared to develop on various features of environmental relief (creek banks, basins) and in various developmental series and stages of vegetation (first on the creek banks and later on in the basins).

Long-term studies on the vegetation can be carried out in three ways: (1) by anticipating the appearance of unpredictable, but afterwards still detectable environmental changes (e.g. severe winters, dry summers); (2) by utilizing predictable environmental changes, mostly the result of planned technical interference (e.g. dredging, damming); and (3) by deliberate experiment. To illustrate dynamic changes in salt-marsh vegetation some

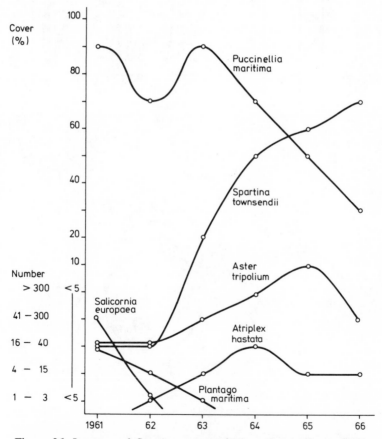

Figure 36. Increase of *Spartina townsendii* in a *Puccinellia maritima* community growing on a curved sand-spit in the Voorne dune system, as a consequence of rapidly increasing deposition of silt from dredging. (Estimation of cover–abundance according to Doing Kraft, 1954)

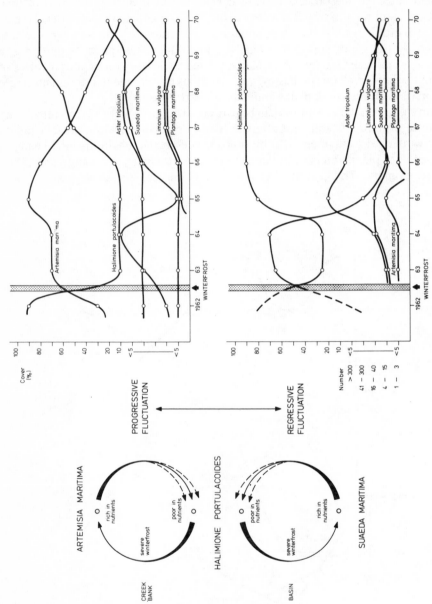

Figure 37. Progressively and regressively deviating fluctuations in *Halimione portulacoides* communities induced by severe winter frost in 1962–63 on the Springersgors, a salt-marsh on the south Coast of Goeree, Netherlands. (After Beeftink, 1973, reproduced by permission of Pudoc, Wageningen; estimation of cover-abundance according to Doing Kraft, 1954)

results are given of studies made in south-west Netherlands in connection with the so-called Delta Plan (Hydro Delft, 1971) for coastal protection in the region of the Rhine–Maas estuary (see also Beeftink, 1975b). Figure 36 shows the rapid increase of *Spartina townsendii* in four years at the cost of *Puccinellia maritima* on a marsh where the deposition of silt increased after 1963 as a result of dredging operations and dike building in the neighbourhood. *Atriplex hastata* and *Aster tripolium* could profit temporarily from organic substances released by dead *Puccinellia* material. (Hubbard (1968) distinguished between the sterile hybrid *S. x townsendii* and the fertile *S. anglica* derived from the primary hybrid by doubling of the chromosomes. Which species occurs at different localities along the European coasts is not certain, probably both are involved. Within the scope of this chapter, however, this question is not relevant as the début of both species interfered ecologically in the same way: they are referred to here under the single name, *S. townsendii*.)

In the case of Figure 37, anticipation has been remunerated by the severe winter frost of 1962–63 during which *Halimione portulacoides* was damaged locally. The temporary outbursts of *Suaeda maritima* and *Aster tripolium* in the *Halimione* communities growing in the basins, and that of *Artemisia maritima* in the creek-bank communities are examples of deviating (non-cyclic) fluctuations. The first phenomenon must be considered as regressively deviating, the second one as progressively deviating fluctuations.

ENVIRONMENTAL MASTER-FACTORS

Water regime

By its nature, the salt-marsh ecosystem involves both above-ground and underground water relations. The first ones are effected by the tides and by rainfall. The underground water relations are influenced by the above-ground ones and by drainage and seepage conditions.

Along coasts, tides vary from place to place in their period, amplitude and phase. Each of these properties will have an influence on the zonation pattern. According to Tutin (1942), the time at which low water of spring tides occurs, as well as the occurrence of double tides, influence the ultimate height to which *Zostera marina* can grow. Inversion phenomena in the zonation of the Puccinellietum and Halimionetum are probably linked to the magnitude of the tidal range (Beeftink, 1965). Telescope phenomena in the zonation pattern may perhaps also be ascribed to the extent of the tidal range, or to differences in height between the high-water levels of neap and spring tides (Beeftink, 1965). However, quantitative data from comparative research are scarce and incidental.

Compared with tidal influence, rainfall is of minor importance in the water regime, except where the latter wedges out into fully terrestrial areas, such as is the case in beach plains partly cut off from the sea. However, the temporary desalination induced in the topsoil layer seems to activate germination of many halophytes (Adriani, 1958; Waisel, 1972).

Soilwater relations are obviously of considerable importance, as many halophytes react strongly to groundwater and aeration conditions (Waisel, 1972). A distinction can be made between the groundwater in the subsoil of a marsh, and the moisture and air conditions in the upper soil-layers bearing the roots of the vegetation (Beeftink, 1965, 1966). In the first soilwater system, pressure conditions depend on external factors such as the tides, the drainage of adjacent agricultural land, and seepage from dunes and other land forms rising above the marsh's level. This pressure system upholds the second one in which the input of water floods and precipitation is more or less balanced by outflow via the creek system, the subsoil, and through evapotranspiration (cf. also Chapman, 1938, 1960b).

Salinity

The salinity of the soil is closely related to its water conditions. Since in most places along west European coasts annual precipitation exceeds annual evapotranspiration, submergence mostly results in an input of salts. Submergence will only have a leaching effect where soilwater salinity exceeds that of the flooding-water in summer.

At higher levels, the lower frequency of submergence does increase the influence of the mutually opposing effects of precipitation and evapotranspiration on salinity. These influences are maximal if the soil salinity is still relatively high; in other words, with respect to the tides there is a level above which these opposing effects on the fluctuation of the soil salinity decrease again because the average salinity level is too low. In south-west Netherlands, maximum fluctuations are found (15)25–35 cm above MHW (Beeftink, 1965, 1966). Under estuarine conditions, this zone shifts down to lower tidal levels

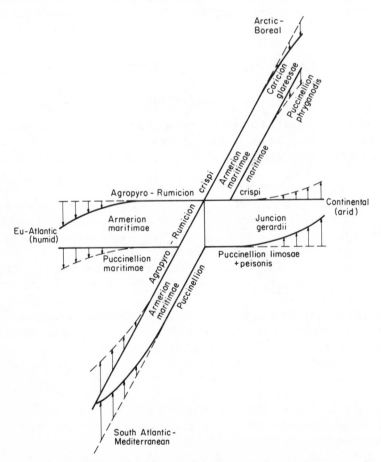

Figure 38. Shifting and telescope phenomena in the zonal patterns of European halophytic vegetation, with the higher salt-marsh communities (Armerion maritimae) in the central position (data from Beeftink, 1965, 1968)

upstream, at the same time dropping to lower values. Similar shifts are found in the zonation pattern of the vegetation (Gillham, 1957; Beeftink, 1965, 1966).

On a European scale, the vertical range over which saline soils extend depends mainly on climatic conditions. Here, too, zonal species combinations show telescope phenomena. A good illustration (Figure 38) is given by the distribution of the alliance Armerion maritimae (alliance is a grouping of similar associations of plant species and is indicated by the suffix '-ion' added to the scientific name of a characterizing genus), the associations of which develop optimally in the upper supralittoral zone of the North Sea region (Beeftink, 1968; 1972). From this region to the west coasts of Ireland and Scotland, the humid and temperate EuAtlantic climate enables more terrestrial species, and even such species as *Cirsium palustre*, *Molinia coerulea*, and *Iris pseudacorus*, to penetrate into the Amerion (Gimingham, 1964 and author's own observations). At the same time, the Armerion species *Armeria maritima* and *Glaux maritima* are well represented at lower levels (i.e. in the Puccinellietum) and may even behave almost as pioneer colonists. In the continental parts of Europe, the Armerion (here called Juncion gerardii) is again lost in more terrestrial communities (cf. Wendelberger, 1950). To the north, the cold Boreal–Arctic climate wedges out the Armerion, broadly in the same way as EuAtlantic climate conditions. Finally, the shifts in a southern direction resemble those as a response to continental conditions, at the same time leaving room (niches) for other groups of communities (Salicornion fruticosae, Limonietalia) which are mainly distributed in the Mediterranean (Beeftink, 1968, 1972).

Sedimentation and erosion

The salt-marsh ecosystem exists as a result of a continuous external supply of inorganic material. The process of maturation, however, is promoted if accretion proceeds slowly and gradually. The sedimentation rate can be considered low if it amounts to a maximum of 3 mm accretion per annum, dependent on the vegetation density (Richards, 1934; Ketner, 1972 from Hollmann, 1962). Rapid sedimentation, amounting to 10 cm and more annually, is injurious to species richness, especially to the algal component (cf. Ranwell, 1964a; Beeftink, 1965). A high accretion rate of heavy mud, as often occurs in *Spartina townsendii* fields, also seems to limit the possibilities of succession towards maturity.

During genesis of salt-marsh, sedimentation of clay and silt takes place especially in the basins, whilst the coarser particles are more and more deposited on the creek banks (see p. 95). The more sand that takes part in the accretion process, relative to the finer particles, the more pronounced the surface relief will be. A prevailing supply of clay and silt, on the other hand, has a levelling influence and decreases the environmental relief. Diversity within the ecosystem is, therefore, favoured when the fine portions in the total amount of the supplied material are low. As the higher accreted elements of the marsh are at the same time built of more easily desiccated and leached deposits, the contrasts with the lower ones will be reinforced.

Erosion occurs in different ways. Sheet erosion is more common on mud-flats than in salt-marshes. Normal scour and suction scour occur in both, although the latter is rare in salt-marsh. Creeks show lateral erosion in the side-walls of their outer bends, and headward erosion at the end of their stream bed. As erosion is normally local, it can be an important feature for the ecosystem. Sheet erosion leads to arrest of the vegetational development or even to regression. Scarp erosion at the edge of the salt-marshes—very common in many parts of the west European coasts—destroys the zonation pattern in the upper eulittoral part and prevents its natural development.

Mineralization and humification; transport of organic matter

Salt-marshes also seem to answer Odum's concept of a young ecosystem in the relation between mineralization and humification processes (Odum, 1971). Mineralization of organic matter probably predominates over humification. The reverse is only observed in the permanent wet to moist coastal dune-slacks situated under peripheral tidal influences. Although the organic content of the soil is higher in brackish-marshes than in salt-marshes, no data are available to decide how far this difference is related to the higher rate at which biomass is produced in the brackish-marshes (Beeftink *et al.* 1977). The intense and often hardly predictable transport of minerals and organic matter by the tides interferes with both mineralization and humification processes.

In the salt-marsh ecosystem, the introduction of *Spartina townsendii* resulted in considerably higher levels of organic matter occurring. Compared with times before its introduction, mass development of this halophyte contributes to both a higher amount of drift deposition and a higher quantity of disintegrated detritus, which only partly recedes with the ebb tide into the adjacent aquatic system. The relatively higher proportion of *Spartina* debris in the drift deposition probably also has a levelling influence on their composition.

TYPES OF PRESSURE ON SALT-MARSH AND THEIR EFFECTS

Pressure or stress can be defined as any burdensome force or influence acting upon or within a salt-marsh. This definition covers not only the different types of, e.g., pollution, but also stresses conditions inherent in the nature of the ecosystem itself. In the latter case, however, the term pressure should be reserved for stressful influences exceeding those inherent in the normal developmental stages through which salt-marshes are passing. Pressure, therefore, has to have a measurable detrimental effect upon the ecosystem.

Natural (non-human) stress is rare and most of it will, on second thoughts, be induced by man in an indirect way, e.g. catchment activities (Inglis and Kestner, 1958; Ranwell, 1972a). Under this restriction, a sudden increase in sedimentation or erosion rate, owing to displacements of tide-runs or increment of shingle-spits, may belong to this type of pressure. Another example may be an intensive eutrophication of the higher marshes as a consequence of masses of detached seaweed washed ashore or owing to bulk transport of plant materials from fluvial catchment areas. Floods caused by earthquakes and tropical hurricanes penetrating the hinterland must also be classified under 'natural stress'.

The types of pressure induced by man are manifold and varied. They can be classified into the following categories (cf. *Nature Conservation at the Coast*, 1969).

Agriculture

The most common use of salt-marsh is grazing: usually by sheep, but also by cows and horses, and even by domestic geese (in north-west Germany). On European salt-marshes, this practice seems to be most common along shores from Scotland and Denmark to north-west France, and less in south-west Europe. Over the past 25 years, however, grazing has been much reduced for economic reasons (cf. Tyler, 1969). Although mostly limited to the higher marshes and coastal dune-slacks, grazing may take up vast areas. Transitions to ungrazed areas may vary from sharp (creeks, fences) to gradual.

Grazing includes selective cutting and removal of plant parts, local manuring, and trampling of the vegetation. The latter also leads to compaction of the topsoil and, more intensified, to destruction of the turf-layer. Extensive grazing seems inherent in the nature of the salt-marsh ecosystem, promoting grass and rush species among other herbs and dwarf brushwood. Over-grazing causes regression phenomena; e.g. increase of *Salicornia europaea* and *S. perennis* in *Puccinellia maritima* stands, and settling of *Puccinellia maritima* in *Festuca rubra* and *Juncus gerardii* stands. *Halimione portulacoides* and *Limonium vulgare* cannot withstand intensive grazing. If over-grazed, there is a decrease in diversity of species and communities. The same, although probably to a lower degree, occurs if the vegetation is totally ungrazed (Westhoff, 1971).

Hay-making is practised very locally on the highest parts of brackish-marshes in estuaries and in Wadden areas, although it has also been reduced in the last decade. Although mowing is less selective with respect to the species affected than in extensive grazing, the soil will be less compacted. Mowing techniques may be a substitute for grazing.

The responses of salt-marsh species to nitrogen and phosphorus fertilizers differ, although the experimental results agree in an over-all increase in biomass on the higher marshes. Tyler (1967) found that a supply of 0·5 mol P as $H_2PO_4^-$ or 2·5 mol N as $NH_4^+ m^{-2}$ resulted in an increase both in the production and in the P or N contents of the shoots of *Plantago maritima* and *Juncus gerardii*. According to this author, this increase in production indicates that these elements are deficient in Baltic salt-marshes, at least for some of the more important species. Pigott (1969), working with *Salicornia* species, *Suaeda maritima*, and *Halimione portulacoides*, found similar results with the addition of nitrogen and phosphate fertilizers. He concluded, however, that the apparent nutrient-deficiency symptoms found in high-marsh plants might be due to a deficient water supply and mechanical restrictions to rooting in the soil, since high marshes may dry out considerably in summer. On the isle of Terschelling, Ketner (1972) noted a decrease in the production of biomass of salt-marsh communities with a decrease in flooding frequency. He also cited Alberda (1970)—see Ketner, 1972), who found that after fertilizing with N, P, and K the biomass of a *Festuca rubra* community amounted to about twice that of unfertilized vegetation of that kind. In the lower *Spartina townsendii* marshes, Ranwell (1964c) found a high level of essential nutrients in a marsh with a high accretion rate. These results agree with the hypothesis brought forward by Feekes (1936), who stated that the available nitrogen content acts as a limiting factor as soon as the flooding frequency decreases.

Gathering plants and animals for food is an old custom which is still regionally practised. In south-west Netherlands, for instance, cutting of young *Salicornia* and of leaves of *Aster tripolium* for cooking is popular, as is gathering of periwinkles (*Littorina littorea*). In north-west Germany, *Triglochin maritima* and *Plantago maritima* were once used for food. Damage caused seems to be slight.

Open-cast mining

Cutting and removal of turf for laying lawns and reinforcing sea-walls is the first step to open-cast mining in order to win natural resources from the soil. In England, turf-cutting for sale is practised. Careful control with fertilizers, selective weed killers and repeated mowing may be required to suppress species diversity (Gray, 1972). Ranwell (1972a) noted that 5 cm (2 in) strips are usually left between cuts to improve regeneration, and within five years the same areas may be cut again.

From Denmark to the Netherlands, turf-cutting is practised for covering and reinforcing sea-walls (Kamps, 1962). On the German North Sea coast, seed stocks of *Festuca rubra* and *Agrostis stolonifera* are also propagated for use on embankments (Wohlenberg, 1965). Repercussions of turf-cutting on salt-marsh vegetation are unknown. But when carried out incidentally and without extensive pretreatments, secondary succession towards natural development of the vegetation can proceed within the period of one decade, perhaps after a longer period. Ranwell (1972a) mentions experiments in progress to determine the botanical changes of cutting and regeneration. In south-west Netherlands, Beeftink and his collaborators have started experiments with trampling, mowing, spraying, turf-cutting digging and excavating.

Clay-winning for building sea-walls, etc. is a more drastic form of mining. In the Low Countries, large amounts of clay are sometimes needed, e.g. after flood disasters have broken sea-walls and attacked their embanked areas. Clay pits in the generally low-lying salt-marshes may be valuable habitats for waterfowl, but they also often provide favourable conditions for the mass establishment of *Spartina townsendii*. Moreover, the interference brought about by excavation on the natural geomorphological and vegetational patterns is generally disastrous for several decades, perhaps for more than a century, since it leaves the sites in total disorder.

Land-reclamation and improvement for agricultural purposes

Measures for reclaiming salt-marshes involve different techniques (cf. Beeftink, 1975a). Resulting from experience in England and on the Wadden coasts, methods were developed using a system for trapping silt and clay by means of sedimentation fields. These fields were built by a system of squares made of brushwood and earthen groynes, intersected by grips of ditches (Carey and Oliver, 1918; Jakobsen and Jensen, 1956; Kamps, 1962). This technique starts on low sand- and mud-flats of sheltered shores, reinforced by *Spartina* plantations (Figure 39). After the Second World War, this practice was almost abandoned owing to the increased cost of labour.

As vegetation may have a considerable effect on the rate of accretion, techniques were also developed for promoting vegetation by means of the sowing and planting of *Spartina townsendii*, *Salicornia stricta*, and *Puccinellia maritima* (Wohlenberg, 1938; 1939; 1965; Kamps, 1962). The first species, especially, appears to be successful in promoting accretion. However, whether it serves the purpose ultimately is still an open question: experience in south-west Netherlands indicates that the heavy clay of old *Spartina* fields may turn out to be difficult to cultivate.

Moreover, if they are not embanked there are indications that in the long run *Spartina townsendii* communities are incapable of fitting into the natural succession series of the salt-marsh vegetation, at least under euhaline and polyhaline conditions (Figure 40). In the mesohaline Bridgwater Bay, Ranwell (1961, 1964b,c) found that grazing pressure and accumulation of *Spartina* litter at the top of the *Spartina* marsh facilitate the invasion of higher marsh species. In the first case, *Spartina* is succeeded by *Puccinellia maritima*; in the case of drift accumulation, it is replaced by *Scirpus maritimus*, *Phragmites communis*, *Aster tripolium*, and *Puccinellia maritima*. Apparently, *Spartina townsendii* fits better into ecosystems under estuarine conditions than in those which are in a closer connection with the sea. However, the wholesale die-back phenomenon in the oldest *Spartina* fields in Southampton Water (Goodman *et al.* 1959, 1961; Goodman, 1960) may point to incompatibility under polyhaline conditions between the soil conditions developed by this *Spartina* species, and the salt-marsh species which 'ought' to have settled instead of *Spartina* considering the level to which the soil had been accreted with respect to the tides. Generally speaking, the recent genesis of *Spartina townsendii* seems to be in disfunction with the existing, internally equilibrated ecosystem (Figure 40).

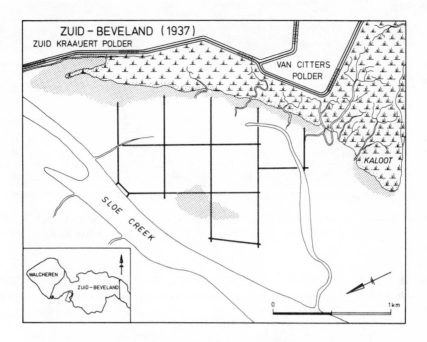

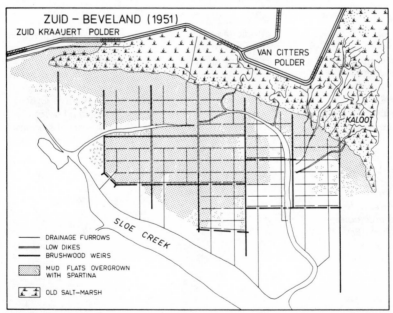

Figure 39. Settling-fields in the Zuid-Sloe east of Vlissingen, Netherlands, promoting sedimentation of silt and increase in *Spartina townsendii* from 1937 to 1951 (maps reproduced by permission of the Dienst der Domeinen)

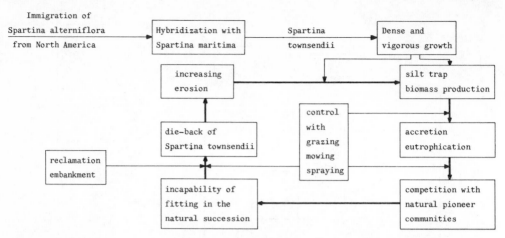

Figure 40. Disfunction of the introduction of *Spartina townsendii* into the salt-marsh ecosystem. (After Beeftink, 1975a; reproduced by permission of the British Ecological Society)

Table 10   Preference of halophytes for tidal and non-tidal saline habitats in south-west Netherlands

	Preference for:	No preference
Tidal habitats	Non-tidal habitats	
Exclusive		
Spartina maritima	*Ruppia maritima*	*Zostera marina*
Salicornia stricta	*Puccinellia fasciculata*	*Salicornia europaea agg*
Halimione portulacoides	*Hordeum marinum*	*Aster tripolium*
Artemisia maritima	*Hordeum jubatum*	*Puccinellia maritima*
Atriplex littoralis		*Triglochin maritima*
Armeria maritima		*Glaux maritima*
Carex extensa		*Juncus gerardii*
Scirpus rufus		*Festuca rubra*
Cochlearia anglica?		*Parapholis strigosa*
		Scirpus maritimus
Selective		
Zostera nana	*Salicornia brachystachya*	
Suaeda maritima	*Puccinellia retroflexa?*	
Spartina townsendii		
Cochlearia officinalis		
Althaea officinalis		
Alopecurus bulbosus		
Oenanthe lachenalii		
Preferent		
Spergularia marginata	*Puccinellia distans*	
Plantago maritima	*Spergularia salina*	
Juncus maritimus	*Halimione pedunculata*	
Sagina maritima	*Juncus maritimus*	
Elytrigia pungens		
Atriplex hastata		
Carex distans		
Ranunculus sardous		
Apium graveolens		

Drainage of marshes is another form of land-reclamation. It results in a lowering of the water-table and thus in changes of habitat, flora, and fauna. Traversing with ditches is harmful to the geomorphological development, for the natural watercourses may lose their function (Figure 41).

Embankment is one of the most serious impacts on salt-marshes (Beeftink, 1975a). Poldering includes isolation from the tides, drainage through sluices, drying and leaching of the soil, and agricultural exploitation. Studying the St. Lawrence *Spartina* marshes, Reed and Moisan (1971) came to estimate their ecological, recreational, educational, and

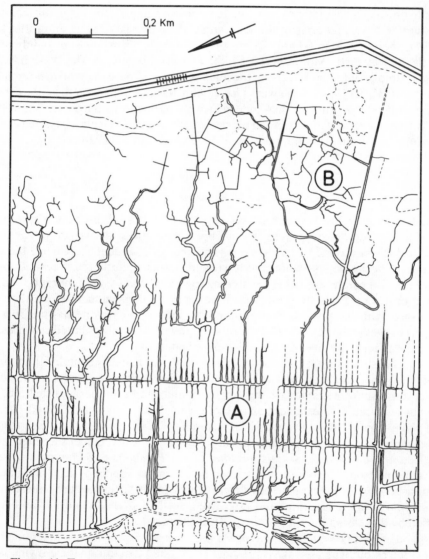

Figure 41. Two patterns of artificial creek systems in the Zuid-Sloe near Vlissingen. (A) Drainage furrows within settling-fields on the mud-flats (cf. Figure 39); (B) An unusual drainage-furrow system in a grazed salt-marsh (derived, by permission, from Map WS 68 A of the Survey Section of the Department of Roads and Waterways)

tourist values higher than the agricultural values for which they are currently being exploited. Halophytes can only survive in former bare depressions (creeks, bottom land, etc.), albeit in other species combinations than on the tidal marshes. Experience in south-western Netherlands shows that some halophyte species prefer tidal habitats but others non-tidal ones (Table 10). Therefore, in addition to tidal salt-marshes, well-managed saltings in the polderland contribute to the species diversity of the coastal habitat as a whole.

Large-scale impoundment of coastal waters

Recent hydro-technical methods have made impressive improvements in our abilities to dam estuaries, bays, and arms of the sea. An example of a large-scale enterprise still in progress is the Delta Plan (Hydro Delft, 1971). In England, schemes to build barrages across the Solway Firth, Morecambe Bay, the Dee estuary, and in The Wash have been proposed; storage of freshwater being the main purpose in these cases (*Water resources in England and Wales*, 1974). Such works have a much wider impact on the coastal habitat than embankment of salt-marshes for exclusively agricultural purposes. Some of the biological implications of the proposed barrages enclosing these English bays and estuaries are outlined by Gilson (1966), Gray (1970, 1972) and Barnes (1974).

Pollution

Especially in industrialized estuaries, bays, etc., salt-marshes are threatened by a complex array of man-made detritus, litter, and chemical substances, derived from agricultural wastes, fertilizer outwash, urban sewage, and industrial effluents. Among the latter, spillage of oil, heavy metals, radioactive waste and thermal pollution can be mentioned. As Ranwell (1972a) emphasized, it is frightening to realize that we are ignorant of both the nutritive or toxic effects of these substances and of their influences on the life of salt-marshes, with the possible exception of oil. Up to now no one has measured, for example, changes in the deathline for salt-marsh growth in heavily polluted estuaries. Only some single effects have been incidentally investigated.

In older industrial coastal centres, sediment profiles of marshes contain a high proportion of man-made detritus (McCrone, 1966). Urban sewage, agricultural wastes and fertilizer outwash from arable farming areas accumulate in estuaries and sheltered bays. These nutrients cause an extensive growth of green algae (*Enteromorpha* and *Ulva* spp) and of phanerogamic life in low marshes. Ranwell (1972a) reported that these masses of algae smother salt-marsh growth and appear to replace it locally. Under such high nutrient levels, *Spartina townsendii*, *Scirpus maritimus*, and *Phragmites communis* also show a heavy standing crop, which in turn results in an enormous supply of dead plant material washing ashore on the higher marshes during winter. In the Scheldt estuary, these bulks of drift material promote luxurious and extensive growth of *Elytrigia pungens* and *Atriplex* spp, thereby decreasing the species diversity of the marsh.

Wass and Wright (1969) noted that in Virginia *Spartina* species grow taller and have a darker green colour when sewage is present. They stated that during the growing season, salt-marsh is able to absorb some of the excess nutrients which otherwise would create algal blooms in the estuaries. Thus, as natural estuarine conditions already act as a nutrient trap creating a sort of 'self-enriching' system (Odum, 1971), sewage may induce 'over-enriched' conditions with collapse tendencies.

Thermal pollution may have similar effects. Anderson (1969) noted in Chesapeake Bay that *Spartina alterniflora* and *Phragmites communis* can both tolerate temperatures up to 35 °C. *Spartina* grew about twice as tall in the heated water.

Teal (1969) described the vulnerability of East American salt-marshes to pollution by heavy metals and D.D.T. residues, which accumulate in the sediments and are ingested by marsh invertebrates. Wass and Wright (1969) cite a study of the role of *Spartina alterniflora* in introducing iron, zinc, and manganese into estuarine food chains.

The effect of oil-spillage is most evident and has been well-documented by both close observation and experimental work in Great Britain. Salt-marsh and mangrove swamp act as traps for drifting oil (Nelson-Smith, 1972a). As such they are a valuable aid in purifying the adjacent aquatic ecosystem, and contribute towards mitigation of the damage of waterfowl and fish (Ranwell, 1968b; Stebbings, 1970). Using this fact as an extenuating circumstance to alleviate the serious crime of unjustifiable oil-spillage, however, is relying too heavily on the capacities of salt-marsh.

Heavy but isolated oil contaminations may be tolerated without long-term damage. Death of oiled shoots is a short-term effect; in the case of perennials, followed by new growth from the plant bases. During the recovery period, there may be reduced germination and flowering, a reduced population of animals, and even growth stimulation of some plant species (Baker, 1971). However, most vegetation is rapidly killed by the emulsifiers used to remobilize and disperse oil (O'Sullivan and Richardson, 1967; Ranwell, 1968b; Baker, 1971), and the use of these substances often spreads the damage to previously unaffected areas (Ranwell, 1968b).

Chronic oil pollution, even in low concentrations, will ultimately kill the vegetation and prevent recolonization. The oil content is mainly built up in the creeks, where the vegetation (if any is present) then consists of blue-green algae (Baker, 1971; König, 1968). Recovery probably occurs after the pollution has ceased but major changes in erosion and deposition patterns may have affected the marsh, and *Spartina townsendii*, if present in the neighbourhood, will probably be the first to recolonize large parts.

Outstanding contributions have been made to elucidating the effects of oil and emulsifiers on salt marshes and salt-marsh plants by Baker (1971, 1973). She demonstrated the greater capacity of deep-rooting perennials and *Puccinellia* turf to survive as compared with annuals, the promotion of anerobic soil conditions by the oil cover and a possibly indirect nutritive effect of oil on salt-marsh plants.

The effects of air pollution on the salt-marsh ecosystem are probably minimal, owing to tidal flooding. However, McCrone (1966) found that algae and silt can accumulate radioisotopes carried by air.

Recreation

Excepting beach plains and coastal dune-slacks, salt-marshes are not very attractive for recreation and tourism. Amateur inshore fishing, wildfowl shooting, and some aquatic sports generally make only a moderate use of these silty and muddy habitats. The sandy marshes, however, stand a good chance of being involved in other recreational activities, such as walking and rambling, picnicking, horseriding, camping and caravanning, etc. These activities can easily result in disturbance of wildfowl habitats, and in damage to the vegetation through trampling, followed by loss of the humic topsoil layer and erosion. Other threats are litter accumulation, fire, and trespass. Trampling kept within bounds with respect to space and intensity may, however, promote some species diversity, provide pressure does not cause any significant erosion.

Industrial and urban sites

The oldest forms of industry in salt-marshes are salt-making and oyster-culture. For salt-making by evaporation of seawater, embanked basins are cut in the marshes and filled through sluices. This very old technique is still employed from south Brittany to the coasts of Portugal. The middle and higher marshes are destroyed thoroughly by this industry, leaving only small dams between the basins. The same holds for oyster-culture sites established in salt-marsh areas.

The establishment of modern harbour-works, industrial and power plants, or urban sites in salt-marshes completely destroys the ecosystem and is often associated with further disturbance of the neighbouring coastal habitats by road and pipeline works, subsidiary buildings, air and water pollution, litter, increased public pressure, etc. These human activities generally cause large-scale biodegradation of the coasts (*Nature Conservation at the Coast*, 1969).

Education and field studies

Scientific field-work and organized activities for out-door natural history may lead to interference and disturbance, if not in wildfowl relations then at least in the vegetation. Collecting flowers of *Limonium vulgare* may ultimately damage its populations and can change the species composition of the vegetation.

Scientific work may also leave its mark upon the vegetation in the shape of footpaths, and barren and bare spots attracting rabbits. Penetrating impervious soil-layers in the sediment with a soil auger can change soil–water relations. Ecologists should always be aware of the possibility of describing 'changes-in-time' which they induced themselves!

USES OF THE COASTAL MARSHES

In enumerating the kinds of use of the coastal marsh ecosystem, all forms of evidently harmful use will remain outside our consideration. Use, here, is interpreted as utilization which includes preservation of the nature of the ecosystem as well as giving maximum protection. As a borderline case we will consider a long-term multiple-use policy for the purpose of preservation of the ecosystem's natural resources (Odum, 1971). From this starting point, the uses are considered in the fields of education and scientific need, of recreation and of economic utilization.

Education and scientific need

Coastal marsh is eminently suited to the education of the interested layman and amateur ecologist. Compared with other ecosystems, its vegetation is simply structured and the ecological requirements with their internal and external relations are easily surveyed. The dynamic aspects of salt-marsh are particularly suitable for the study of field natural history; for example, repeated vegetation mapping, birdwatching, detecting the historical development of marshes, making inventories of the degree of contamination, damage and erosion, etc. The large expanse of many marshes enables individual and organized youth activities to combine field studies with their need for conservation, rehabilitation schemes and roving spirit (*Nature Conservation at the Coast*, 1969).

Within the scientific scope, the tidal border ecosystem, coastal marshes included, is of basic importance for taxonomic, physiological, and evolutionary studies of halophytes, as well as for ecological and geobotanical research with emphasis on 'pattern and process' or 'biomass–energy-cycling' approaches. The linear extent of tidal marshes on a worldwide scale, as well as their situation between the marine and terrestrial environments, make this habitat an object of fundamental scientific interest.

In the field of the applied sciences, coastal marshes are of interest for varying reasons:

(1) For estimation of their production rate in order to determine the agricultural poten- tialities of natural ecosystems of 'young' (immature) nature (Odum, 1971), and to delimit their part in the productivity of the adjacent near-shore and estuarine environments.

(2) As gene-reservoirs for further hybridization and selection of new crops (e.g. *Festuca rubra, Elytrigia pungens, Beta maritima*), and of new ornamental flowers (*Limonium* spp, *Armeria* spp); for detection of useful plant substances; and as seed-stocks for salt-tolerant species suitable for use in salt-affected roadside environments (Ranwell *et al.*, 1973).

(3) As a buffering environment between the marine or estuarine and terrestrial zones— firstly for defence of the land against tidal and wave action, secondly for trapping flotsam, and marine or estuarine pollutants, some of which are actually destroyed or absorbed in this ecosystem, and thirdly for signalling the harmful effects of marine pollution on plants and animals (excess of nutrients, oil, chlorated hydrocarbons, heavy metals such as mercury, pesticides, and other factory effluents).

Recreation

Salt-marsh is not suited to mass recreation. The mostly muddy soil and the frequent difficulty of access allow only the enthusiast to neglect these objections. Only coastal dune-slacks and beach plains are comparatively attractive.

In tidal marshes, recreation may have the character of education (see above), of gathering food for private use, or simply of leisure. Gathering of 'samphire' (*Salicornia* spp), sea-aster (*Aster tripolium*), bivalves (*Mytilus edulis, Cardium edule, Ostrea* spp) and periwinkles (*Littorina littorea*), as well as inshore fishing, are spare-time practices on many European coasts. Walking and rambling, picnicking, ponyriding, birdwatching, boating, and shooting wildfowl can only be allowed within very limited bounds and under supervision.

Economic use

The economic significance of salt-marsh is very wide but not very great. Marshes are a source of turf, and of minerals (clay, salts, sand) trapped from the floods which will be resupplied after removal. They are also a source of plant materials, which can be utilized for agricultural purposes. A part of these materials may be transported, as detritus, towards the near-shore waters, contributing to their fertility. Indirectly, tidal marshes have an economic value as a habitat for the already mentioned educational, scientific, and recreational uses.

Commercial cutting of sea-washed turf for lawns may be a lucrative job (Ranwell, 1968a; Gray, 1972), but the excavation of clay on a large scale is considered a predatory short-term policy with respect to other uses.

Like other natural grassland, salt-marshes are best suited for grazing and hay-making. By far the most widespread use is open-range grazing for domestic animals. However, in the last decade, first hay-making and then grazing have ceased in many areas of Europe (cf. Tyler, 1969). At the same time, grazing pressure has been continously reduced as the number of cattle per unit area decreased. The costs of sheep and cattle breeding and of hay-making have become and are still becoming too high in many European countries to be economically attractive. Consequently, the area of salt-marsh pasture abandoned nowadays is gradually increasing.

Grazing intensity depends on the number of days in which the marshes are free from tidal cover. For eastern American *Spartina* and *Distichlis* marshes, Williams (1955, 1959) mentions one cow for every 2–4 acres (0·8–1·6 ha) during a six month grazing season. According to Ranwell (1972a), European saltings composed of *Agrostis stolonifera*, *Festuca rubra*, and *Puccinellia maritima* support two to three sheep to the acre (0·4 ha), grazing most of the year. Gray (1972) discusses the results of comparative case-studies of eleven farms in the Morecambe Bay region with salt-marsh grazing of sheep. In spite of the higher costs of labour and of loss of stock by drowning, etc., the quality and weight of lambs from the marshes are generally favourable. Stocking rates are calculated to average about 4·5 sheep ha^{-1} raising to 6·5 sheep ha^{-1} (lambs calculated as 0·5 ewe units) with an average annual loss of 80–100 grazing days owing to high tides (see Table 11).

The new vigorous *Spartina townsendii*, introduced by man for sea-defence and reclamation purposes, is one of the most widespread salt-marsh plants on the European coasts today, often colonizing vast coastal mud-flats. Especially in Britain, this has stimulated ecologists to evaluate its economic significance.

From the viewpoint of coastal protection and reclamation, *Spartina townsendii* has undoubtedly an economic value (Ranwell, 1967), be it a relative one. For sea-defence, die-back phenomenon in *Spartina* and the consequent erosion of the marshes (Goodman *et al.*, 1959; Hubbard, 1965), makes the use of this species only a short-term solution.

Although coastal *Spartina* marshes have a high production rate, it is difficult to assess to what extent economic utilization of this extensive and potentially valuable protein resource can be realized (Ranwell, 1967). The most obvious use in grazing by sheep and cattle—mainly practised in the British Isles, to a lesser extent on coasts of the European continent, and quite extensively on North American coasts. The soil weakness and the long periods of tidal submergence, however, are limitative for grazing, especially with cattle, and may easily induce diseases (foot-rot in sheep).

The results of harvesting for silage and fodder are encouraging, according to Ranwell (1967), but the weakness and topography of the soil has so far limited the extent to which *Spartina* marshes can be cropped. Ranwell (1967) noted that *Spartina* provides suitable fodder for the older breeds of beef-cattle and sheep, but not for some modern ones.

With inadequately controlled introduction of *Spartina townsendii*, however, problems arise in many salt-marshes with respect to their economic use. Often the quality for grazing is reduced: the increasing capacity of trapping mud becomes inadequate and drainage is impeded; and the increasing eutrophication with *Spartina* debris stimulates hard grasses (*Elytrigia pungens*) at the cost of soft ones (mainly *Festuca rubra*).

METHODS OF GAINING AN ECOLOGICAL BASIS TO MANAGEMENT

In the study of pattern and process in vegetation and environment, a distinction must be made between methods aiming to elucidate the difference in space between structural patterns and those endeavouring to understand structural changes in time. In the spatial approach, one has to chose suitable moments to obtain a picture of the spatial variation. With this method, characterized by the term 'mapping', developmental and regressing processes remain uncertain. On the other hand, the approach in time involves a selection of marked habitats limited in space. This type of method can be characterized by the term 'monitoring'. In a limited scale, both approaches can be united in iterative studies of selective areas by means of accurate descriptive techniques, such as transects across the shoreline (e.g. Shimwell, 1971).

Investigation of structural patterns in the distribution of vegetation in space can be carried out on different levels: firstly, the distribution of indicator species and communities over the marsh; and secondly, the distribution of basic species combinations, life forms and structures of vegetation. In all cases, remote-sensing and aerial photography (Grimes and Hubbard, 1969, 1972; Hubbard, 1971) at different scales are obvious means of mapping. These forms of vegetation mapping find their ultimate expression in studies of diversity patterns both within and between habitats (e.g. Whittaker, 1965). The concept of habitat can then be viewed in terms of environmental selection mechanisms within a marsh, as well as of a marsh as a whole compared with other marshes.

Mapping the variety in the vegetation should be carried out at different scales. Repeated mapping of large areas is more suited to long-term control, whilst for short-term management decisions, an annual study of vegetation samples by means of permanent quadrats and transects is required.

For both spatial research and the study of dynamics in vegetation, the descriptive technique of the Zürich–Montpellier School of Braun-Blanquet (1964) (see also Shimwell, 1971) can be recommended, especially the refined estimation scales of Doing Kraft (1954) and Barkman *et al.* (1964). With these latter codes, minor annual changes can be traced and more information is provided for a choice of management measures. Besides the traditional estimation of cover–abundance (supplemented with a sociability figure giving a more complete picture of the community structure), phenology, vitality, and fertility data may be helpful for the interpretation of tendencies in vegetation dynamics (Barkman *et al.*, 1964).

The most important environmental factors can mostly be mapped or diagrammatized in a similar way to the vegetation units. The relation of species and communities to the tides (frequency of flooding, duration of inundation) can be determined by tide-gauges, combined with surveying of the soil-surface. In lonely areas, stakes planted into the mud with a row of cups at 2 cm intervals above one another to catch the floodwater, may be helpful for comparative measurements.

Groundwater relations can be measured by means of plastic tubes with perforated lower ends at different depths. To prevent ingress of soil particles, the perforated ends are covered with jute or nylon stockings. These tubes are also well suited to groundwater sampling. Salinity is an excellent tracer of groundwater movements, vertical as well as horizontal, in relation to flooding, precipitation, drainage, and seepage.

Soil samples for the analysis of physical properties (silt, water, humus, carbonate) as well as chemical ones (salinity, pH, nitrogen, phosphate, etc.) must be spaced systematically, in quadrats other than those marked out for vegetation analysis, because sampling generally disturbs the vegetation too much.

Without disregarding the quality of other ecological work, the publications of the following authors give much information on the organization and methods of ecological investigation in salt-marshes: Chapman (1934, 1938–1959), Gillham (1957), Gillner (1960), Ranwell (1961, 1964a, 1964b), Ranwell *et al.* (1964), Tyler (1967, 1968, 1969, 1971a, 1971b), and Dalby (1970). Investigations of the dynamics of salt-marshes are still rather scarce but have been carried out by Christiansen (1937), Feekes (1936, 1943), Voderberg and Fröde (1967), Westhoff (1969), and Beeftink *et al.* (1971).

The strategy for investigating salt-marshes for management purposes differs according to the situation at hand. The simplest approach is to follow the impact upon the vegetation of environmental changes in, e.g., grazing intensity, recreation pressure, degree of pollution, withdrawal of fresh groundwater from adjacent resources, accretion rate caused by dredging, etc. In most cases, however, it will be very difficult to counteract harmful effects, once they are detected. Therefore the information thus gained must generally be applied to similar threats later on and elsewhere.

Besides this approach, a more active management policy can be mapped out, first isolating the major impending management problems in the region or country, and then setting up experiments to investigate their impact on the ecosystem. Simulation techniques for field and laboratory experiments should be developed. The work of Baker (1971, 1973) on the effect of oil and its emulsifiers on the vegetation serves as a combined example of these approaches.

Much bigger problems arise if large coastal areas are reclaimed by embankment, or will be impounded behind barrages. These engineering works will interfere with salt-marshes to such an extent that the original ecosystem can never survive and quite a different one will develop. In such a situation, it is of paramount importance to decide which policy will be pursued with respect to the area. For nature conservation, the choice of that policy will depend on different considerations of which are the most important: (1) the need for conservation of species and communities, for instance rare species combinations of plants or rare breeding birds; (2) the ecological possibilities of the area with respect to its geomorphology, present and potential conditions of soil and climate, and internal stability; and (3) the possibilities of managing the area in view of the environmental conditions chosen. In this respect, it may be worthwhile to develop efficient techniques to construct new arrangements of topographical features adapted to natural situations.

CRITERIA FOR AN ECOLOGICAL EVALUATION

Human influence on the salt-marsh environment is rapidly changing whilst still increasing. The ecologist and the physiographer are burdened with a severe responsibility, as it is their task to advise the authorities on policies of planning and deciding future destinations. For that purpose, they will have to gather a good deal of data pertaining to the situations at hand, as well as to evaluate criteria for the assessment of scientific value in the fields of botany, zoology, geomorphology and geology. These include the following (cf. *Nature Conservation at the Coast*, 1969).

Extent

The size of the area can be important. Usually, the larger the area, the more likely it is to function as a self-perpetuating and self-supporting unit, and the more valuable it should be from the conservational and the scientific point of view. The critical size, however, is difficult to assess; the actual situation of the area with respect to, e.g., future shifting and magnitude of tide-runs is of paramount importance.

Diversity

One of the most important criteria is the diversity of species and communities. Diversity is related to the numbers of species and communities per unit area. Comparisons should be made in a limited region, preferably within a geographical floral element. Large numbers of species and communities and high diversities are usually related to large habitat diversity. Salt-marshes in contact with quite different environments (dunes, shingle beaches, bogs, acid substrates, etc.) will have high diversity values. In the botanical field, the concept of γ-diversity of Whittaker (1965)—the species diversity of a large community complex based on α-diversity of the component communities and the β-diversity of environmental types—is recommended. The species richness of an area as a whole can be measured by comparing its species number with the average number for an area of that size, broadly in the same way as introduced by Adriani and van der Maarel (1968) for the dunes of Voorne (cf. also van der Maarel, 1971). As the species richness of salt-marshes is rather low compared with other ecosystems, the rarities in species, communities and subhabitats must be proportionally highly estimated.

Rarity

The presence of rare species and communities gives the marshes an 'excess' value depending on the number and degree of the rarities. Rare species and communities can be expected in the contact zones with other geological formations, or simply in habitats with exceptional combinations of salinity with other environmental factors, such as are realized in stable contacts of tidal marsh soils with other types of substrate. The occurrence of rare species or communities must be judged both on the national and on the international scale (including feeding grounds for migratory birds). Within this field, the location of coastal marshes with respect to boundary zones of coastal climates or of geographical elements of the coastal flora is of great importance (e.g. Beeftink, 1965; Ranwell, 1968a). In these zones, the areal distribution limits of species are not only concentrated, but they have a relatively high proportion of endemics. According to van der Maarel (1971), rarity can be determined from a frequency class division in which both the number of inventory squares (e.g. Perring and Walters, 1962; van Rompaey and Delvosalle, 1972) and the average number of individuals or 'plant units' are taken into account. In countries where a square-grid floristic investigation has not yet been carried out, one has to content oneself for the present with a much rougher estimate; e.g. data from floras or estimates of the habitat rarity.

Naturalness

Human impact may have damaged the natural topography of the salt-marsh as well as its soil profile and prevented their natural development. Habitats which have been only slightly modified are often of higher quality than those which have been considerably modified. On the other hand, however, a certain period of stability afterwards does reincrease the ecological value of such young ecosystems.

Replacement

The degree to which a salt-marsh could be replaced in the event of its destruction or occupation for industrial or civil–technical purposes is a rather theoretical criterion for the time being. In low countries such as the Netherlands, salt-marshes could theoretically and technically be rather easily replaced by abandoning embanked agricultural land, as naturally achieved by flood disasters in former ages. Other possibilities include creating near-shore catchment areas with breakwaters and thus promoting sedimentation of sand and silt. In several cases, however, such nature engineering will not be possible owing to lack of space, or may be undesirable on account of unfortunate ecological side-effects in the surrounding areas.

There is urgent need for an internationally established databank for species lists, community descriptions (*relevés*, *Aufnahmen*), habitat data, etc. to gain a better founded evaluation of coastal habitats on a European scale. For such a work, subdivision of coasts into a grid of squares or into coastal units of a fixed running distance will be of basic importance.

MANAGEMENT FOR CONSERVATION OF SALT-MARSH BIOTA

As is demonstrated above, it is of outstanding importance to distinguish between the external and the internal aspects of management activities. With respect to external human influences, management should mainly have the character of protection, whilst internal human influences principally require regulation measures.

Management as a protection against external human influences

Pollution

External human influences emanate from the sea as well as from the land, and air pollution must also be taken into account. Direct protection of tidal marshes against pollution from the sea seems to be practically impossible, as an open relationship with the tides is a prerequisite for the survival of the marsh ecosystem. The only possibility left is to strive for purification of the estuarine and coastal waters themselves. The same holds for pollution from the air.

Indirectly, the influence of excess nutrients supplied by the tides may be partly counteracted by a higher grazing pressure, although danger that the highest parts of the marshes will be over-grazed exists. Occasionally oiled salt-marshes are best left to recover naturally. Baker (1973) found recently in field experiments that recovery from up to four oilings was generally good, but considerable damage resulted from eight to twelve oilings.

Pollution from the landward side into the marsh is usually more local and therefore ought to be more easily stemmed: the mostly local input of rubbish dumps, drains for sewage, agricultural wastes, chemicals and factory effluents might be prevented by more concentrated storage and, in the case of fluid pollution, by purification before discharge. However, special attention must be paid to the 'shock biotope' of small freshwater streams and trickles discharging into the littoral zone. These biotopes are inhabited by a very characteristic fauna and flora. The South African composite *Cotula coronopifolia* has invaded these biotopes along the west and south-west European coasts.

Introduced species

Another powerful human influence from outside has been the deliberate introduction of *Spartina townsendii* to several salt-marshes, and its subsequent uncontrolled spread (Ranwell, 1967). As Ranwell (1972a) has pointed out, the rationale behind many of these introductions is questionable. Many attempts have been made to eradicate it locally, especially in Great Britain, but most of them have failed owing to the high reproductive potential of this species. Ranwell (1972a) mentioned new efforts to determine the population level at which effective control can be achieved at reasonable cost.

Regulation measures in salt-marsh management

The internal influences in vogue vary from very mild forms such as gathering salt-marsh plants and wildfowling—or even no interference at all—to profound impacts on the vegetation and its physical and chemical environment, such as reclamation schemes and salt panning. Although there are many traces of past human activity in nearly all salt-marshes, most of them will be absorbed afterwards into the ecosystem. Here, only those management measures are considered which keep diversity up to the existing level or increase it to an acceptable value for nature conservation.

No interference

Sound management of whatever type of ecosystem needs zero levels to measure the effects of human influence. No interference at all can only be effective in salt-marshes which were abandoned a long time ago and on which pollution, including the excess eutrophication caused by *Spartina townsendii*, is minimal. In these cases, the chance that diversity will be lowered to a level under that of, e.g., extensive grazing, is minimal. In the case of excess nutrients, the more eutrophicated the coastal marsh is, the more regulating measures are required.

Grazing

As salt-marshes are for the most part natural grassland, grazing will approach most closely the natural circumstances prevailing before human occupation, in which differentiation of the vegetation was induced by the natural adaptations between herbivores (mammals, wildfowl) and herbage. According to this viewpoint, there are two aspects of outstanding importance in assessing a grazing policy for nature conservation: (1) open-range grazing and (2) a moderate grazing pressure linked with a prolonged stay of the flocks or herds.

Open-range grazing is preferable to shifting grazing within fenced areas. Fenced regions will be totally cropped in a short space of time, and tillering grasses, such as *Puccinellia maritima*, *Festuca rubra*, and *Agrostis stolonifera*, will be stimulated at the cost of other salt-marsh species. Shifting grazing also easily disturbs the breeding of wildfowl. Open-range grazing can easily be combined with long-term moderate grazing pressure, resulting in the development of gradients in the degree of grazing influences (selective removal of the overground parts, trampling, manuring). For that purpose, stocking densities have to

Table 11 Recommended grazing pressures (animals per hectare) in salt-marsh

References	Domestic animals	Salt-marsh	Brackish-marsh	Beach plains
Williams (1955, 1959) (E. American salt marshes)	Cows	0·62–1·25		
Gray (1972) (Morecambe Bay)	Sheep	Average c 4·5, maximum 6·5		
Ranwell (1972a) (European saltings)	Sheep	5·0–7·5		
Beeftink and Daane (unpublished) (SW. Netherlands)	Sheep	2	3	0·5–1·0
	Cattle	0·33	0·5	0·25

be chosen to accord with conservation of the ecosystem, rather than with the short-term production of domestic animals. In Table 11, a comparison of the numbers for such production with those given by Beeftink and Daane (unpublished) shows that for conservation purposes about half the density for domestic purposes is required. The Dutch numbers exclude the use of soil fertilizers and the addition of fodder, and suppose a grazing period as long as possible.

When conservation is intended, three factors tend to make grazing pressure too heavy:

(1) The biomass produced yearly is maximal in the lower marshes, and here more grazing days are lost owing to tidal cover. In other words, the higher marshes tend to be over-grazed and the lower to be under-grazed.
(2) Sandy salt-marshes generally produce less food for domestic animals than clay ones; at the same time, sandy marshes are frequently an important food resource for rabbits, but a precarious one owing to myxomatosis.
(3) During dry summers, the increased soil salinity suppresses food production considerably.

Therefore, grazing tends to reinforce differences in standing crop between higher and lower marshes, and between clay and sand saltings.

Cropping and spraying

Mowing of grass-rich marshes can be an important addition to management by grazing. Besides reed (*Phragmites*) and rush (*Scirpus lacustris*) cutting which could be and often is lucrative, hay-making in brackish-marshes and beach plains seems still to be just possible at reasonable cost. There are indications that mowing without grazing afterwards will impoverish the diversity of the vegetation. Extensive grazing afterwards might alleviate the threat of environmental equalization evoked by mowing, adding diversity through selective grazing, cattle tracks and accumulations of manure. The hay is usable as additional fodder for cattle and sheep.

On the north-east coast of North America, *Spartina* species are harvested and are under investigation for use as hay and pasture (Ranwell, 1972a). In England, Hubbard and Ranwell (1966) demonstrated that it was possible to cut *Spartina townsendii* marsh in dry weather, using a light tractor, and to make palatable and digestable silage for sheep. The reaction of the *Spartina* to cutting for several years, however, is still uncertain, and, in addition, it is problematical whether it will be technically possible to cut every year. As management measures for conservation purposes, both cutting and grazing seem to be the only acceptable measures for forcing back the introduced *Spartina* species. The use of herbicides like Dalapon (sodium salt of 2,2-dichloropropionic acid) is not very effective in marshes washed daily by the tides. More effective herbicides, such as Fenuron (3-phenyl-1,1-dimethylurea), may be harmful owing to side-effects on invertebrates (Ranwell and Downing, 1960). Another risk is that spraying on a wide scale may promote erosion of large areas (Ranwell, 1972a). Experience in south-west Netherlands has taught that on plots where *Spartina* was killed by Gramoxone [paraquat(dimethyldipyridylium)-dichloride], resettling of new individuals of both *Spartina maritima* and *S. townsendii* was still reluctant after eight years. Spraying, therefore, seems only acceptable as an ultimate expedient in saving the last northern *Spartina maritima* sites and possibly other rare communities.

Turf cutting and excavating

Removal of the upper parts of the soil-profile is a means of converting vegetation in an advanced stage of succession to younger stages. From the viewpoint that all salt-marsh communities should be conserved, turf cutting and even excavating clay can therefore be an obvious way of achieving this. Ranwell (1972a) stated that experiments are in progress to reverse relatively unpalatable *Festuca rubra* marsh into *Puccinellia* marsh, which is more palatable for wildfowl.

For application on a wider scale, two conditions seem to be important.

(1) The impact of turf cutting, and still more of excavating, must be adapted as far as possible to the topography of the marsh, or be carried out in accordance with a new arrangement of topographical features of the environment—contrary to the usual way of leaving it in a disturbed and derelict state.
(2) The area must not be threatened by pollution, nor infected with *Spartina townsendii*.

The first condition safeguards maximal re-establishment of a renewed natural vegetation; the second one will prevent vigorous growth of *Spartina townsendii* on the bare mud. Under suitable conditions excavations attract masses of waterfowl.

Drainage

Whilst turf cutting and excavation rejuvenate salt-marsh vegetation, drainage induces a forced succession. This effect is mostly not beneficial to the natural diversity pattern, as ditching usually reduces spatial variation and introduces changes in both the pattern of flooding and drainage, and in the process of sedimentation. For conservation purposes, drainage techniques may therefore only be helpful in promoting spatial diversity in monotonous salt-marshes, such as *Spartina townsendii* fields. In fact, this example of a management technique—as with the others—should be applied with great care, and accurate preparatory study of the topography and the existing drainage pattern of the marsh is essential.

CONCLUSION

As van der Maarel (1971) has stated for the dune grasslands of Voorne (Netherlands), the natural and semi-natural salt-marsh vegetation should be managed in such a manner as to leave their standing biomass on a relatively low level and their stability on a high level. For conservation, grazing seems to be the principal form of management, although moderate cropping and no interference at all are useful. Turf cutting, excavation and drainage may only be means of introducing some differentiation into the monotonous character of some marshes, either by 'rejuvenating' or by 'ageing'.

The experience that nature conservation is in fact not so much a question of ecological knowledge of nature, as of ethics, notion and co-operation in the world of man, gives the ecologist's work an evident social dimension.

ACKNOWLEDGEMENTS

The author is much indebted to Drs K. F. Vaas and R. S. K. Barnes for their valuable help and advice in formulating the text in English.

Chapter 7

Estuaries

A. NELSON-SMITH

BASIC PHYSICAL AND ECOLOGICAL FEATURES

It is difficult if not impossible to devise a precise definition of an estuary which would satisfy workers in all of the wide variety of disciplines concerned with such regions. Essential elements in such a definition have been discussed by many of the contributors to an invaluable and exhaustive volume edited by Lauff (1967). The loose picture of an estuary which emerges is the more or less enclosed region at the mouth of a river where freshwater from land drainage mixes to a greater or lesser extent with saline water from a tidal sea. What makes estuaries a unique habitat is the continually changing cycle of salinity, water level, and—to a lesser extent—such characteristics as temperature, within a region typified by shelter, abundant soft sediments, and a constant supply of detritus or dissolved nutrients (Barnes, 1974). Freshened non-tidal inlets and coastal lagoons, in contrast, tend to acquire stable values or gradients which change only slowly; they are also excluded on etymological grounds (*aestus* = tide).

Typical estuaries do not always form where a freshwater flow reaches the sea; on a mountainous coast, saline water may barely penetrate a river flowing fast down a steep bed while, at the other extreme, the marshy creeks fringing a low, flat coast may receive such sluggish flows that they are fully saline almost to their head. The limits of an estuary are often defined for administrative purposes by the end of tidal influence at its head and the line of the open coast at its mouth, but this has little practical use. At the head, the tide may be moving freshwater only—for example, the Elbe is brackish only in the lower quarter of its tidal reaches (Figure 42)—while the sea may be appreciably diluted by the flow of a major river for many miles offshore.

Pritchard (1967) has classified estuaries, as understood here, into four basic types. The classical form is the ria or drowned river valley which he calls a 'coastal plain estuary', and to which the major part of the discussion which follows will refer. The fjord is a product of glaciation; in Norway it usually has a shallow rocky sill and behaves as an estuary to sill depth, although the inner waters below this depth often remain stagnant for most of the year. On the Pacific coast of North America, however, the entrance may also be deep. Bar-built estuaries are shallow and are often fed by more than one river; water movement may not be very vigorous, so that wind-mixing becomes important. The fourth group, of estuaries formed by tectonic processes, is a miscellaneous one and includes indentations produced in the coastline by faulting, subsidence and so on. Various schemes have also

123

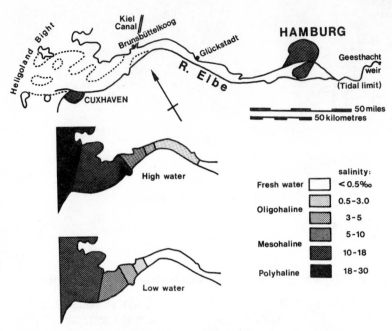

Figure 42. The Elbe estuary, showing the salinity in its lower reaches at low
and high water (partly after Caspers, 1959, see also Kühl, 1972)

been proposed to classify sections of an estuary according to their salinity; they have been
reviewed by Segerstråle (1959) amongst others. The simplest and most recent defines
freshwater as having a salinity less than 0·5‰ (parts per thousand); between 0·5‰ and
5·0‰ the habitat is oligohaline, at 5–18‰ it is mesohaline, at 18–30‰ polyhaline, and over
30‰ (a reasonable lower limit for coastal sea-water) euhaline. In considering the ecology
of an estuary, however, topography and other local factors are so important that schemes
of subdivision must be devised individually, as for Milford Haven (Figure 43), if they are to
have any practical significance. Table 12 lists a selection of such studies.

In the typical coastal plain estuary, the mixing of fresh and salt water is not complete, the
less dense freshwater tending to float so that there is a gradient in salinity from surface to
bottom as well as from head to mouth; the estuary is said to be stratified. Disregarding the

Table 12   A selection of some of the more general studies of estuaries in north-western Europe.
(Many of the papers listed make extensive reference to more specialized work.)

Estuary	Scope of study	Reference
Bantry Bay (W Ireland)	B	Crapp, 1973
Blackwater (SE England)	B H	Coughlan, 1966; Hawes (undated)
Burry Inlet (S Wales)	B H P S	Nelson-Smith and Bridges, 1976
Clyde (SW Scotland)	P S	Porter, 1973
Dee (N Wales/NW England)	B	Stopford, 1951; Buxton, Gillham and Pugh Thomas, 1976
Dee (E Scotland)	B	Milne, 1940a

Estuary	Scope of study	Reference
Dovey (Dyfi) (W Wales)	B H	Hayes and Dobson, 1969
Elbe (NW Germany)	B H	Kühl and Mann, 1962; Riemann, 1966; Kühl, 1972
Exe (SW England)	B	Allen and Todd, 1902; Holme, 1949; Gillham, 1957
Forth, Firth of (SE Scotland)	B H P	Rattray, 1886; Royal Society of Edinburgh, 1972
Galician Rías (Camariñas, el Barquero, Vigo) (NW Spain)	B	Ardré *et al.*, 1958; Fischer-Piette and Seoane-Camba, 1962, 1963
Hardangerfjord (W Norway)	B H	Saelan, 1962; Jorde and Klavestad, 1963; Brattegard, 1966
Loch Linnhe/Loch Eil (SW Scotland)	B P	Pearson, 1970, 1972
Mersey (NW England)	B H P	Fraser, 1932; Bassindale, 1938; Corlett, 1948; Bowden and Sharaf el Din, 1966; Porter, 1973
Milford Haven (SW Wales)	B H P	Nelson-Smith, 1965, 1967; various authors in Cowell, 1971; Gabriel, Dias and Nelson-Smith, 1975; De Turville, 1975
La Rance (NW France)	B	Fischer-Piette, 1931
Severn Estuary/Bristol Channel (SW England/S Wales)	B H P	Bassindale, 1941, 1943a, 1943b; Haderlie and Clark, 1959; National Parks Commission, 1967; Institute for Marine Environmental Research, 1971–75; Natural Environment Research Council, 1972; Boyden and Little, 1973; Welsh Office, 1974; Wigham, 1976
Solway Firth (SW Scotland/NW England	B H P	Perkins and Williams, 1963, 1964; Perkins, Bailey and Williams, 1964; Williams, Perkins and Hinde, 1965; Williams, Perkins and Gorman, 1965; Perkins, 1972; Perkins, 1976
Southampton Water/Solent (S England)	B P S	Raymont, 1972; Dartington Amenity Research Trust, 1973; Barnes, 1973; Barnes, Coughlan and Holmes, 1973
Tamar (SW England)	B H	Percival, 1929; Hartley and Spooner, 1938; Milne, 1938, 1940b; Spooner and Moore, 1940
Tay (E Scotland)	B H P	Buller, McManus and Williams, 1971; Royal Society of Edinburgh, 1972
Tees (NE England)	B H P	Alexander, Southgate and Bassindale, 1936; Bassindale, 1943b; Porter, 1973
Thames (SE England)	H P	Inglish and Allen, 1957; Gameson and Barrett, 1958; Ministry of Housing and Local Government, 1964; Barrett, 1972
Tyne (NE England)	H	Fisher Cassie *et al.*, 1962
Ythan (E Scotland)	B H	Leach, 1971; Milne and Dunnet, 1972; Chambers and Milne, 1975a, 1975b

B = biology; H = hydrography and physiography; P = pollution; S = socio-economic studies

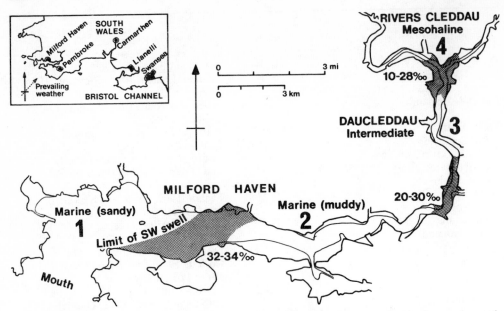

Figure 43. Ecological divisions of Milford Haven on the basis of topography, hydrography, and the distribution of sedentary marine species; shaded areas indicate critical regions forming boundaries between zones (from Nelson-Smith, 1972a)

reciprocating movement of the tides, a net circulation is set up in which the movement of freshwater seawards at the surface is to some extent balanced by a movement of saltwater upstream along the bottom (Figure 44). This intrusion is often called the 'salt-wedge', hence the alternative name 'salt-wedge estuary' for one of this type. In the northern hemisphere, the Coriolis effect due to the earth's rotation causes the wedge to tilt so that surface water flows out towards the right-hand side while the inflowing saline water moves towards the left. The salt-wedge is best defined when there is a relatively large river flow

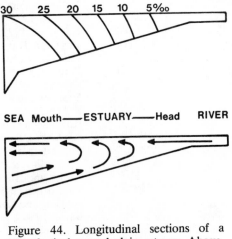

Figure 44. Longitudinal sections of a hypothetical coastal-plain estuary. Above, isohalines showing stratification; Below, simplified pattern of net circulation showing intrusion of the salt-wedge

and the bed is straight, deep, and narrow. With a greater tidal range and a wide or shallow channel, stratification is less apparent although, of course, changes in the degree of stratification can also occur within the same estuary following seasonal changes in river flow.

Shores around north-western Europe experience semidiurnal tides (high tide occurs every 12·5 h) whose rise and fall typically follow a symmetrical sine curve, passing through a fortnightly cycle in which spring tides (reaching greater extreme levels above and below mean water level) alternate with the lesser neap tides. Estuarine tides may deviate from this pattern. Where the channel is deep and the tidal range large (for example, in the Severn Estuary) more water may enter on a spring-tide flood than can move down again on the ebb so that, although high-water levels are considerably greater, low water reaches slightly lower levels during neap tides. The Severn Estuary is, incidentally, also noted for its bore, a related tidal phenomenon well explained by Bassindale (1943a). Where there is more than one connection with the sea, there may be double tides—for example, in Southampton Water, whose mouth is divided by the Isle of Wight. The tidal wave passing up the English Channel enters first through The Solent, giving rise to the main high water, but about two hours later a secondary surge passes up Spithead. Between these times, the water level remains more or less static, but then falls particularly rapidly. Whatever its detailed pattern, tidal rise and fall results in a 'piston effect' as a block of water corresponding to the tidal volume moves up and down. Obviously, much of the water which ebbs from the upper reaches will return on the next flood unless the estuary virtually dries out at low water, so that the resistance time of a particular sample of water may be lengthy—as much as 150 days in the Severn Estuary, for example (see Institute for Marine Environmental Research, 1971–75). The net movement of any solid object or diffuse material in such a sample will depend on its buoyancy or density; an object or immiscible substance which floats may escape into the open sea quite quickly although, if it is dense enough to sink to the bottom, it might be carried a considerable distance up-channel by the intruding salt-wedge.

There is still disagreement about the main origin of sediments in estuaries—see, for example, Guilcher (1967). It might seem obvious that most of this material must have been brought down by river but, in the main, this is true only where a delta is formed, as for example at the mouth of the Rhône. Sediments in many of the large estuaries of north-western Europe (such as those of the Elbe, Weser, Rhine, and Schelde) are said to be derived entirely from the sea, while the Seine receives about three-quarters from this source and the Thames a smaller proportion. The load of suspended materials and the newly deposited mudbanks in some smaller estuaries arise from the reworking of surface sediments in the estuary itself. Turbidity depends not only on the amount of sediment carried into the estuary but also on the availability of energy to keep it in suspension; the Severn estuary, for example, is very turbid because of the great range of its tides. At the head of an estuary, the rate at which suspended matter settles depends largely on river flow; the nature of the intertidal banks tends to be that of the lower reaches of the river, often a fairly firm gravelly mud. Around the mouth, shores are exposed to the action of incoming waves, so they are usually of clean sand, shingle or rocks, resembling those of the open coast to each side. On passing into shelter up the estuary, the settled sediments become finer just as in sheltered coastal inlets (see Figure 43). However, at a particular region in the upper reaches, there are usually large banks of a particularly soft, glutinous mud whose settlement is due not to sheltered conditions or diminished water velocity but to flocculation of colloidal particles, whose negative charge is neutralized by the positively charged ions of the intruding seawater.

Turbidity due to suspended sediments is of ecological importance because it limits the penetration of light required by plants for photosynthesis; the larger seaweeds may be restricted to a depth as little as 0·5 or 1 m below low water, although in clear seas they extend 60 m or so below this level. Pathogenic bacteria discharged in untreated sewage are rapidly killed by sunlight, while some toxic chemicals are absorbed and thus concentrated on suspended particles, so their typically high turbidity also makes estuarine waters less suitable to receive such wastes. A high rate of deposition or erosion may make it difficult or impossible for animals to colonize sand- or mudbanks, while their nature (especially the particle size) will determine the abundance and variety of this fauna. Intertidally, few animals live on the surface of soft sediments. Those which move about at high tide dig in or retire to a prepared hiding-place when the water leaves the banks, while the majority occupy some sort of burrow throughout their adult lives. With the exception of coarser sediments at the top of the shore, sands and muds retain water, so there is no danger of desiccation. This retained water remains fairly static except where local seepage from the land occurs; thus salinity and temperature are far more constant a few centimetres below the surface and the distribution of the infauna, both down the shore and up the estuary, depends greatly on the nature of the sediment (see Figure 46). However, stable mud and sand also become anaerobic beneath the surface, so their inhabitants have to depend on the overlying water for respiration, either by extending fine gills into it or drawing it through the burrow. Most immobile animals use these respiratory currents for feeding as well, sometimes filtering particles including plankton and larvae from the water (suspension-feeders), sometimes sucking or shovelling in the abundant detritus which has already settled (deposit-feeders). The extensive water movements of an estuary are constantly bringing fresh supplies of food within range, but filter-feeders usually select particles primarily by size; too many inedible particles may thus overload the system and lead to slow growth while, if they are sharp-edged or chemically active, they may also damage the delicate filters. By filtering large volumes of water, they also risk accumulating harmful doses of toxic pollutants even when these are at very low concentrations.

When they are actively sampling the water, even safely buried animals are obviously subject to salinity changes. Those which cannot tolerate low levels may nevertheless penetrate far up the estuary if they are able to obtain sufficient food in a short time; amongst bivalves, for example, deposit-feeders like *Scrobicularia* can usually take in food more rapidly than the suspension-feeding cockles. The horizontal distribution of the epifauna (animals living on the surface of either soft or hard shores) between the mouth and head of the estuary is, in contrast, determined more by salinity than any other factor, so that it is possible to plot a curve of diversity against salinity (Figure 45). It is noteworthy that freshwater species are, in general, intolerant of all but the lowest salinities. A few marine species are able to penetrate into water that is almost completely fresh although their numbers, too, fall off quite rapidly. In Milford Haven, half of the marine species of rocky shores fail to penetrate into the region where salinity may fall below 20‰ (zone 3 in Figure 43) while three-quarters fall short of the mesohaline region beyond (zone 4) where it may be lower than 10‰; a similar decline has been recorded in the Severn estuary (Figure 46). Even though a few specialized forms occur in the intermediate zone between the range of more typically fresh- and salt-water species, diversity is still at its lowest there, but some of its inhabitants may occur in very large numbers. In soft substrata amongst the worms alone, there may be up to $1,750,000 \text{ m}^{-2}$; the marine polychaetes *Nereis diversicolor* and *Pygospio elegans* can reach densities of $100,000 \text{ m}^{-2}$ and $30,000 \text{ m}^{-2}$, respectively.

The vertical distribution of sedentary plants and animals, particularly on rocky shores or hard artificial surfaces, is largely determined by tidal level, as on the open coast; but the

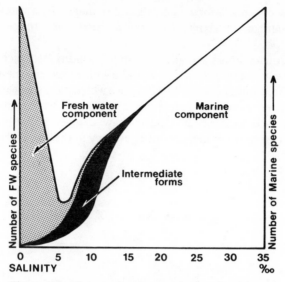

Figure 45. Changes in the number and origin of species along the salinity gradient of an estuary, originally from Remane's work in the Baltic Sea (1934). The intermediate forms are often termed the 'brackish component', but Barnes (1974) holds that they are merely specialized inhabitants of sheltered muds, not necessarily restricted to brackish conditions

pattern is confused by changes in salinity at any given point on the shore resulting from tidal movements of water up or down the estuary. In a temperate zone estuary, salinities will always be lower at its head than at its mouth, at any tidal level; as the tide ebbs, the shore will be lapped by water from progressively farther upstream. Thus the lower shore will be exposed to more extreme changes. Similarly but ecologically less important changes will be experienced with water temperatures, as the sea lags behind the land and its rivers both in warming up and cooling down. In temperate regions generally, rivers are warmer than the sea in early summer but cooler in early winter.

To illustrate how much more demanding is this rapidly changing environment than the fairly stable freshwater or fully marine habitats, we might consider more closely how the salt content of each affects the animals living there. The composition of their body fluids needs to be under fairly rigid control, but they must also maintain continuity with their environment, for example in the exchange of dissolved respiratory gases across gills. Gill membranes also permit the movement of ions by diffusion and the passage of water under osmotic pressure into the more concentrated solution. Fresh water contains much lower concentrations of salts than the fluids of animals living there, so these require an outer covering as impermeable as possible, together with very efficient excretory organs, to avoid both the loss of valuable salts and the possibility of becoming so bloated that their physiological processes break down. Lower vertebrates such as fish are believed by several authorities to have evolved in brackishwater and thus have body fluids rather more dilute than seawater, unlike marine invertebrates which have few osmotic problems. Sea fish thus face the paradoxical danger of desiccation by osmotic loss of water. Most replace inevitable losses by swallowing the seawater and rejecting unwanted salts by active secretion across the gills. The necessity of drinking large volumes of water makes marine fish, like the filter-feeding bottom animals mentioned above, liable to accumulate toxic pollutants from low concentrations. Fish gills are damaged by a wide range of pollutants and their involvement in salt control as well as respiration makes this damage more serious.

Although numerous animals can adapt to stable conditions which differ substantially from their norm, many cannot respond both to the wide range experienced in estuaries and the rapidity with which values change there. Those which have succeeded in colonizing the estuarine environment overcome their osmotic difficulties by various degrees of conformity, regulation or avoidance. Several species have evolved physiological specializations enabling them to achieve some measure of constancy in their body fluids, in the face of a

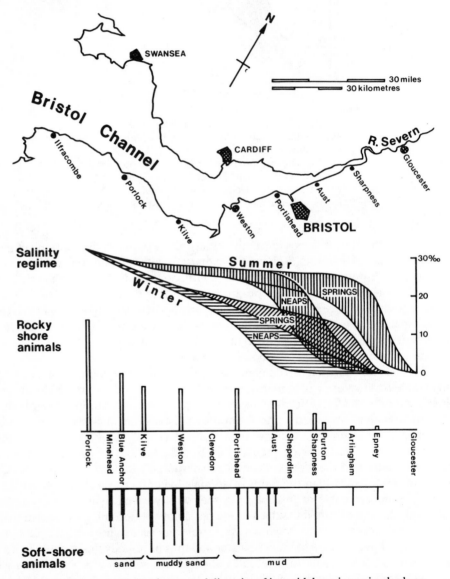

Figure 46. Range of salinity change and diversity of intertidal marine animals along the southern shore of the Severn Estuary (partly from Bassindale, 1943; soft-shore data from Boyden and Little, 1973)

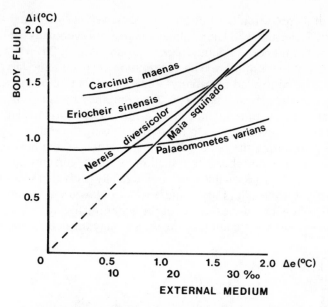

Figure 47. Changes in the concentration of the body fluid of various animals following changes in the surrounding water, measured as freezing-point depression (the approximate equivalent in salinity is also shown); the broken line indicates complete conformity (from various sources, particularly Beadle, 1943). The spider-crab *Maia* is unable to control the concentration of its body fluids; *Carcinus* (shore-crab) and *Eriocheir* (mitten-crab) exert strong control, as does a brackish shrimp, *Palaemonetes*, at a lower level; the ragworm *Nereis* cannot control its body fluids well, but tolerates their dilution

fluctuating external medium, as indicated in Figure 47; others rely on behavioural adaptations.

Highly mobile animals, like fish, may penetrate the estuary from the sea on a flood tide but retreat again on the ebb. Sedentary forms on hard substrata, lacking mobility, can often still avoid periods of intolerable conditions just as well as burrowers in soft sediments by closing down the defences which they have to deploy, even in fully marine conditions, against desiccation at low tide, buffeting by rough water or attack by predators—although this also means a reduction in the time available for feeding or other vital activities. For example, limpets graze the slime of small algae from the rocks, returning to a home-scar where the conical shell and rock surface have worn so as to fit each other tightly. At the point of their farthest penetration up Milford Haven, common limpets *Patella vulgata* on the upper shore close down only when the tide leaves them, at a salinity of about 26‰; lower down the shore, the salinity falls to as little as 12‰ before they are uncovered by the ebb, but they can tolerate dilution only as far as 24‰. Through having to close down early, they thus lose all the advantage of extra feeding time which is available to lower-shore limpets on the open coast. In spite of this, they feed well enough in that productive estuary to grow faster and larger than most limpets do in a more typical habitat. Smaller creatures may take more passive advantage of water movements to stay within the limits which suit them best; the tiny snail *Hydrobia ulvae*, which occurs in high densities on intertidal flats of muddy sand, goes through a cycle of floating on the rising tide and sinking during the ebb which keeps it at its preferred level on the shore. Newell (1964) found that in an Essex estuary, this activity declines with diminishing salinity until eventually the snails fail to float when covered by the tide or, if floating, sink at around 2‰. In this way they avoid being carried into lethally low salinities. The distribution of planktonic larvae of sedentary animals may also be controlled by salinity—for example, those of the worm *Phyllodoce maculata* swim less strongly at less than 12‰ so that they remain in the lower layers of intruding seawater and are thus retained in the estuary. Others use tidal movements: older oyster larvae tend to drop near or on to the bottom during the ebb, but rise into the main force of the current during the flood tide, thus remaining in the headwaters (Carriker, 1951). Some inlets are too short, too shallow or have currents too weak for such mechanisms to maintain stocks in this way; commercially cultivated species such as oysters then have to be laid artificially. Larval tolerance of unfavourable temperatures, salinity and other physicochemical characteristics of the water (including its load of pollutants) is usually lower than that of the adult (Figure 48) and may thus be the main factor limiting

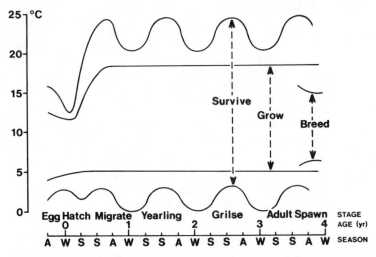

Figure 48. Temperature requirements for survival, growth, and breeding of Pacific salmon at various stages in their life history (from Nelson-Smith, 1972a, after Brett, 1956). Note that breeding and growth of the young can take place only over about half the range in which adults can survive

colonization of apparently suitable regions. Similarly, these factors may interact; high temperatures, low oxygen levels, or the presence of toxic wastes may reduce the ability to tolerate low salinities while, conversely, animals existing towards the limit of their salinity tolerance will be more susceptible to pollution.

SPECIAL FEATURES AND USES

The outstanding feature of estuaries, from the aspect of applied ecology, is the high productivity which results from the constant supply of nutrients to a sheltered but dynamic habitat. Indeed, estuaries and tidal marshes are regarded as amongst the most fertile natural areas in the world, being up to seven times as productive as a typical wheat field and 20 times more than the open sea. Primary production (that is, by photosynthesis) probably lies in the region of 500 g of organic matter per square metre per annum in an estuary of average richness; more food is brought in by the tide (in the form of plankton) and from the land (as organic detritus), to be trapped in filter-feeders and others low in the food-web. Some of these riches are exported; at one end of the scale, estuaries are an important source of vitamin B_{12} (which is elaborated by micro-organisms but essential to the growth both of higher plants and animals) for coastal waters as a whole—see Stewart (1972). At the other end, they form the breeding and nursery grounds for a variety of sea fish. Estuarine flats also provide food and winter shelter for large numbers of waders, ducks, and other waterfowl. After a serious oil-spill in the Medway, the small number of over-wintering birds was at first thought to reflect the magnitude of casualties amongst them; only when the customary numbers reappeared the following year was it realized that most had been forced to move elsewhere by heavy mortalities amongst their food organisms. Milne and Dunnet (1972) have summarized the extensive investigations by workers from the University of Aberdeen on the activities of birds and fish in the nearby Ythan estuary, which illustrates this function very clearly.

The greatest immediate commercial value of this high productivity, however, lies in fisheries within the estuary, mostly for bivalve molluscs. At the same symposium, Walne (1972) gave examples of the standing stock of some commercial shellfish beds in northern Europe, recording values of nearly 2,000 g dry meat per square metre for mussels, 200 g m^{-2} for oysters and 120 g m^{-2} for cockles. A reasonably productive British estuary should yield about 1,000 kg live weight per hectare of either cockles or muscles over all; a properly cultivated mussel-bed ought to provide eight times this, which is equal to between 50 and 100 times the meat yield from fatstock raised by traditional methods on grazing land. Cockles live at a shallow depth beneath muddy sand and are able to burrow again if disturbed, unless taken by a predator first; they are left to fend for themselves apart from regulations limiting the size and quantity which may be taken. They usually occupy intertidal flats, where many are eaten by fish at high tide and birds at low water. Mussels attach themselves to solid objects with short, strong threads; they can move slightly on these to avoid overcrowding or smothering but, if they become too dense or distant from their original anchorage, waves or currents may wash them away. Transplanting experiments on shallow Danish beds have not been successful, but raft culture has now become common in the deep rías of north-west Spain. Each raft supports hundred of long ropes, hanging vertically; the young mussels attach themselves naturally to these ropes and can utilize the full depth of water, out of reach of bottom-feeding predators such as crabs or starfish. Oysters also cement themselves to solid objects, but they are thereafter immovable and thus at the mercy of abnormal conditions or mechanical disturbance. In the main oyster fisheries of France, Japan, and the United States, the shells or tiles offered for natural settlement are then racked or suspended in areas specially suitable for good growth. This mollusc now commands such high prices that artificial rearing of larvae is economically feasible. A number of bivalves utilized elsewhere are ignored in Britain—for example, the deep-burrowing soft-shell clam *Mya arenaria*, which is highly prized in the United States. However, *Mercenaria mercenaria* (known as the hard-shell clam or quahog in North America) has become naturalized on the French coast and around Southampton Water, where it is farmed experimentally with the aid of warm-water effluents from power generation.

With the exception of local races of herring, living for example in the Clyde estuary or the upper reaches of Milford Haven, and the salmon or seatrout which pass through to breed in the rivers, the finfish of estuaries are not generally greatly valuable. They include dabs and flounders, the flatfish which are easiest to farm in enclosures. Growing marine fish to marketable size is not yet economic and the most desirable species would require cleaner water than can be obtained from most European estuaries; experiments have so far been confined to remote sea lochs, or to breeding young fish for release. Unfortunately, when this is done with plaice (a more valuable flatfish) the released young move off to deeper water, where their contribution to fisheries is problematical. Nevertheless, some estuaries (e.g. parts of the Wadden Zee) are the regular nursery-ground for plaice and other subjects of valuable coastal fisheries.

Human settlements have been established on the shores of estuaries from time immemorial, most often either at the first place within the mouth offering a sheltered boat-landing, or the farthest downstream crossing-point. Apart from organized fisheries, the early settlers must have exploited the biota in such individual pursuits as angling, bait- or clam-digging and wildfowling, later adding less utilitarian activities such as birdwatch-ing or recreational walking. Even though interest in such leisure pursuits has increased enormously in recent years, together with the use of various habitats in educational field studies, they still have little impact except in the immediate vicinity of large towns, popular holiday areas or the less well-run field study centres. This is far from true of the utilization of the physical properties of estuaries as shown, for example, by a study of the recent history of four industrialized British estuaries (Porter, 1973). To many laymen, an estuary must seem to be merely the lowest section of the river. This is true in the sense that it takes over the frequent role of a river as sewer and main drain; waste which enters a river and is not neutralized there by settlement, evaporation, chemical reaction, or biological degrada-tion must pass into its estuary. There, it joins further wastes discharged either deliberately, to take advantage of the notional greater dilution and dispersion, or accidentally in the course of transhipment, manufacturing or other operations. Domestic sewage probably represents the greatest volume of effluent containing tangible additions to reach most estuaries. A recent survey of the discharges of sewage to the coastal waters of England and Wales (Department of the Environment, 1973) covered 333 'principal outfalls', which emit a volume of nearly one million cubic metres per day, and revealed that in only ten cases (discharging $10,500 \, m^3$ per day) was the sewage fully treated. These serve a population of about 75,000 people, from a total of nearly five million. Cooling water is discharged in enormous volumes from modern plant—for example, $90,000 \, m^3 \, h^{-1}$ from the 360 MW electricity-generating nuclear station at Hunterston in the Clyde estuary or $230,000 \, m^3 \, h^{-1}$ from the 2,000 MW oil-fired station at Pembroke in Milford Haven, in each case at 8–10 °C above intake temperature. Iron and steel manufacture draws nearly $60,000 \, m^3 \, h^{-1}$ from British estuaries, while I.C.I. chemical works use nearly $40,000 \, m^3 \, h^{-1}$ from the Tees estuary alone, mostly for cooling. Although the behaviour of a given waste can be predicted only from a detailed knowledge of its composition and the hydrography of the receiving estuary, enough has been said of stratification, salt-wedge intrusion, tidal retention, flocculation, and the reworking of sediments to indicate that this is unlikely to be a simple matter.

A more extreme use involves physical interference with the estuary; this may take place at any point from a dam in its headwaters to a breakwater across its mouth. By retaining water, dams tend to reduce the scouring and freshening effects of spate flow—which is probably, on balance, favourable to the sedentary biota—but they also retain silt. Where the river is the main source of sediment, the existing balance between deposition and erosion must be disturbed, so that the shoreline around the mouth may recede—as has happened to the Nile Delta since the construction of the first Aswan barriers, not to mention the new High Dam. Conversely, a barrage across the lower reaches of a more typical estuary will accumulate sediment to the seaward side, because this is the main direction of its origin and tidal scouring will be reduced. The purpose of such a barrage is usually to impound river water; cut off from the sea, the reservoir so formed will rapidly freshen, so that the barrage becomes the effective head of the estuary. A more complicated situation arises with a tidal barrage used for power generation, as in La Rance, especially if the turbines can be reversed to provide pumped storage and so further confuse the 'tides' experienced by upstream biota. The lower reaches of many estuaries used as commercial waterways have channels bounded by training-walls or breakwaters, often erected by civil engineers in the later stages of the industrial revolution and frequently not producing the desired results; for example, the decline of Llanelli docks in South Wales dates from the erection of diversion walls intended to improve the approaches. Such works do, however, provide hard substrata in what were previously expanses mainly of soft sediments. Estuarine marshes and flats have long been regarded as land of little value, so that it has been cheaper to 'reclaim' shoreline areas there for heavy industry such as steelworks, oil refineries, chemical works, and power stations than to site these inland. Once these areas are cut off from the tide, it can neither deposit its silt load amongst the marsh plants nor carry away the nutritious organic detritus. Redirected currents and changes in the pattern of sediment deposition or erosion may then again lead to unpredictable changes in navigation or drainage channels. Elsewhere, the natural tendency of salt-marshes to trap silt and thus rise in level has been enhanced by the deliberate planting of colonizers like the hybrid *Spartina anglica* and its subsequent escape to other areas. This rapidly spreading plant provides good grazing for sheep and horses but, once well established, it is hard to control, as at the once picturesque seaside resort of Parkgate in the Dee estuary which still has its promenade, although this now faces an extensive muddy marsh rather than a sandy beach.

The ultimate in reclamation has, of course, been carried out by the Dutch—but not without cost. Before it was filled in, the Oosterschelde produced 30 million oysters per annum. From the point of view of a marine scientist, such reclamation can be regarded as permanent pollution. Intrusion of holiday-home/marina developments (common in North American estuaries and bays, particularly in Florida, but in Europe restricted mainly to the Mediterranean coast of France) not only blocks or diverts tidal movements but also adds pollution of various sorts. Domestic and garden chemicals may spill directly into the water; house sewage is likely to be treated but that from boats is only now beginning to come under control. Pleasure boats eject, in total, large volumes of oil from their exhausts and bilges, while it has been pointed out as a curiosity of human behaviour that dumping garbage from a motor car is generally regarded as antisocial, whereas doing so from a boat is not. In the affluent industrial nations, boating is an increasingly popular leisure activity—for example, more than 4,000 boats are now kept in Southampton Water and its tributary estuaries (Dartington Amenity Research Trust, 1973); over half of these have living accommodation and nearly three-quarters are motor-powered.

For far longer than boats have been used for pleasure alone, estuaries have provided a safe anchorage for working vessels. For comparison with the intensity of pleasure boating, there were over 700 commercial shipping movements per week in Southampton Water during 1970, ranging from cross-channel ferries to the largest oil tankers. Many of these vessels also discharge sewage, garbage, waste oil, and dirty ballast into the water, together with further fallout from the air. Because of the concentration of shipping and the lack of space for manoeuvre, there is a greater likelihood of collision or stranding, leading to the occasional spillage of hazardous bulk cargoes and frequent breaching of fuel-oil tanks. Dredging to maintain or deepen navigation channels disrupts bottom communities, increases the turbidity of the water and may resuspend settled polluting wastes. Finally, of course, commercial shipping demands the construction of wharves, docks and storage or transhipment sites, replacing varied shores of gentle gradient with vertical walls.

MAJOR POLLUTANTS AND THEIR EFFECTS

It is convenient, even if not accurate in terms of the ultimate physiological effect on individual organisms, to divide water pollutants broadly into those having physical and those having chemical effects. Physical pollution in estuaries often exaggerates conditions which are already present to some extent. For example, particulate wastes create turbidity and may damage gills when suspended, or blanket the bottom and may smother plants and sessile animals when settled. It they are organic, their decomposition may also deplete the oxygen supply. Some of the larger discharges of sewage or industrial effluent, as well as the undesirable materials which they contain, may represent a sizeable addition of fresh water. Chemical additives in discharged cooling-water are rarely sufficient to alter it permanently from its condition when it entered the plant; however, its elevated temperature not only accentuates the effects of other pollutants (especially their oxygen demand) but also alters the local environment sufficiently to justify the term pollution in its own right. Some organisms already present are likely to be able to survive these exaggerated conditions, so the general effect of such pollution will be a further reduction in species diversity but not necessarily in biomass. An existing community may be well adapted to a particular pollutant—for example, the annelids and small crustaceans of Lochs Linnhe and Eil which naturally inhabit leaf-litter but readily colonized wood-fibre deposits from a pulp-mill discharge (Pearson, 1972). In other cases, species which are naturally rare, or occasional exotic introductions, are favoured by the changed conditions above those which were previously dominant. This is particularly true of thermal pollution; Naylor (1965a,b) has described the largely exotic fauna of a warmed dock and the changes which followed its reversion to more normal temperatures.

Detergents are largely physical in their effects; they reduce aeration at the water surface, facilitate the entry of other pollutants into living organisms and themselves damage cell membranes at the sites where respiratory and ion exchange takes place. Oil occupies an intermediate position between these broad categories; crude petroleum which has been spilt at sea, together with some fuel-oils even when fresh, behave as inert materials which may float or sink but ultimately are usually accumulated along the high-water mark as encrustations which smother existing attached animals and deter the settlement of others. Oil reaching estuarine shores from collisions, strandings, or spillage from shore installations is, by contrast, still fresh enough to contain its toxic elements, although, in bulk, its first effect may still be a smothering one. In making this distinction it should not be forgotten that over half the oil reaching the marine environment originates from landward sources, most of it via rivers. Estuaries containing refineries also carry numbers of tankers taking away petroleum products, which may be markedly more toxic than the incoming crude oil. For example, the tanker *Dona Marika* lost many thousands of gallons of high-octane gasoline when she was stranded on rocks in Milford Haven during August 1973, killing practically all the intertidal gastropods in neighbouring bays. Such spillages provide a clear demonstration of the dynamic nature of rocky shore communities; when grazers such as limpets and winkles are killed in sufficient numbers, the algae which they normally control take over the rock surface. After the *Torrey Canyon* spill (when the grazers were killed by toxic solvent–emulsifiers used to remove the oil, which was itself well weathered) a summer bloom of green algae was replaced in the following year by a heavier and more permanent cover of brown fucoid seaweeds (see Nelson-Smith, 1968a,b; Smith, 1968) which persisted for a number of years. The increased weed cover discourages the settlement of barnacles which must, in turn, affect their predators such as dogwhelks *Nucella lapillus* and the polychaete *Eulalia viridis*.

Shores well populated by a varied flora and fauna recover from single catastrophic oil-spills quite well; Cornwall has not been subjected to another tanker wreck during the recovery period and may avoid further disasters for many years. Milford Haven has been the scene of serious spills, on average, once every two years; of course, these have not all occurred in the same place, but there is a much greater possibility that a shore will become polluted again before it has recovered, so that this process may receive a severe set-back. Continual pollution at a low level can also take its toll; a large refinery in Southampton Water discharges oil in its cooling- and process-water at a concentration measured in parts per million, but the total volume of water is such that some 5,000 l of oil are lost in it per day, flowing across a salt-marsh. The vegetation, trapping this oil, was killed off over a large area so that the marsh surface began to erode away (Figure 49). Fortunately, after recent improvements the area is becoming recolonized by plants (Dicks, 1973). A much smaller discharge across a rocky cove in Milfor Haven has brought about changes similar to a catastrophic spill (but over a much smaller area), with a decline in grazers and a

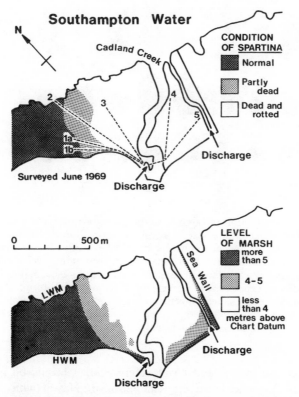

Figure 49. The effect of chronic discharges of oily water on a salt-marsh in Southampton Water, recorded in June 1969 (from Nelson-Smith, 1972a, after Baker, 1970). The broken lines numbered one to five indicate the positions of the survey transects from which the maps were plotted

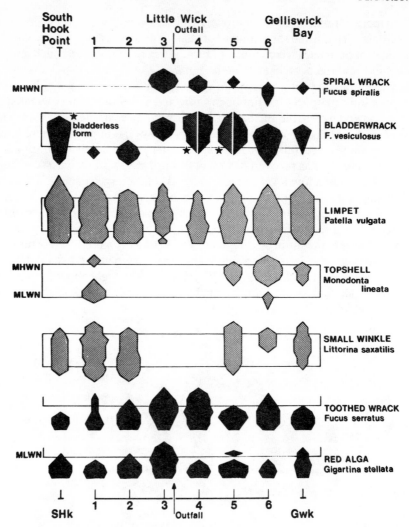

Figure 50. Distribution and abundance of some grazing molluscs and common seaweeds on a series of transects along the north shore of Milford Haven to either side of a small oil-refinery outfall. The width of each histogram is proportional to abundance; horizontal lines represent the level of either mean high or mean low water level of neap tides. At stations 4 and 5, the bladderless form of *Fucus vesiculosus* is mixed with the more usual form, which is not typical of unaffected shores; it is the only form at South Hook Point, which is slightly more exposed than the other sites. (Reproduced by permission of Applied Science Publishers from Nelson-Smith, 1975).

corresponding increase in seaweed cover (Figure 50). Chronic oil pollution can have ecological effects over a wider area, although in Milford Haven (which is well flushed and carefully controlled) no obvious decline has been detected during the first decade of large-scale oil industry operations among rocky shore organisms (Crapp, 1971a) or plankton (Gabriel, Dias and Nelson-Smith, 1975). In less well favoured areas, low levels of oil pollution may depress photosynthesis in plants and bring about the faulty development of animal larvae, deter the more sensitive or delicate amongst mobile animals, and weaken those which are sedentary. Even where the fauna shows no ill effects, edible molluscs and fish may be unsaleable because of an oily taste.

A wide variety of chemical wastes enters estuaries, ranging in toxicity from the very lethal cyanides, chlorine, or ammonia to salts which are almost inert. Highly active biocides also enter from agricultural, industrial and domestic use, discharges from the manufacturers, occasional spillages in transit, and atmospheric fallout. Such pollutants may be very selective in their action—for example, crustaceans are highly sensitive to pesticides used against the closely related insects, whereas molluscs may be relatively quite tolerant (Table 13). They may also become concentrated particularly in one part of the environment—for example, phenols are absorbed onto mobile mud particles—but the most important distinction is between those which are readily degraded and the persistent

Table 13 The toxicity of the major pollutants of water to the main groups of aquatic animals, generalized from a wide range of publications. The greatest number of symbols indicates the highest toxicity: +++++ represents an $LC_{50}(24 \text{ h})$ of less than 0·3 p.p.m., while + indicates a median lethal concentration of over 1,000 p.p.m. Toxicities will vary according to environmental conditions and species which are either specially sensitive or specially tolerant have been ignored. Simplified from Nelson-Smith (1971, 1972a).

Pollutant	Plankton and invertebrate larvae	Lower invertebrates	Crustaceans	Molluscs	Fish
Heavy metals (salts)					
Copper	++++		++++	++++	++++
Lead			++		++++
Zinc	++		+++	+++	+++
Mercury	+++++		++++	++++	++++
Cadmium					+++++
Chlorine		+++	++++	+++	++++
CNCl					+++++
Cyanide			++++	+++	+++++
Fluoride					+++
Sulphide					++++
Mercaptan					++++
Phenol	++++		+++	++	+++
Cresol			+++	+++	+++
Formaldehyde			++		++
Herbicides					
Paraquat, simazine			+++		++++
Pentachlorophenate	++++		++++	++++	++++
2,4-D					++
Pesticides					
Rotenone				+++++	+++++
Chlorinated HC, PBC			+++++		++++
Organophosphorus			+++++		+++++
Typical crude and fuel oils	+++	+	+	+	++
Low aromatic hydrocarbons			+++		++++
Light oil products	++				++
Oil-spill cleansers					
Early (e.g. BP1002)			++++	+++	++++
Modern (e.g. BP1100)			++	+	+
Surfactants (anionic and non-ionic)	++++	++++	+++	+++	++++

pollutants. Cyanides and ammonia, being compounds which are readily broken down chemically or by various bacteria into harmless constituents, may present only a temporary or local problem; toxic heavy metals such are mercury, cadmium, arsenic, lead, copper, or zinc are chemical elements and can only be converted from one salt to another, while chlorinated hydrocarbon insecticides are extremely resistant to change and may break down into compounds which are even more stable and toxic. Because of their persistence, they may accumulate in the environment itself (for example, in the sediments around an effluent outfall) or within the tissues of plants and animals, which are able to concentrate some ions many times over from trace quantities in the water. Extreme cases are scallops containing 2·25 million-times the cadmium, or winkles 1·5 million-times the chromium, present in the surrounding medium, although such concentrations may have occurred partly in the smaller organisms forming their food. At each step up a food-chain, only about 10% of material from the body of the prey is incorporated into the tissue of its predator, the rest being expended as fuel—but persistent chemicals of this sort remain within that 10%. In a Long Island estuary, algae and plankton were found to contain 0·04–0·08 p.p.m. of DDT or its derivatives; molluscs and small fish, feeding on these, had 0·26–0·42 p.p.m. Carnivorous fish contained over 2 p.p.m., while the birds feeding on them generally had between 4 and 26 p.p.m.—some gulls had acquired up to 75 p.p.m. (Woodwell, 1971). These figures are calculated on a whole-body basis; chlorinated hydrocarbons accumulate in fatty tissue and have their greatest effect in times of unusual stress or during reproductive stages, when these are used up quickly. The toxic heavy metals are also often accumulated in specific organs. Humans are not only at the head of most food-chains but may select specific organs or tissues as food; depending on circumstances, this may protect them from toxic materials or greatly increase their risk.

Human health may be affected by bioconcentration in other ways. Although some authorities maintain that bathers rarely become infected from swallowing pathogens discharged in sewage, bivalves may readily accumulate sufficient bacteria to become dangerous if not properly cleansed before eating, as seems to have happened during a recent cholera outbreak in the Mediterranean. Large quantities of nutrients, especially nitrates and phosphates, are also discharged in sewage; a human adult excretes, on average, 9 g of nitrogen and 2 g of phosphorus per day, amounts which are not greatly affected by subsequent treatment of the sewage. The middle portion of the Severn Estuary/Bristol Channel (between the suspension bridge and Lundy Island) receives 15,000 kg of nitrogen and 3,500 kg of phosphorus daily from the inhabitants of its coastal towns alone, disregarding the contributions brought down by the Severn and tributary rivers. To this must be added increasing amounts of nitrogen from the intensive farming both of arable land and livestock; domestic detergents add about another 20% to the phosphorus in sewage. High concentrations of such nutrients, flowing into coastal waters, are reflected by an abundance of common planktonic plants; they sometimes stimulate a dense bloom of dinoflagellates, producing a 'red tide' so called from the colour which is imparted to the sea. These organisms may cause trouble merely by clogging the gills of finfish and shellfish but, usually, they also produce a paralytic poison which can kill either the fish or the humans and birds feeding on them. In the worst outbreaks, even seaspray can irritate exposed skin. In more enclosed waters, a massive temporary bloom of planktonic algae or even attached plants may occur; the plants are usually short-lived and their subsequent decomposition uses up much of the available oxygen. This is similar to a longer-term, natural process in lakes called eutrophication; but it might be less confusing to adopt the broader expression 'secondary organic pollution', offered by Aubert and Aubert (1973), for conditions such as the recent decline in oxygen and thus macroscopic life in the deeper parts of the Baltic and the Oslofjord resulting from the continual discharge of sewage and other organic wastes there.

It has already been mentioned that discharges of warmed water may favour the spread of exotic plants or animals, and that such effluents might be of use in seafood production. They may also have the undesirable effect of stimulating the growth of animals which bore into or encrust ships' hulls and harbour structures, such as tubeworms, the ship-'worm' *Teredo*, barnacles, and gribbles (*Limnoria*); some species are not greatly active, or cannot breed, in normal north European temperatures. A recent general account has been published by Jones and Eltringham (1971). A further group of troublesome creatures which might loosely be described as 'biological pollution' may be introduced by ships or when shellfish from overseas are laid for seeding or fattening; they include the slipper-limpet *Crepidula fornicata*, which competes all too successfully with oysters; an American whelk *Urosalpinx cinerea*, which eats them; and, more recently, the Japanese seaweed *Sargassum muticum*, which has become established along the north-east coast of the Isle of Wight at the mouth of Southampton Water. From experience in British Columbia, it is feared that this seaweed may spread as rapidly and uncontrollably as the cord-grass *Spartina anglica* (see above; Jones and Farnham, 1973).

METHODS

To determine the effects of human intervention on the ecology of an estuary, the nature, distribution and abundance of its flora and fauna should ideally have been recorded beforehand. Of course, this is rarely possible, although in some cases there has been an abrupt change in the scale of developments since ecological studies began (as with the growth of Milford Haven into a major oil port since 1960); in others, long-standing pollution has been quite rapidly abated (for example, the introduction of proper sewage treatment before its disposal into Biscayne Bay, Florida—see McNulty, 1970) so that changes can be recorded in reverse, as it were. Even when they exist, ancient faunistic surveys are rarely helpful; they are mostly lists biased towards the collector's speciality and with little ecological detail. The comparison of such lists, whether between different localities or at various dates in the same locality, is of little use unless they were made by the same worker under similar conditions and show the ecological importance of each organism. At almost any locality, further search will reveal greater variety, but the freshly discovered species are unlikely to be of great ecological significance. A well studied shore in the United States was resurveyed after a serious oil-spill and found to show a greater variety of seaweeds than before—but only because the latest investigator was a specialist in the smaller marine algae. Diversity indices may nevertheless be useful, especially in comparing various sites along a gradient in environmental conditions; the numerous ways of arriving at and interpreting such indices are explained by Warren (1971).

The flora and fauna of rocky shores are relatively easy to survey, since they are visible, easy to enumerate and, in most circumstances, do not show great fluctuations with time. In Milford Haven, a baseline was established by assessing the abundance of some 50 common species on belt transects down a series of typical shores throughout the estuary (Moyse and Nelson-Smith, 1963; Nelson-Smith, 1967). These surveys have since enabled the effects of catastrophic spills or effluent discharge and the general health of the estuary to be assessed (Figure 51). Similar transect surveys helped to elucidate the effects of and recovery from *Torrey Canyon* spillage in Cornwall (Nelson-Smith, 1968a,b). On soft-sediment shores, the fauna is not only hidden from sight and distributed in a less predictable manner but also tends to fluctuate more, both seasonally and from year to year. At each sampling station,

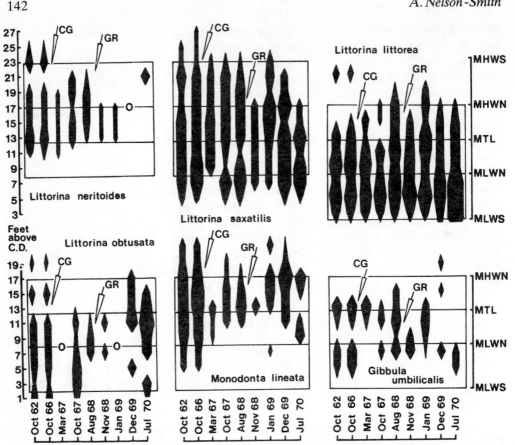

Figure 51. Effects of two successive spills on the distribution and abundance of shore gastropods at Hazelbeach, Milford Haven, after spillage from the tanker *Chryssi P. Goulandris* (CG) in January 1967, and the Gulf refinery (GR) in November 1968 (from Nelson-Smith, 1972a, using data published by that author in 1967, 1968a, b, and by Crapp, 1971a)

the sediment has to be dug and sieved to reveal its inhabitants; to provide adequate information, numerous stations should be sampled both down and across the shore, while sampling must be repeated frequently (perhaps bimonthly) to establish the range of natural variation before any special significance can be attached to observed changes. Results of some typical studies have been illustrated by Crapp, Withers and Sullivan (1971) or Boyden and Little (1973). Meaningful results are even more difficult to obtain below tide-marks, where 'blind' sampling from a boat can give quite different data from selective sampling by a diver. Such surveys require time and expert personnel, particularly on soft substrata. It is therefore not surprising that investigators have sought indicator species whose presence or abundance might be used to diagnose ecological health or the intensity of pollution without the necessity of a full survey. On rocky shores, it seems that the size–frequency distribution of limpets (which gives a crude indication of their growth-rate and recruitment) might be a good indicator, at least of ill effects from oil pollution (Crapp, 1971a), although Lewis (1972) has demonstrated considerable natural variations amongst mussels, barnacles, and limpets on the harsh British north-east coast. In shallow water offshore, Bellamy and his colleagues (Bellamy, 1968; Bellamy *et al.*, 1972) have shown that growth of the kelp *Laminaria hyperborea* provides a history of water quality, in the short-term by size of the lamina (which is shed each year) and over the lifetime of the plant by that of the stipe or stem, which shows its age in clear annual rings. On soft bottoms, Reish (1960, 1972a) has found that the polychaete *Capitella capitata* is diagnostic of organic pollution from food-processing wastes or domestic sewage. He reported (1972b) the use of another polychaete (*Syllides edentula*) to establish polluted areas in the port of Livorno, and a more direct indication of spread from sludge-dumping beds off New York in the incorporation of aluminium foil, bandages, and seeds into tubes built by similar worms. Tomato seeds remain undecomposed for long periods and are widely recognized as indicators of sewage-polluted sediments, while stranded filter-cigarette ends or rubber condoms in large numbers serve as warnings of possible beach contamination. Faecal bacteria are also direct indicators of sewage pollution, of course, but water samples need culturing before numbers can be estimated, while salinity and turbidity (both very variable in estuaries) have an influence on bacterial survival.

If one is seeking the effects of a specific outfall, or the distribution of a particular parameter, water movements must be taken into account; in most estuaries, a full understanding of these can be obtained only after extensive measurement. Any adequate schemes requires a combination of expensive boat-work and elaborate instrumentation. Dispersion tests can be carried out in the field, using such markers as buoys, drift-cards, oranges, plastic beads, dyes, exotic bacteria or radioactive particles to simulate the waste discharge, but the scale and thus expense of such investigations usually limit them to a few tidal cycles at most, during which weather conditions or river flows may not be typical; certainly neither physical nor biological accumulation can be demonstrated in this way. If the proposed development (which may be structural works rather than a polluting effluent) is sufficiently important, a hydrographical model may be built in which various states of weather and water can be simulated in a scaled-down waterway, but this too is expensive. Now that computers are widely available, it is more usual to develop a mathematical model with which an infinite variety of changes can be investigated very quickly (see e.g. Taylor and Gear, 1972). It is important to remember, however, that the conclusions printed out can only be as good as the data fed in! Although simple systems like the middle reaches of rivers have been widely simulated, few models of estuaries are sophisticated enough for ecological studies; perhaps the most elaborate and successful has been that of the Thames (Barrett, 1972).

Ecological surveys, hydrographical observations and the manipulation of mathematical models are most often utilized to give an over-all picture of the workings or present condition of an estuary. Plant operators or regulating authorities frequently face the more practical but limited problem of the probable safety and permissible levels of given constituents in an effluent. This problem is usually met by carrying out toxicity tests. Much has been written about the proper design and interpretation of such tests, to which Warren (1971) gives a good background; the main difficulties lie in the selection of suitable organisms and conditions for the test. Traditionally, convenient animals such as young trout have been exposed for one or two days to a range of concentrations of individual chemical compounds; the concentration which kills half of them within the period of the test is noted (as the LC-50 or median lethal concentration) and some fraction of this—often one hundredth—was then assumed to be 'safe'. Clearly, the best organisms to use should rather be typical of the habitat, in their most sensitive life-stage, and tested under the chemical and physical conditions experienced in that habitat; while the possibility of bioconcentration and of sublethal effects on growth rate, condition or reproductive success should also be taken into account in interpreting the tests. The more enlightened industrial firms now go a long way towards meeting these requirements (see

Carter, 1973). The components of an effluent may act differently in combination, too; the practice in chemical industry, where effluents are not only complex but variable, is simply to test a bulked sample of the actual effluent against a selection of typical inhabitants of the receiving water.

MANAGEMENT

Estuaries, then, are valued for different reasons by those engaged in commerce, manufacturing industry, fisheries, leisure pursuits, and the conservation of landscape or wildlife; not surprisingly, their activities or demands are often in conflict, since the effects of one may be disadvantageous to another. Satisfactory management demands a full knowledge of the physical and biological processes taking place in a given estuary, as well as its major uses, the demands they make and the effects they create. It is probably safe to say that such detailed knowledge does not yet exist for any major estuary, in spite of the profusion of published accounts of which only a few have been mentioned in this chapter. An attempt is being made to remedy this in Britain by the Natural Environment Research Council, by setting up research groups amongst those interested in each of the main estuaries to collate existing knowledge, identify 'blind spots' and, it is hoped, to initiate or guide investigations into these areas of ignorance. A single manufacturer or statutory authority can scarcely be expected to carry out more than an *ad hoc* study of the probable effects of their own proposed developments, although large-scale schemes, such as the one to dump sewage sludge containing trade wastes from several large Midlands towns into Liverpool Bay at the mouth of the Mersey, have been the subject of extensive multidisciplinary studies (see e.g. Department of the Environment, 1972).

Effective management implies one over-all authority or, at worst, a small number of closely-linked ones. The main bodies concerned with any British estuary might include the river authority (now the regional water authority, usually having a tidal order extending its interest below the limit of the river proper), and sea fisheries committee, both of which are involved in imposing standards for waste discharges; several port authorities might deal with the maintenance of channels or regulate the navigation and mooring of small leisure craft as well as commercial shipping. Waterside developments in general are regulated by the planning departments of major local authorities; since rivers often act as county boundaries, there will probably be at least one for each shore. Sewage treatment and disposal was, until recently, provided by individual local authorities, but is now under the control of the regional water authority. Marshes, dunes, cliff tops and other backshore areas are protected by the Nature Conservancy Council, although this body does not itself own land. Areas of special landscape or wildlife value may be preserved by non-statutory bodies such as the National Trust (who have acquired much of Britain's remaining unspoilt coastline under 'Enterprise Neptune'), county naturalists' trusts or the Royal Society for the Protection of Birds; these and many other organizations will often present objections to schemes for development, or may make positive proposals for conservation measures themselves.

An example of the problems which can arise from this patchwork of authorities followed a leakage of heavy crude oil from a tanker berthed at Dundee during 1968. The oil at first formed a coherent slick which could have been skimmed using available equipment, or, at worst, dispersed chemically; but the local authorities on the shores of the Tay estuary could not finance such operations beyond their own beaches. The Harbour Trust was responsible only if the oil was a hazard to navigation, while the River Purification Board (the Scottish version of a river authority) had not obtained its tidal order—but, anyway, was empowered only to prosecute polluters, not to deal with the pollution itself. While legal powers and responsibilities were being determined, the oil spread for many miles on successive spring tides. The cost of inaction was not only that of cleaning shores over a much wider area, but also the loss of a substantial proportion of the eider-duck resident in Britain—duck which, furthermore, had been carefully ringed and were under observation as part of a long-term research project. At a less catastrophic level, plans to discharge untreated coke-oven wastes from a steelworks extension near Newport into the Severn Estuary were delayed by objections from the Bristol Docks Authority on the opposite bank, who discovered that the outfall would open in their waters only from a newspaper report a day or two before the date set for dispersion tests on site.

Some local authority, water quality, fisheries, and conservation interests have been nominally united in the Department of the Environment, although in practice they frequently still function in isolation. Regional water authorities have taken over from the old river authorities administering water resources (the quality of natural waters, sewage disposal and the regulation of waste discharges from other sources, impoundment, and water supply) over a wider area. It seems inevitable that difficulties will arise in reconciling these conflicting responsibilities. Reed (1972) has suggested that the sea fisheries committees should, in a similar way, be reconstituted as coastal sea authorities with powers to administer the maritime zone (including estuaries) and coastal waters as a total resource. They should not suffer from quite such severe conflicts of interest, although Reed envisaged that their costs should be partly met by fees charged for waste disposal. He compares this with port dues, royalties for offshore gravel extraction or fishing licence fees and, indeed, the suggestion is consistent with the emerging philosophy that 'the polluter should pay'. Admittedly it is a practical solution to the problem of financing environmental protection, but it could easily give rise to unfortunate attitudes; the manufacturer acquires what must appear to him as a 'right to pollute' by paying a fee to do so (and, if it is a modest one, loses any incentive to find a better way of dealing with his wastes), while the authority, if dependent on the income from discharge licences, could hardly be expected to pursue vigorously any policy of reducing the number or volume of discharges. In fact, once it had been persuaded of the relative harmlessness of further outfalls, it might begin to justify approval of them by reference to the good use which could be made of the resulting income.

In contrast to the situation in Dundee (which, to be fair, is an old-established general port unaccustomed to bulk oil movements), the Milford Haven Conservancy Board took advantage of their recent and sudden emergence as a major oil-port authority to insist on the participation of each oil installation in a master plan whose basis is the appropriate treatment of all oil-spills as soon as they occur. A spray-boat and other equipment are maintained jointly by subscribers to the plan. Responsibility for meeting the cost of each operation is apportioned after the spillage has been dealt with and, if the culprit is not immediately obvious, all parties are agreed to accept the decision of the harbour-master; if he is unable to assign the blame, the cost is shared. In Milford Haven, as elsewhere after the *Torrey Canyon* disaster, shores were assessed according to their amenity value or biological importance. Recommended methods of treating oil-spills were laid down for each category; shores fronting seaside towns, boatyards and so on might be cleansed without further ado while, for most others, prior consultation is required with the Nature Conservancy Council, fisheries authorities or advisers from local marine laboratories and university departments. Few if any treatments are regarded as suitable for shores of the highest biological value. Such a classification of seashores obviously has wider value. Glude (1972) proposed the establishment of a computer-stored Coastal Resources Atlas of the United States, constantly updated and containing not only details of the uses and value of shores sensitive to oil-spillage or cleansing, but also information on fisheries, spawning periods, seasonal changes in current patterns and so forth, so that data essential to decision making might instantly be available at the time of any disastrous event. Estuaries, as we have seen, are among the most complex of environments both physically and biologically. As accumulating data emphasize this complexity, such an information bank would seem particularly appropriate for these regions, providing up-to-date summaries of all relevant material not only in emergency but as a basis for decisions on proposals for any significant development.

Chapter 8

Rocky foreshores

J. R. LEWIS

PHYSICAL BACKGROUND

Rocky shores occur most extensively where the coastal region is mountainous or at least rugged. Topographically they are more variable than other coastal habitats. Depending on the geological character of the coast they range from steep, inaccessible cliffs to wide, gently-sloping platforms; from fringing islets to long narrow inlets; from smooth, uniform slopes to highly dissected, irregular masses or extensive boulder beaches. They may extend virtually unbroken over great distances or alternate with sandy bays, or be represented solely by the artificial substrata of harbours and breakwaters. Other physical conditions are also variable. The majority of rocky shores, on open coasts, experience the relatively stable conditions of fully marine situations, but some experience the regular or intermittent low salinities and high turbidities of estuaries. Those in long, narrow inlets may be subject to wide, seasonal changes of temperature and salinity, and to relatively little exchange of water with the open sea.

Superimposed upon these local or regional variations of topography and water quality are major differences in the amount of wave action experienced. To the layman the terms 'rocky coast' and 'exposed coast' are often synonomous, for many rocky areas are subjected to the continuous violence of oceanic swell and waves, but on much-indented coastlines conditions may be perpetually calm. On open coasts themselves the amount of wave crash or surge is highly variable locally and largely depends upon the very local topography. Offshore islets or reefs naturally protect the main shoreline; but in their absence the shallower the water, or the more extensive the shore from high- to low-water mark, or the more broken and irregular it is, the more will the violence of the sea be dissipated at the seaward edge or on projecting peaks. Thus while it is natural to expect that oceanic shores will have a characteristically 'exposed fauna and flora' the intimate interaction between waves and topography produces areas supporting populations that are more typical of sheltered, land-locked situations.

Additional sources of local variation in the physical environment are the geological nature of the rocks (and especially their surface structure, and their suitability for boring-organisms); the presence of freshwater outflows across the shore; the occurrence of sand or shingle adjacent to the rocks; and the natural turbidity of the adjacent waters.

147

BIOLOGICAL BACKGROUND

The conspicuous and typical species of open rock surfaces are either attached (e.g. barnacles, mussels, oysters, tube-worms, lichens, and the many types of algae) or else, if mobile, are nevertheless capable of holding tightly to the surface of the rock or of retreating to protective clefts as occasion demands (e.g. limpets and the many types of more typical snails, i.e. periwinkles, topshells, dogwhelks). Much less conspicuous are small crustaceans and worms that live in the secondary habitats provided among closely packed mussels and barnacles, or among the algae. The organisms of the open rock surface live in one of the most stressful of marine habitats. Tidal ebb exposes them to desiccation or rainfall, to high and low temperatures and so forth, but the relatively few species adapted to such conditions may occur in very large numbers.

Towards low-tide level, where desiccation is less, and especially on shaded gully walls, not only does the diversity increase but there is often a marked change in the types of organisms. Barnacles, mussels, and large brown algae give way to small red algae and to attached sponges, hydroids, pendulous anemones, polyzoans and compound sea-squirts, together with their associated microspecies. Most of these attached animals are unknown to the layman. They appear as a 'fur', or as brightly coloured encrustations or gelatinous blobs. Such forms, often referred to as the 'cryptofauna', also characterize the under-surfaces of stable boulders, and where the latter stand in pools, they are accompanied by mobile and tube-living worms, chitons, saddle-oysters, crabs and other crustaceans, and by several types of littoral fish.

Relatively few rocky shore species are ubiquitous. The majority occupy only parts of the gradients from high- to low-water mark and from full wave-exposure to full shelter. Then within these major limits, further, more local variation in distribution and abundance stems from the influence of substrate, freshwater outflows, aspect and, most importantly, from the interactions of the species themselves. The distribution of rocky shore populations in European waters is probably better known than for any other marine habitat, and there is an almost overwhelming literature describing the many and subtle local variations. As a general guide the broad characteristics may be summarized as follows.

(1) Astride and above the high-tide level, the terrestrial lichen vegetation (usually grey-green and orange in colour) gives way to a 'black zone' of lichens and microscopic algae that also supports many small snails (up to c 5–8 mm long). This upper fringe to the shore is many metres deep on exposed, spray-swept headlands, is very narrow in extreme shelter and may be poorly represented on the artificial substrata of harbours and breakwaters.

(2) The main intertidal area is richer in species and therefore potentially more variable. Its upper levels are characterized by large numbers of barnacles in conditions of exposure, and by brown algae (Fucaceae, or 'fucoids') in shelter. The upper limits of both types of organisms often form conspicuous biological horizons. This exposure–shelter difference then persists in the middle and lower levels. Thus the barnacles may be joined, or even subordinated to mussels, together with limpets, dogwhelks, and short robust algae (usually reddish in colour). Towards low water there is often an increase in red algae and in limey- or sandy-tubed worms, and the cryptofauna appears on the gullies and overhangs. Here too, essentially subtidal species such as starfish and sea-urchins may intermittently make local upshore migrations and occur in considerable quantities. In sheltered positions, the high-level fucoid algae are succeeded downshore by the larger 'bladdered' or 'serrated' types. The ultimate in sheltered conditions is marked by free-living algae that simply rest on stable, muddy shingle.

Subordinate species show similar 'exposure tolerances'. Some snails, for example, are often most abundant under intermediate conditions where they are neither dislodged by waves nor displaced by masses of large algae. In areas of intermediate exposure—especially those with wide and broken foreshores—the barnacle–mussel and brown algal populations may intermingle or occur in bewildering mosaics. And whatever the exposure conditions, all the species present will demonstrate a vertical pattern of distribution in accordance with their individual desiccation tolerances.

(3) The lowest tidal levels are characterized by the upper fringe of essentially subtidal populations, dominated by various types of large brown algae (the kelps) and by the sudden elimination of most of the midshore species. In wave-swept sites, this level supports few macroscopic species apart from kelps and its most distinguishing features are the pinkish encrustations of calcium-depositing red algae. In sheltered areas free from mud, and specially stable boulder beaches, the kelps are joined by a wide variety of other algae and by a greatly increased range of cryptofaunal species, some on the rock, some on the algae.

In the functional structure of rocky shore populations, we may recognize several sources of primary production or of input of organic matter. Most obvious are the many types of algae which, unlike much of the vegetation of salt-marshes, derive nothing from the substrate except anchorage. All their nutrient salt requirements are met from the seawater itself. Although a few animals feed on the macroscopic algae directly, the latter are principally utilized in animal food-chains either at their microscopic sporeling or young stages (when they sustain the grazing limpets, snails, chitons, and sea-urchins) or at the organic detrital level, when their contribution joins that derived from terrestrial runoff and mixes with the phytoplankton material of coastal waters. To these three sources of plant material in suspension in tidal waters must be added the microzooplankton that includes the eggs and larvae of littoral and sublittoral animals. All these forms of organic material constitute the food of the other basic animal level—the suspension or filter-feeding barnacles, mussels, oysters, tube-worms, anemones, and the majority of sedentary, cryptofaunal species. Then preying upon either the grazing or suspension-feeding animals are various types of carnivores, including dogwhelks, starfish, crabs, nudibranchs and, finally, sea birds and man.

The direct dependence of part of the shore population upon food material in suspension demonstrates at once that rocky shore communities are not discrete, self-contained and self-sustaining entities. Furthermore a high percentage of species shed reproductive stages into the sea, where growth and/or dispersal phases lasting from a few hours to several weeks or months may precede the final re-establishment on the shore. Recruitment in one locality may thus be dependent upon the reproductive output of another locality and upon the growth and mortality occurring during the drifting phase. Thus rocky shores are part of a wider inter-related system. They contribute massively to the coastal pool of basic production; they are much dependent upon the overlying water for suspended nutrients and food; they interact with the planktonic environment during the reproductive phase; and they contribute to some extent to the food cycles of subtidal and avian predators.

SOURCES OF BIOLOGICAL DAMAGE

Damage to rocky shore habitats and communities derives from two main and markedly dissimilar sources; chemical and physical changes of the sea water, i.e. pollution in all its various forms; and the over exploitation of their natural attributes for economic recreational and educational purposes, i.e. direct 'people-pressures'.

Since rocky shores tend to occur mostly on the more rugged coastlines, many are, inevitably, distant from centres of population and industry. This isolation is little or no protection against drifting oil nor against any generally distributed and persistent pollutants of coastal waters, but it does mean that the rocky shore is generally the least at risk of all coastal habitats between the tide-marks. Nevertheless there remain many stretches of rocky coastline adjacent to populous areas, and the people-pressures are tending to become less and less local.

Pollution

Of the various types of pollution damage, that which may be arising from generally distributed chemicals is not only difficult to demonstrate and assess, but also raises wider issues of control that are beyond both local jurisdiction and the scope of this book. Oil pollution is a subject that has a vast literature, both emotive and scientific, the latter being most ably summarized by Nelson-Smith (1972a). For present purposes it suffices to emphasize that although spillage, especially of crude oils into enclosed bays or directly onto the shore line, causes mortality to barnacles, limpets, snails, etc., most stranded oil has usually lost its toxic components by evaporation by the time it is washed ashore (Cowell, 1971). Heavy strandings may then have suffocating or mechanical effects on rocky shore organisms, but the bulk of the recorded damage has come from attempts to remove the oil rather than from the oil itself. It is worth pointing out that oil is degraded and removed by biological agencies of which marine bacteria are probably the most effective. The spraying of oil-slicks by emulsifiers not only aids mechanical dispersal of the oil, but by creating very small droplets and thereby increasing the ratio of surface area to volume, actively increases the bacterial degradation rate. Large lumps or encrustations of stranded oil on the shores are degraded more slowly, but grazing molluscs, especially limpets and chitons, scrape away encrusted oil during browsing. It then passes through the gut to emerge in a form more readily usable by bacteria.

The outfalls of large urban–industrial areas carry a wide range of potentially damaging substances: chemical effluents, oily runoff from land drains, domestic sewage with its high solid and organic content, and in some cases water of high temperature. The combined effects of such components are seen most strongly where they are discharged into confined waters such as harbours and docks (Persoone and de Pauw, 1968). Two opposing trends are recorded. Firstly destruction, which may be attributed to the chemicals, sediment, oil or heat, and secondly a selective promoting effect due to enrichment by nutrients and organic matter (e.g. Sawyer, 1965; O'Sullivan, 1971; Johnson, 1972). In the most severe cases there is an extreme paucity of life, with, at best, only slimy films of a few highly-tolerant diatoms and microscopic algae. Further away and with greater dilution, the diversity of species increases, although the actual components will depend upon the bias of the effluent; high sediment content, for example, being detrimental to the cryptofauna. At this stage the enrichment effect may now be reflected in an abundance of the few tolerant species, a further contributory factor being the reduced competition and predation. Organisms reported to prosper include diatoms, blue-green algae, the green alga *Ulva*, mussels, some species of barnacles and tube-forming polychaete worms (Smyth, 1968; Tulkki, 1968). With further improvement in water quality, the range of species increases and the effects of pollution will become increasingly difficult to detect, as they will be reflected only by particular species sensitive to particular effluent components, and may well be indistinguishable among the natural background variations (see pp. 153–4).

This type of sequence has rarely been recorded on rocky shores, perhaps because such habitats are rare in estuaries where most discharges are concentrated, perhaps also because most attention has been directed to either the planktonic or soft-bottom spheres where enrichment and sedimentation processes respectively appear to create more severe conditions and responses than in the intertidal area.

People-pressures

Man's other activities and uses of the rocky intertidal zone act more directly and often specifically upon adult stages of organisms, although there is also some unwitting general destruction. To this extent they are potentially more capable of immediate control, although this may require education and self-discipline on an unlikely scale.

Economic exploitation involves the collection of mussels, winkles, crabs, and sea-urchins for food, and of limpets, mussels, and anemones as bait for line fishing. A variety of red algae (*Porphyra, Dilsea, Gigartina, Chondrus, Rhodymenia*) may still be used for food or medicinal purposes, and the large brown algae are dug into the land for manurial purposes (Boney, 1965). Such activities on the scale of small isolated fishing–farming communities have probably existed for centuries and were largely self-regulatory. Local depletions resulted in diminishing returns, cessation of collecting, or transfer of activity to other areas, and thus an opportunity for recovery, either by growth of juveniles or by the inflow of new recruitment stages. Many red algae, the fucoids and the kelps are, however, now used for the commercial extraction of alginates, and stacks of dried algae awaiting transportation are a familiar feature on the Atlantic coasts of Europe. Much of the alga used in alginate extraction is even now washed up naturally, but hand cropping of the intertidal species (*Ascophyllum* especially) and the mechanical cutting or dredging of living kelps are now increasing and may result in locally severe denudation. Fortunately there are large reserves of inaccessible kelps not amenable to mechanical cropping, and both *Ascophyllum* and the red algae regenerate, provided enough of the holdfast and lower thallus is left untouched. Nevertheless the slow growth of *Ascophyllum* and its low establishment rate suggest that stocks are not as inexhaustible as the long-fronded festoons (how many tens of years old?) superficially suggest. Cropping any algae unwittingly changes the entire local biota, being especially destructive of the small forms of animal life attached to the weed itself, and to the underlying and shaded plants and animals on the rock. The large white encrustations that characterize former *Ascophyllum* habitats are the remains of calcareous red algae that have been subjected to unusual desiccation stress.

Just as the alginate industry is increasing its demands on natural stocks (and fortunately turning to culture methods), so improved transport and access to markets is starting to increase the scale on which other species of edible or souvenir value are being collected. Much sought species include the sea-urchins *Echinus* and *Paracentrotus*, the ormer or abalone *Haliotis*, and the coral *Eunicella*, while even the common winkle (*Littorina littorea*) and the laver weed (*Porhyra*) are sent to distant markets. These commercial activities could also be self-regulatory in that exhaustion of local stocks could lead to cessation of activities, but conducted on a 'clean out and move on' basis, using power boats, SCUBA-diving techniques, etc., they could well be placing some strain on the local regenerative powers of the species concerned.

This danger is being augmented by a considerable increase in amateur collecting by holidaymakers. Indeed, since such people are not motivated by profit and collect at leisure they are likely to be much more thorough searchers than commercial collectors. Their thoroughness is reflected by the mounds of urchin shells that characterize picnic sites on Mediterranean shores, and by the virtually complete absence of edible winkles on many suitable parts of the Britanny coast. (Limpets and topshells, although prized by gourmets, do not yet find general favour!) Holidaymakers also cause much unwitting damage. The general and usually unfounded belief that there are crabs and lobsters of edible size in every tide pool or cleft, or under every large stone in the lower shore leads to much physical damage to the habitat of the cryptofauna.

The final and increasing stress on accessible rocky shores comes from the unprecedented increase in field studies in recent years. Traditionally this was an activity confined to university parties, but now it involves in Britain at least, all age groups in the educational system and too often these are under the supervision of other enthusiasts with little training in biology. While the more serious and disciplined groups utilize the superb opportunities of rocky shores to study the ecology of the common species, too many 'leave no stone unturned' in the search for either crabs and fishes (the only marine animals they know), or at the other extreme, for novel or rare species. By autumn the more heavily-used beaches are often marked by stones and boulders with strangely white upper surfaces—the dead remains of encrusting algae and animal colonies that were formerly face down, and about whose existence the eager amateurs were largely ignorant.

It would be unduly alarmist to suggest that all these activities constitute a serious threat to rocky shore populations in general. Extensive areas are protected by their remoteness or inaccessibility (and a few because they are on military firing ranges), so that for those species with planktonic phases—and these fortunately are the majority—there should always be reservoirs from which new recruitment can take place. Unfortunately we do not have as much data as one would wish on reproductive and recruitment rates, on the drifting phase, and on lifespans and the onset of sexual maturity to feel confident that the entire system is in balance. Indeed for many species and localities we do not have enough information on population levels to be sure that the known declines of some species in some localities are not considerably more general. Certainly the increasing use of more of the coastline must lead us to ask when the still largely local deteriorations will become matters of wider concern. Conservationists in northern latitudes may perhaps take comfort from the fact that the degree of risk will probably be directly related to the length of the holiday season, and to the warmth and clarity of the sea!

ASSESSING BIOLOGICAL CHANGES

Before considering particular types of change and their possible interpretation and implications, it is necessary to appreciate some of the general problems of detecting significant changes and of making assessments. The permanent and drastic reduction in the number of species that is the most certain indication of severe stress presents no problem of detection. Fortunately this condition is rare or highly localized on rocky coasts. More probably, deterioration will manifest itself in the following ways.

(1) Modest changes in species diversity.
(2) Changes in numbers of individuals of particular species (increase or decrease is possible because the physical conditions or ecological balance or both may have changed).
(3) Changes in the 'performance' of particular species, i.e. they may grow at faster or slower rates, or have a different pattern of reproduction.

These lesser scales of response do not lessen their importance for, ideally, the aim should be to detect incipient changes before the situation deteriorates too far. Unfortunately the greater subtlety makes detection more difficult and could involve professional biologists in several years study.

Taking numerical change as the superficially more amenable criterion, two approaches exist: observation of changes with time, or comparisons with control shores that appear to be free from stress or pressure. Observation of temporal changes necessitates regular quantitative surveys, and several years might elapse—depending upon the severity of the stress—before a definite trend emerged. Unless this trend were very strong (i.e. severe decline; dominance by one or two new species) and clearly correlated with observable pollution or over-use, there would be no certainty that the change had anything but natural causes.

Few members of the general public, and not all biologists, appreciate that abundance and local distribution are not constant. They fluctuate seasonally, annually, and irregularly in response to natural factors, but the scale of the fluctuations and relative importance of various possible causes is barely understood for even the common species. Climatic variation is the most obvious and widely operating factor. The severe winter of 1962–63 caused heavy mortalities—now largely made good—on the western coasts of France and Britain. But smaller variations of sea and air temperature can differentially influence fecundity, growth, and general efficiency of species competing for the same space or food resources, and so bring about temporary but striking changes in local abundance. Even when reproductive output appears similar from year to year, the final recruitment rate depends upon events during the planktonic phase, including such chance matters as wind direction and surface drift when the larvae are ready to settle.

Having successfully settled, the young stages have entered a habitat where competition, predation, and grazing may produce highly local but striking variations in numbers over periods of a few months or several years. For example, a decrease in limpets (because of several years low recruitment, or bird predation, or human collecting) will permit the fucoid algae to increase. These in turn destroy existing barnacles, and prevent barnacle settlement until increased numbers of limpets reduce the establishment rate of the algae. Similarly, a sudden incursion of starfish into the lower shore may remove all the mussels and lead to barnacle dominance—until a successful new mussel settlement swamps the barnacles. Even highly successful recruitment has its pitfalls, for the resulting intense competition for space or food among individuals of the same species may result in mass mortality.

This local instability, which obviously complicates the problem of detecting trends, is most marked in the lower half of exposed and semi-exposed shores. In the former it largely reflects the dominance of animals with planktonic larvae and the abundance of predators; in the latter it arises from the ability of some exposed-shore and some sheltered-shore species to intermingle and to ebb and flow as the ecological balance shifts slightly. Only in full shelter where the large fucoid algae are very abundant is there a natural long-term stability—yet even here removal of the algae at once allows barnacle-dominated communities to develop.

Another source of temporal variation stems from the accidental introduction of species from other geographical areas. They may arrive on the undersides of boats (Nair, 1962), through insufficient care when commercial species are transported live, or simply by drifting on floating objects. If they are able to survive the local conditions they may then become established and, as happened between 1940–60 with the Australasian barnacle, *Elminius*, may displace the local species over much of the shoreline.

It may be the case that a locality was the scene of a survey made many years ago, even during the last century. These records can, with discretion, provide useful information for comparison with the present day. The data in the old surveys will usually take the form of species lists, and not only may the names have changed but checking upon current presence or absence is a skilled task often requiring intimate experience of the habitat preference of the species concerned. A negative result today may do no more than reflect the ignorance or impatience of the investigator.

The second approach to the detection of biological change, that is comparative surveys, may therefore appear more suitable, especially if the sites are along a gradient of decreasing stress or use. At once, however, one encounters the many local variations arising from differences in shore topography, wave action, substrate, aspect, etc., so that ideally all such comparisons should be between sites that are as physically alike as possible. This is a requirement rarely to be met on natural rocky shore lines, so unless the biological changes are dramatic they may need to be interpreted by scientists fully experienced in the subtleties of local distribution. Finally, it should be appreciated that temporal fluctuations, since they are partly caused by local biological factors, may be out of phase from one part of the gradient to another.

These cautionary paragraphs are necessary as an antidote to the widely held assumption that ecological deterioration is easily detected by 'instant ecologists'. The more severe the stress or the greater the other forms of activity the more this will be so, but in the majority of cases considerable difficulty will be encountered. Pessimistic or alarmist diagnoses may err on the 'right side' from an outright conservation viewpoint, but it is irresponsible to cry wolf unnecessarily and could be counter-productive in the long run. But balanced judgements and predictions are difficult and those responsible for local control should seek whatever advice and assistance is available from experienced biologists and hydrographers—appreciating in advance that without long-term data or new, intensive investigations this advice may be cautious and limited.

BIOLOGICAL OBSERVATIONS AND POSSIBLE INTERPRETATIONS

The two forms of stress—pollution and people-pressure—clearly overlap in their ultimate biological effects and it is necessary if possible to try to distinguish between the two. Initially this problem is simplified because areas subject to severe pollution are not likely to be used for leisure and educational activities. The overlap is most likely to occur in the vicinity of large resort areas discharging effluent on to open coasts; yet these are the rocky habitats where natural fluctuations are most likely to occur. Because of the resultant interpretational difficulties there is little point in trying to implicate either type of man-made influence unless there are strong *a priori* grounds, i.e. proximity to a large outfall; discoloured, oily, or highly turbid water; floating and stranded sewage; observed and frequent use by commercial or amateur collectors and educational parties. Furthermore, because of the apparently greater effect of sewage enrichment and solids upon the phytoplankton and soft-bottom communities, corroboration should if possible be obtained from these habitats rather than relying solely upon deductions based on shoreline observations.

The following paragraphs list types of observations and their probable or possible interpretations.

(1) *Extreme scarcity of macroscopic flora and fauna*
    This is the extreme manifestation of pollution. It occurs most extensively in the very enclosed waters of harbours and docks, but may be seen on a limited scale in the immediate vicinity of outfalls discharging just at low-tide level of rocky coasts. See also (2).
    Caution: not to be confused with (4) or (5) or (6).

(2) *Hard surfaces abundantly coated with slime*

This condition arises from the high sediment and nutrient content of sewage outfalls. The fine silt and organic solids are colonized by tolerant diatoms and microscopic algae (Cyanophyceae). Usually accompanies (1).

Caution: (a) Slimy films of diatoms occur in spring on bare rock where the water is naturally turbid.

(b) Slippery films of Cyanophyceae characterize the high-level black zone (see p. 148) in spring.

(3) *Scarcity of sedentary suspension-feeding animals* (e.g., barnacles and oysters on open rock; sponges, hydroids, polyzoans and ascidians on under-surfaces)

This condition arises when an increase in fine sediment smothers existing animals or prevents larval attachment. It may be deduced from the presence not only of sediment, but also of dead, silt-filled barnacle shells, possibly some development of slime (2), and other replacing organisms [see (10)]. This is probably the most common form of 'pollution' on rocky shores.

Caution: Not to be confused with temporary or permanent absence due to natural causes (4) and (6).

(4) *Patches of bare rock*

Unless accompanied by silt and slime, and near to an outfall, patchiness will almost certainly have a natural explanation, e.g. severe predation or grazing; mass elimination due to overcrowding; storm damage; abrasion caused by seasonal movement of sand or shingle; friable rock surface unsuitable for attached organisms. Confirmation may be obtained from the possible presence of a few healthy individuals that escaped destruction, from slime trails of migratory grazing-molluscs across the bare areas, or by observing phases of reoccupation and later denudation.

(5) *Absence or scarcity of algae* (animals present)

This has natural causes. It may reflect excessive wave action (see p. 148) or grazing by limpets and snails (intertidally) and by sea-urchins (at low-tide level and below). The grazing control exerted by limpets can be deduced from the algal colonization that follows experimental removal of limpets from areas several metres in extent.

Caution: Since sea-urchins are migratory their absence when observations are made does not mean they are not responsible.

(6) *Absence or scarcity of rock animals* (perennial algae present)

In the absence of sediment or slime, (2) or (3), this condition is natural. Barnacles, mussels, oysters, tube-worms and even limpets and some snails have difficulty competing with perennial turf-like carpets of red algae or long-fronded masses of fucoid algae that usually predominate at low levels or in shelter respectively.

(7) *Scarcity or absence of newly settled spat and juveniles*

Since reproductive stages are generally more susceptible to pollutants than adults their absence, when sediment and slime are also lacking, could be the first indication of a deterioration of water quality. This is however a difficult and specialized criterion to use, for it necessitates accurate knowledge of reproductive periods and of the natural annual variability in recruitment rates, together with the skill to locate and identify the settling stages.

(8) *Scarcity of large individuals* (small forms present)

If this occurs over a wide area of coast it could reflect recruitment failures several years earlier. On a more limited scale, and especially if it involves species used as food, bait, or souvenirs, or which are much-prized biologically it will indicate over-collection. On a very localized scale scarcity will probably stem from natural predation or will represent a stage in the recolonization of areas previous denuded.

Caution: Small size is not always an indication of young animals. The small closely-packed mussels in the upper shore, or the small limpets living among dense upper shore barnacles are small because feeding and growth are difficult for them in these habitats. They may well be much older than large animals living elsewhere on the same shore.

(9) *General increase in algae, and especially of green algae*

This could be a response to the increased nutrients associated with sewage discharge. The green alga *Ulva* is regarded as a good sewage indicator. However, an increase in algae, with the green algae as the first stage colonists, is an almost automatic result of a decrease in grazing animals, especially limpets, which may be natural or due to over-collection [see (5)].

Caution: Green algae (*Enteromorpha* especially) are short-lived, opportunistic species capable of rapid colonization and growth; and in some years during late spring and early summer they may blanket all other species on the shore in localities far removed from pollution. They also occur naturally in the lower spray zone, especially in spring, and are common where fresh water flows across the foreshore.

(10) *Increase in mat-forming organisms*

Some small, filamentous red algae and polychaete worms (*Polydora*) bind fine sediment to form a spongy layer 2–3 cm deep on rocks in the lower shore. Thus a local and new development of this mat-formation may indicate a new high level of suspended solids.

Caution: Mat-formation occurs naturally in non-polluted areas, Therefore its use as a pollution indicator should be in conjunction with other sediment indicators, i.e. (2) and (3).

(11) *Local appearance of 'exotic' species*

As a pollution indicator, this phenomenon relates to the sudden appearance of warm-water species in enclosed areas receiving the heated discharge of power stations. The most conspicuous example on hard substrates is the replacement of native barnacles by *Balanus amphitrite*. (Additional species in this and other habitats are given in Naylor, 1965a). Further manifestations of thermal pollution are changes in growth rates and in the reproductive cycles, but these require specialist study.

(12) *Appearance of whitish encrustations on boulders*

This results from the death and bleaching of calcareous red algae and encrustations of colonial animals on upturned surfaces. It may result from storm action, but usually reflects ignorant collecting [cf. (13)].

(13) *Appearance of whitish encrustations on open rock surface*

This results from the death and bleaching of calcareous red algae and encrustations of colonial animals when natural or artificial removal of the larger algae exposes them to severe desiccation [cf. (12)].

(14) *'Abnormal' coloration of attached algae*

Bright pink colours in normally dark red algae, or green patches in brown algae can indicate death due to adverse water quality. However the range of natural colour variation in red algae is considerable and so professional experience may be required, while bleaching and green patches in kelps especially is also associated with death due to desiccation when unusually low tides coincide with hot weather.

Numerical criteria have been deliberately excluded above, because the range of normal variabilities is so wide that to give suggested guideline densities could well lead to erroneous conclusions locally. For example, limpet densities can range up to 400–500 m^{-2}, yet numbers as low as 20–50 m^{-2} can still be 'normal' at another site on the same shore; and the latter population could still prevent dense growth of algae if the individuals were large and the nearest source of reproductive stages of the algae were some distance away. Whatever the initial density, it is a progressive or abrupt and severe decline in numbers or in maximum size (and/or an increase in perennial algae) that should serve as the warning signal and direct attention to possible causes. Numerical pitfalls also arise from seasonal changes at the same site. Barnacle densities may reach 40,000–60,000 m^{-2} just after spatfall, but could fall, for entirely natural reasons, to 5,000–10,000 m^{-2} or even to zero before the next recruitment period. It is an unfortunate fact that before significant trends can emerge from the monitoring of temporal changes, it may be necessary to have several years familiarity with the ranges of densities and fluctuations that are 'normal' for that particular shore with its own distinctive set of conditions.

In gradient surveys, the absolute numbers matter much less. Provided one is aware of the sources of misinterpretation suggested on p. 155–6, it is the existence of persistent, *relative* changes along gradients away from the area of suspected stress or pressure that should immediately justify concern.

PROTECTIVE AND CONSERVATION MEASURES

The alleviation of damage that is attributable to local sources of pollution is a matter about which little can be done immediately or individually. It involves the entire process of waste treatment and disposal by coastal townships, and in spite of the increasing biological awareness, seems likely to be a side-benefit from remedial action that will be taken for aesthetic and health reasons. Nevertheless, the careful and systematic compilation of data relating to biological damage along the shoreline must surely serve to strengthen the case for more enlightened policies.

Oil pollution, by contrast, invokes immediate decisions and actions for which elaborate plans and chains of responsibility are now being established locally, regionally, and nationally in many countries. The many options of treatment and equipment for use, both before and after stranding are discussed in many reports and are summarized by Nelson-Smith (1972a). Accordingly there is little point in further summary here except to reiterate a few general recommendations of principles.

(1) Since removal or treatment is easier and preferable at sea rather than on shore, and because freshly spilt oil is the most toxic, every effort should be made to ensure the presence of permanent or portable booms and barriers etc. to prevent oil reaching shores in the vicinity of oil terminals and transhipment points.
(2) Once oil is stranded hasten slowly, for the treatment may be more harmful than the oil.
(3) On little-used or remote shores, give favourable consideration to 'no action'.
(4) If the nature and quantity of the strandings necessitate cleaning action use mechanical means of dispersal or collection in preference to emulsifiers and dispersants.
(5) If amenity considerations justify rapid cleaning, use only recently developed emulsifiers of low toxicity, and always under the direction of experts from the oil industry or Government laboratories. Indeed, all who might be involved in such action should have the techniques and the reasons for them explained in advance, not during a crisis. Thus we might avoid repetition of the case in which an order not to spray detergents on a particular beach was interpreted as 'apply by any other means except spray'!

Increasing awareness of the need to regulate human activities on the shore line has led in some countries to the establishment of coastal or marine reserves, to limitation of access, or to restrictions on the collection of specimens. While it is possible that such practices will become more common in future, the most valuable contribution to future conservation can only come from enhanced awareness and education on the part of the many and diverse users of the shoreline. This must inevitably be a long term objective that will most likely be achieved only via the present generations of schoolchildren and through the co-operation of teachers, especially those responsible for field courses and excursions.

The current upsurge of interest in field studies and conservation could, unwittingly, be detrimental to the cause of conservation. To prevent this paradox it is to be hoped that teachers who are not biologically trained will increase their own awareness of the less obvious organisms on the shore (e.g. the cryptofauna under stones) and that all will attempt to inculcate some simple code of conservation practice. This should place the emphasis upon observation rather than collection and ultimate destruction (especially of species that are hard to find), on returning free-living specimens to their correct habitat, and on carefully replacing all stones and boulders in exactly the same position as before.

In order to get the same message across to the general public it is perhaps worth suggesting that local authorities or naturalists' associations might place notices in car parks and on the approaches to beaches explaining the importance of conservation and inviting public co-operation.

PART III

PREDOMINANTLY TERRESTRIAL ENVIRONMENTS

Chapter 9

Sand-dunes

L. A. Boorman

INTRODUCTION

Geomorphological history and general character

Sand-dunes and salt-marshes are not infrequently found in close association since both are formed from marine sediments stabilized by vegetation. Salt-marshes however are formed from fine sediments brought in by water transport, while sand-dunes are composed of coarse sediments transported by wind. Thus there are two prerequisites for dune formation, an adequate supply of sand and sufficient wind. The stabilization of the mobile sand by vegetation is the second phase in the process. Details of this aspect will be considered later. It is sufficient to mention here that vegetation plays two roles in dune formation; it stabilizes existing sand surfaces and it speeds further accretion by reducing wind speed over the surface of the sand. The stabilization of sand surfaces by vegetation differs from other forms of stabilization in that the plant cover can keep pace with the accreting surface. This contrasts with the action of artificial stabilizing agents which can only work at a single preset level, and therefore for a relatively short period of time.

Supply of sand

Dunes are formed by the wind transport of marine sands and thus a prerequisite of dune formation is the occurrence of sand-flats at a sufficiently high level for the surface layer to dry out between the tides. The lower limit of sand movement will vary with the tidal regime and the interaction of wind and atmospheric humidity. The quantity of sand within this drying zone will in itself be insufficient for extensive dune building, but as the sand is removed from this zone by wind action it will be replenished from the zones below it by water transport during tidal submergence. Krumbein and Slack (1956) divide the beach zone of wind transport into two; the foreshore where there is transport by water currents, by waves, and occasionally by wind; and the backshore where there is transport primarily by wind with breaking waves having only a minor influence. In practice the situation is complex and the zones recognized should perhaps be regarded as forming a continuum up the shore with water having a decreasing effect, and wind an increasing one. It is important to remember however that the beach is the area of transport between the submarine sand deposits and the growing dunes. The sources of these marine sand deposits vary; they can be relatively old (Pleistocene) marine deposits, or from more recent coastal erosion, or of recent fluvial origin. In many areas, e.g. off the coast of East Anglia, there is considerable recycling of beach-sand with exchange between the beach and offshore sand-bars. This sand supply is also augmented by cliff erosion and longshore drift. Cambers (1976) has shown this to be true for the dunes at Winterton, Norfolk. The most extensive studies of the origin of coastal sands have been done on the Dutch coast. The coast of the Netherlands from the Ems to the Scheldt, approximately 400 km long, is predominantly a dune coastline broken only by a number of inlets and estuaries. Eisma (1968) has shown that the coastal sands are mainly reworked early and middle Pleistocene deposits, with a mixture of (Saalian) glacial sands and (Rhine–Maas) river sands; with the former source being more important to the north and the latter to the south of the Netherlands. The reworking of beach sands with a mixing of sand types occurred mainly during spells of coastal erosion in the Holocene.

Wind speed and sand transport

The classic studies of Bagnold (1941) were done on desert sands but they are also applicable to dune sands. He showed that sand moves by saltation; a few grains moved by the wind fall back on the sand surface at an angle and on impact set other grains in motion. He also showed that there was a minimum threshold wind speed for sand transport ($4 \cdot 5$ m s^{-1}) and that above this speed the rate of sand movement was proportional to the cube of the wind speed. In any consideration of sand erosion and deposition the effect of topography on wind speed is paramount. Olson (1958b) described how wind velocity near the surface increases up the windward slope to a maximum on the crown of the dune, then decreases very rapidly down the leeward face. The extent of the zones where $4 \cdot 5$ m s^{-1} is exceeded at the surface will depend on the overall wind speed, but the windward slope and windward side of the crown will be the zones of highest wind speeds, with the possibility of sand movement and erosion, and the leeward face will be primarily a zone of sand deposition. The critical factor is wind speed at the sand surface. This will depend on the roughness of the surface of the sand.

Bare sand has a fairly low surface roughness (Bagnold, 1941) and this means that wind speeds only slightly higher than the critical speed will induce sand movement. Sand surfaces with vegetation have a high degree of roughness and vegetation reduces the wind speed to values well below the critical speed thus inducing sand deposition and dune growth (Olson, 1958b). The wind at the surface will increase as the height of the dune increases and it is this factor, taken with the average wind speed in the area, that determines the overall height of a dune. In an exposed area the dune reaches a height at which the rates of erosion and accretion, even from a vegetated surface, are balanced. In a relatively sheltered area however rate of supply of sand up the windward face may limit the extent of vertical growth. It should be noted here that the supply of sand from the shore by onshore winds may be in a different direction from the strongest winds that produce maximum dune growth. The highest dunes are found some way inland when the prevailing winds are onshore (Willis *et al.*, 1959) but when the prevailing winds are offshore the highest dunes are near the seaward edge (Ranwell, 1972a).

Dune formation and movement

Wind is the transport agent for dune building and it also determines dune form and dune movement. The other important factor is whether the coast is a prograding one with an abundant sand supply or an eroding one with a limited sand supply. In the former case, particularly if the winds are not very strong, a series of ridges is formed in sequence, with the youngest to seaward, and in due course these dunes become stabilized more or less where they were formed. If however the sand supply is limited, and particularly in exposed sites, the seaward ridge grows to a maximum height and then by a process of local erosion and deposition moves landwards, either as a parabolic dune or sometimes as a complete dune ridge. Ultimate stability may only occur after centuries of instability (Ranwell, 1972a). This can happen in the prograding situation if there are sufficiently strong onshore winds. An understanding of the dynamics of a particular system is necessary before any dune stabilization measures are planned. Newborough Warren, Anglesey, is a dune site which is very exposed to onshore winds. Landsberg (1956) showed a good general correlation between wind direction and orientation of the parabolic dunes. The rate of movement of the parabolic dunes and entire dune ridges was measured by Ranwell (1958) who showed that maximum accretion normally occurred at 0–18 m behind the crest of relatively stable dunes but up to 183 m downwind in the unstable dunes. He estimated that a coast dune would take 50 years to reach maximum height and that its mean rate of movement landward would be 6·7 m per year. Thus a dune would only have moved landward sufficiently for the development of new embryo dunes after 70–80 years. The time for further movement landwards sufficient for the acquisition of a degree of stability would be much longer than this. These results were derived from one particularly exposed site but other exposed sites have comparable rates of movement e.g. Kurische Nehrung 5·5–6·1 m per year (Carey and Oliver, 1918). Sites on the east coast of England are rather more sheltered and dune movement is slower. Ranwell (1958) showed a rate of movement of 1·5 m per year on the Norfolk coast. Dune movement has been explained on the basis of erosion of the windward face with deposition to leeward. The process of erosion usually continues downward until the level of permanently wet sand is reached. This produces a flat-bottomed dune valley or dune slack, which becomes colonized by its own characteristic vegetation. The whole process of dune and dune slack formation has been studied in detail at Newborough Warren (Ranwell, 1959, 1972a) and is diagrammatically summarized in Figure 52. The orientation of the dunes in relation to the wind may appear to be parallel or at right angles depending on whether the primary dune ridges or the trailing arms of the parabolic dunes respectively are the more prominent. Dune slacks are also formed in the prograding system between successive dune ridges.

The time sequence in dune processes along the Dutch coast has been described by Jelgersma and van Regteren Altena (1969). They distinguish two types of dune landscape, 'older dunes' and 'younger dunes'. The older dune landscape consists of a series of low ridges more or less parallel to the coast separated by peaty depressions. They are considered to have been formed in the sub-boreal and early sub-atlantic and completed before Roman times. These older dunes formerly occurred all along the western Netherlands but now are restricted to the coast between Hook of Holland and Schoorl to the north. Elsewhere the older dunes have been lost by erosion. The younger dunes were formed from the twelfth century onwards. Excavations for industrial development at IJmuiden and at Haarlem have exposed the sequential deposition of younger dunes over a similar sequence of old dunes. The authors consider that the present day dune system was completed before the end of the sixteenth century with only local redepositions at a later date. The period between the formation of the older and younger dunes was one of dune soil formation. There is a podsolized layer between the younger and older dunes which is the result of a period of afforestation between the first and twelfth centuries A.D. The younger dune landscape with ridges to 40 m above mean sea level contrasts with the

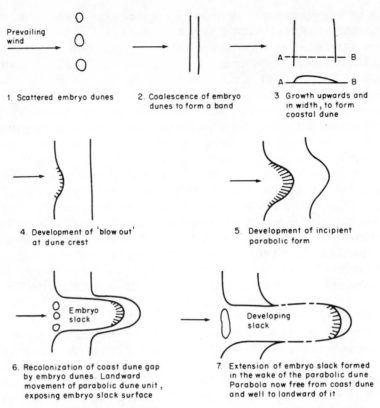

Figure 52. The formation and development of a parabolic dune slack unit (Reproduced from Ranwell, 1972a, by permission of D. S. Ranwell)

comparatively low older dunes and there are parabolic dunes inland (van Straaten, 1961). Coastal erosion of the older dunes, which began about the eighth century, released the large amount of sand required for the formation of the younger dunes. Young dunes frequently contain relatively larger amounts of calcium carbonate, primarily from shell fragments and detrital calcite grains, and as dunes age this calcium carbonate is removed from the surface layers by leaching. The speed of this process depends on climatic conditions. Ranwell (1972a) has estimated that, in temperate climates, dunes with an initial calcium carbonate level of 5% will lose free carbonate from the top 10 cm in about 300 years. Randall (1973) has used changes in calcium carbonate content to study dune development in the Monach Isles (Outer Hebrides) and suggests the use of this technique for studying geomorphological changes in dunes elsewhere.

It is an over-simplification to say that dune development is aligned to the prevailing wind. It has been shown that the dune trends in eight out of twelve parabolic coastal dune fields studied in the northern hemisphere are to the left of the resultant sand movement (average recorded wind direction weighted in relation to its ability to move sand) (Landsberg, 1956). This has been explained by friction on the earth's surface inducing deflection of the ground wind from the geostrophic direction (Warren, 1976). There is a difference of as much as 15° in the deflection between water surfaces and nearby rough surfaces e.g. roof tops (Haltiner and Martin, 1957). Two sites showed no difference (including Newborough Warren, Anglesey) and in the remaining two sites there were differences between the exposure of the wind recording station and the dunes. The correct interpretation of past dune development and the prediction of future trends, particularly in natural dune movements, is important both for dune stabilization work and for long-term dune conservation programmes.

Role of plants in dune growth and development

The growth and stabilization of dunes depends largely on the existence of a barrier to reduce surface wind speed and thus to increase accretion and reduce erosion. Man-made fixed barriers can be used but they rapidly become buried in sand and cease to function. Dune vegetation does not suffer from this disadvantage as it is able to grow with the developing dune, and to a certain extent to regenerate when damaged. However the conditions at different stages in dune development vary and thus more than one species is required to complete the process of stabilization. The detailed interplay of factors will be described in the next section but the main features of the stages in dune succession and the species involved will be considered here. The embryo dunes are still liable to occasional submergence by peak spring tides. Initially, in north-west Europe these embryo dunes are fixed by the perennial grasses *Agropyron junceiforme* and *Elymus arenarius*. Both species are able to spread vegetatively and to develop extensive root systems both horizontally and vertically. Nicholson (1952) has shown that *Agropyron* can tolerate accretion rates of up to 60 cm per year, however it is unlikely that either *Agropyron* or *Elymus* could withstand continued accretion at these rates (Ranwell, 1972a) and they are effectively limited to areas with accretion rates of less than 30 cm per year. *Ammophila arenaria* on the other hand can withstand accretion rates of up to 1 m per year at least for short periods (Ranwell, 1958) and *Ammophila* also has the ability to spread horizontally by rhizome growth.

The two species *Ammophila arenaria* and the American *A. breviligulata* are among the most important dune-forming species in the world. *Ammophila* readily regenerates from broken fragments of rhizome and thus eroded *Ammophila* dune faces can be recolonized naturally as soon as erosion is halted. It has been observed that only when *Ammophila* is growing in an accreting situation it is really vigorous but the reasons for this are not entirely clear. There seems to be some evidence that roots have a short life and that vigour can only be maintained when new internodal growth, and in consequence new root growth, is produced in response to burial (Hope–Simpson and Jefferies, 1966). In the non- or slowly-accreting situation other species less tolerant of accretion than *Ammophila* can come in and competition could have some effect but it is difficult to imagine any other grass or herb species competing successfully with vigorously growing *Ammophila*. The succession from *Ammophila* dune to dune grassland is an interesting one (Hewett, 1970). Recent measurements at Holkham, Norfolk, have shown that large quantities of organic matter are tied up in the *Ammophila* phase in the form of standing dead stems. The transfer of this material back to the soil is a key factor in the development of dune grasslands. The total amount of standing crop of *Ammophila* was estimated to be approximately 1 kg m^{-2} dry matter and of this only about 10% was living material.

The persistence of the dune grassland stage is encouraged by grazing animals but in their absence dune scrub and dune woodland can develop. Dunes have been extensively afforested with *Pinus nigra* spp *laricio* and *P. sylvestris*, e.g. Culbin, Moray and Holkham, but there is little if any natural or semi-natural dune woodland in Britain. In the Netherlands however there is semi-natural dune woodland at Voorne and elsewhere. The dune woodland at Voorne is particular rich and the stages in the succession to the climax *Quercus robur* woodland are well shown (Adriani and van der Maarel, 1968; Blom and Blom-Steinbusch, 1974).

The dunes of north-west Europe contrast with those of Japan (Ishizuka, 1974). Mobile dunes are based on *Elymus mollis* and the sedge, *Carex kobomugi*. The latter is also an important dune builder together with *Wedelia prostrata* (*Compositae*). *Elymus mollis* plays a continuing role in nitrophilous habitats where the original vegetation has been disturbed by human interference. Scrub seems to come in at an earlier stage than in European dunes with *Rosa rugosa* playing a dominant role. The natural succession is to *Quercus dentata* and locally *Q. mongolica* forests. On the coastal dunes to the north-west of Hokkaido pure forests of the conifer *Abies sachalinensis* occur. Frequently however the dunes were planted with *Pinus thunbergii* after the ancient dune forests had been cut down in medieval times. In the warm temperature parts of Japan however *P. thunbergii*, although also extensively planted, is considered by Ishizuka to be occupying its natural distribution.

From this it can be seen that wherever they occur sand-dunes have a range of plant types which can be divided as follows:

(1) Primary colonizers e.g. *Agropyron junceiforme* and *Elymus mollis*
(2) Dune builders e.g. *Ammophila arenaria* and *Wedelia prostrata*
(3) Scrub colonizers e.g. *Hippophaë rhamnoides* and *Rosa rugosa*
(4) Dune forest e.g. *Quercus robur* and *Q. dentata*
(5) Dune plantations e.g. *Pinus nigra* and *P. thunbergii*

The species mentioned above are the dominant species, but frequently there is a rich flora associated with them. Dune management may affect any one species directly or through its effect on one of these dominant species.

SPECIAL FEATURES OF DUNES

Special ecological factors

Dunes are a complex of many very different habitats with contrasting edaphic factors. Perhaps the one dominant factor is the water regime which produces habitats that vary from permanent dune lakes to arid open communities on the exposed dune ridges. The water regime is also a powerful modifying influence on the other factors. Dune temperatures vary between extremes of heat and cold in contrast to other maritime habitats. The build-up of nutrients and organic matter in the soil is a key factor in dune development.

Dune water regimes

Because dune soils are coarse-grained they allow free water movement and the field capacity is relatively low, particularly in young dune soils with little organic matter. Field capacity increases with the increase in organic material as the soils develop and mature. Salisbury (1952) showed that the field capacity increased from 7% in young dune sand to 33% in old dunes. Drainage water moves downward under the influence of gravity to the permanent water-table that exists under most dune systems. This water-table is often floating on a deeper layer of infiltrated seawater. The relation of the water-table to the soil surface varies from standing water (dune lakes), through damp slacks with the water-table near the surface, to the situation on high dunes when the water-table is out of reach of all dune species. Dune plants have been classified according to their requirements in relation to the water-table, and their dependence or otherwise on the permanent water level (Londo, 1975a, 1976). He divided them into three main groups:

(1) Hydrophytes—requiring permanent standing water

(2) Phreatophytes—growing within the sphere of influence of the buried water surface
(3) Aphreatophytes—not dependent on groundwater.

The phreatophytes are further divided into obligate and non-obligate phreatophytes depending on whether or not they can grow elsewhere. The list produced enables the probable water regime of a site to be quickly determined on the basis of the species present.

Once out of the influence of the water-table, in practical terms 2 m or more above the average winter-table level, a plant is largely dependent on seasonal rainfall for its water supply. Dune annuals exploit the higher soil moisture levels in the autumn and spring to complete their lifecycle, remaining dormant as seeds during the summer drought. Dune perennials have root systems adapted to make full use of what water there is during the summer drought. There are two main strategies here; some species e.g. *Carex arenaria* have extensive shallow root systems over a large area allowing these species to exploit summer showers; other species e.g. *Hypochoeris radicata* have a deep tap root system to exploit the rather higher moisture levels at greater depths. The role of internal dew formation has been discussed at some length (e.g. Ranwell, 1972a) since, as Salisbury (1952) showed, rainfall alone appears to be insufficient to maintain dune plants through long periods of drought. Certainly at night the soil temperature gradient is suitable for an upward movement of water vapour from warmer wetter layers deep down (Willis *et al.*, 1959) but their measurements suggested that it was unlikely that the soil temperature would be low enough for the condensation of dew. Dune soil moisture levels critically affect plant germination and establishment and thus small differences in rainfall over the dunes are likely to affect plant distributions. The implications of differences in solar radiation with regard to slope and aspect, and its effect on soil moisture, will be considered in the section on temperature. Except possibly under still conditions the distribution of rainfall will be influenced by variation in topography. Sandsborg (1970) has studied the variations in rainfall over a hillock in Sweden comparable in size to those found in dunes. He showed that there could be variations of up to +12·5% on the windward slope with a wind speed of 3–5 m s^{-1} and −15·0% on the leeward slope. The effect on the vegetation will be minimal under mesophytic conditions but, under the critically low moisture levels that occur on dry dunes, differences of this magnitude could well affect the balance between successful seed germination and plant establishment and complete failure. The extent of this effect will depend on the direction of rain bearing winds; if they come from a number of directions the differences will tend to cancel out whereas if rain is associated with a particular wind direction there will be a marked effect.

Dune slack vegetation obtaining its water from the groundwater-table is less critically affected by variations in the rainfall, but the vertical range of these communities is affected by fluctuations in the water-table levels. On the basis of studies of Newborough Warren, Ranwell (1959, 1960, 1972a) proposed a classification of damp slack plant communities on the basis of an annual fluctuation of the water-table there of 0·7–1·0 m.

He recognized four zones:

(1) Semi-aquatic—flooded autumn to spring with the water-table never more than 0·5 m below the surface
(2) Wet-slack—water-table always within 1 m of surface
(3) Dry-slack—water-table 1–2 m below the surface
(4) Dune habitat—water-table always more than 2 m from the surface. Plants independent of groundwater.

The dry slack category of Ranwell corresponds with the mesosere as defined by Londo

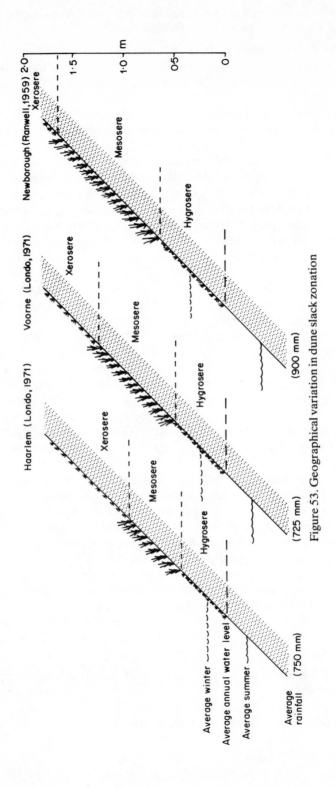

Figure 53. Geographical variation in dune slack zonation

(1971). The vertical extent of the mesosere in the dunes near Haarlem was determined by Londo as being 0·65–1·15 m above the summer water-table. Intermediate values of 0·75–1·50 m have been obtained by van der Maarel (1966b) for the dunes at Voorne (Figure 53). Thus in the dunes at Newborough Warren not only does the mesosere have a greater vertical range, but the mesosere–hydrosere (dry slack–wet slack) boundary is higher above the water-table (i.e. the vertical extent of the hydrosere is greater). These differences are explained by Londo (1971) as being a result of variations in the magnitude of the annual fluctuations in the water level, in the texture of the sand and in the climate.

Jones and Etherington (1971) have studied the growth of certain dune species in relation to the waterlogging of the soil. They showed that dune slack species such as *Carex nigra* and *C. flacca* tolerated waterlogging well, whereas the growth of *Festuca rubra*, a species characteristic of the open dune, was inhibited even under only partly waterlogged conditions. The growth of *Agrostis stolonifera* was also inhibited by waterlogging but the shoot growth that this species made survived, whereas the shoots of waterlogged *Festuca* plants died. However *Agrostis* has a lower overall growth rate than *Festuca*. Thus under dry conditions *Festuca* outgrows *Agrostis* while under wet conditions *Agrostis* can succeed. It is interesting to note that *Carex flacca* did best under half-waterlogged conditions, and that it is classified by Londo (1971) as a non-obligate phreatophyte. The growth of *C. nigra* was however stimulated by complete waterlogging: *C. nigra* is classified by Londo as an obligate phreatophyte. It seems probable that the non-obligate phreatophytes grow in slacks because of a reduction in competition, while obligate phreatophytes have a specific requirement for waterlogged conditions, but more information is needed on this point.

Temperature regimes in the dune habitat

Although coastal climates tend to be more equable than those further inland because the sea exerts a moderating influence, temperatures within sand-dunes show considerable variation. Certainly the influence of the relatively warm sea tends to prevent very low temperatures in the winter but sand surface temperatures in the summer can be very high. Boerboom (1964) made a study of the microclimate of the Wassenaar dunes near The Hague. At 1 cm under the bare sand a maximum temperature of 58·4 °C was recorded, but under a moss layer this temperature reached 63·5 °C: the temperature of the air was only 30·4 °C. The moss layer, because it is dark, absorbs the heat while the sand reflects it; hence the difference in the readings. It does suggest however that the initial effects of dune colonization may be to make the climate more extreme. At the other end of the scale Boerboom also showed that the minimum temperature could be lower in the dunes than on nearby sites. Following the heatwave already referred to, a cooler spell gave a night minimum of 4·9 °C at the nearby meteorological station, while the dune slacks were recorded as having 'heavy frosts'. Similar results have been recorded by the author at Holkham, north Norfolk. A number of temperature probes at different depths on a north-facing and a south-facing slope showed some of the variations within the dune habitat with regard to spring and summer and to cloudy days and sunny days (Figure 54). The differences in soil temperature between different sites can be largely explained on the basis of the effect of slope and aspect on the amount of solar radiation received; and this effect can be estimated empirically (Frank and Lee, 1966). The very high surface temperatures can affect the vegetation directly, e.g. Boerboom (1964) recorded the heatwave of June 1957 as being fatal for *Gentiana cruciata*; but it can also have a marked effect on the water regime. Summer showers falling on hot sand speedily evaporate with

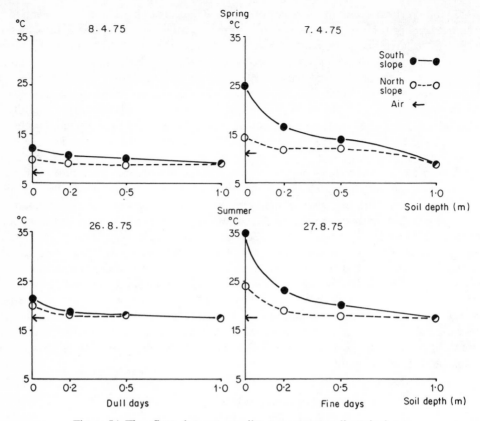

Figure 54. The effect of aspect on soil temperature gradients in dunes

little if any moisture reaching the plant roots and the higher temperatures on south facing slopes magnify this effect. In a study comparing the vegetation on north and south slopes of a sandy (glacial) hillock, Pahlsson (1966) concluded that the differences observed seemed to result mainly from induced differences in soil moisture content. Pahlsson (1974a) showed that the higher soil moisture on the north slope resulted in a significant leaching which was not observed on the southern slope. He suggested (1974b) that the course of colonization on the south slope may be affected by the lower degree of competition that results from the more extreme conditions which favour annual species such as *Cerastium semidecandrum* and *Erophila verna*. While these results were from a sandy hillock on an inland site there is no reason to doubt that they also apply to dune sites.

Organic development of dune soils

The succession from the embryo dunes through to dune woodland with an increasing plant biomass is paralleled by changes in the soil. Boerboom (1963) showed that the humus content of the soil increased from 0·4% on the first coastal dune ridge with *Elymus* and *Ammophila*, and 0·8% on the second ridge with *Hippophaë*, through 1·2% in dune grassland, to 4·9% in dry *Betula* woodland and 5.6% in damp *Betula* woodland. Ball and Williams (1974) have obtained rather lower results for Holkham with 0·2% for the *Ammophila* foredune, 0·9% for dune grassland and 1·9% under planted pines. It is probable however that the Holkham dunes are younger than the Wassenaar dunes studied by Boerboom, with a shorter time for soil development; in addition the higher rainfall at Wassenaar, about 750 mm (Koninklijk Nederlands Meteorologisch Instituut, 1972) compared with about 570 mm per annum at Holkham, will facilitate the organic development of the former. The effects of the gradual build-up in organic material are considerable. The water holding capacity of sandy soils is increased by even small increases in organic content, and this greatly increases the possibility of further successful plant colonization. The initial development of the embryo dunes is greatly influenced by the chance occurrence of organic material along the drift line, e.g. seaweed, water-borne plant debris from salt-marshes, and even litter of human origin. The supplementing of natural low organic levels by artificial means is of great importance in dune stabilization.

Nutrient content of dune soils

The build-up of organic material through the dune succession is accompanied by a general increase in the major nutrients. However there is an unequal distribution of nutrients; dune slack habitats are enriched at the expense of the higher sites as a result of leaching and water movements under the influence of gravity. This naturally favours further vegetation development which in turn increases the difference (Olson, 1958a). Sand in the embryo dunes tends to be nutrient poor. The accumulation of tidal litter provides the basis for the growth of the embryo dunes but as they develop in height the input of tidal debris decreases, and other sources of nutrients become important. Hassouna and Wareing (1964) have suggested that nitrogen fixation, e.g. by bacterial associations, play a part in the nutrition of *Ammophila* and Willis (1965) showed that lack of nitrogen could be a major limiting factor in the growth of *Ammophila*. In the scrub stage of many dunes nitrogen fixation is carried out by root nodules on *Hippophaë* (Stewart, 1967) and also *Alnus* (van Dijk, 1974). The build-up of nitrogen under stands of these two shrub species has important consequences if they are to be cleared to encourage the spread of species-rich grass and herb communities. The higher nitrogen levels often lead to the rapid spread of certain weedy species such as *Urtica* or *Cirsium* at the expense of many other less vigorous species.

The other main source of nitrogen and certain other nutrients is through airborne material. The quantities involved are not very large (Allen *et al.*, 1968) but in a nutrient-poor habitat such as sand-dunes these could have a significant affect. For example it is estimated that nitrogen input per annum amounts to 9–19 kg ha^{-1}. Taking the top 25 cm of the soil with a nitrogen content of 0·01% (e.g. the first dune ridge), the aerial nitrogen washed down by rain amounts to between 1·5 and 2·5% of the total per annum. The quantities of the other major nutrients occurring in rainwater per annum were rather lower e.g. phosphorus 0·2–1·0 kg ha^{-1} and potassium 2·8–5·4 kg ha^{-1}. Nevertheless over a period of years this source could be significant. The aerial supply of nutrients will be particularly important for lichens that are able to take up the nutrients directly from the rainwater. The distribution of nutrients within the dunes varies from place to place and it is likely that the local patterns observed in dune vegetation are at least in part due to differences in nutrient levels. This hypothesis has been strengthened by recent experiments at Holkham where the addition of different combinations of nitrogen, phosphorus, and potassium affected different species and species groups in different ways (Figure 55). The raising of overall nutrient levels however led to a decrease in species diversity, with those species best able to respond to the higher nutrient levels becoming dominant. These changes have considerable implications for dune management.

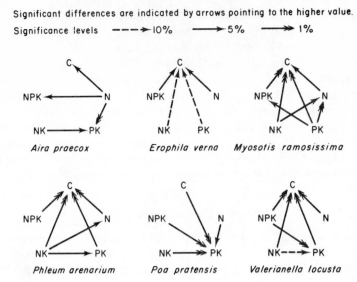

Figure 55. Changes in the frequency of selected species in a dune sward following the addition of nutrients. All are dune annuals except for *Poa pratensis*, a perennial grass. C: untreated. NPK: nitrogen, phosphorus and potassium. NK: nitrogen and potassium. PK: phosphorus and potassium. N: nitrogen

The calcium content of dunes can be very variable, and it affects plant growth both directly and through the effect on pH. Leaching of calcium from the older dunes has already been discussed but there exist mechanisms that lead, at least locally, to the replenishment of calcium in the surface levels of the soil (Salisbury, 1952). These are the wind blow of shell fragments, recycling of calcium via plant roots and leaves, and burrowing activity of rabbits bringing up the calcium-rich soil from the deeper layers. Certain dune species tend to accumulate calcium in their leaves to very high levels, e.g. *Cynoglossum officinale* (Freijsen, 1975), and in the autumn when these leaves die the calcium will be redistributed over the surface layers. The activities of rabbits will tend to create a mosaic of microhabitats and this effect is emphasized by the non-random distribution of rabbit pellets which produces scattered small nutrient-rich areas. Work on *Cynoglossum* (Freijsen, 1975) has shown that there are close links between calcium and nitrogen requirements. Davies and Snayden (1973) have shown that the wide ecological amplitude of *Anthoxanthum odoratum* is at least partly due to the evolution of ecotypes with different calcium requirements and that these differences have evolved over quite small distances (less than 30 m).

These local nutrient variations, which are so important for species and habitat diversity, are emphasized by the very low background levels of nutrients in the dunes and will be evened out by any overall increase in dune nutrient levels. In dune slacks the nutrient picture is complicated by plants rooting down into the water-table, which may have a totally different nutrient spectrum from that of the soil. In old slacks, particularly the drier ones, the surface soils may be rather acid and nutrient-poor but there may be calcareous nutrient-rich groundwater. In this case the flora will be composed of a mixture of shallow-rooting, acid-demanding species and deeper-rooting, calcium-requiring species. Thus it is important not only to know the nutrient requirements of a particular species but also the depth of the active rooting zone of that species.

Types of environmental pressure affecting dunes

Man can affect dunes directly by the introduction or removal of species; or indirectly by changing some environmental factor. Species introduced by man include pine trees, rabbits, domestic grazing animals, and *Hippophaë*. Man affects dunes through trampling and through the industrial and residential development of dune areas. Environmental factors that are affected include various forms of pollution and artificially induced changes in the water-table.

Afforestation of sand dunes

Woodland is the natural vegetational climax for many dune areas but the deliberate planting of tree species, particularly conifers, has occurred extensively. The planting of *Pinus thunbergii* in many parts of the Japanese dunes has already been mentioned. In Britain there are extensive plantations, particularly of *Pinus sylvestris*, at Culbin, Moray and Tentsmuir, Fife. The total area of afforested dune had been estimated at 4,000 ha (Macdonald, 1954) but the area today is greater; the total wooded area of Culbin is 3,100 ha and of Tentsmuir is 1,300 ha with many smaller plantations. The initial plantings were done with the aim of dune stabilization and other species of *Pinus* notably *P. nigra* spp *laricio* and *P. pinaster* were used. The coastal fringe of woodland provides shelter for commercially viable conifer plantations further inland, principally *P. sylvestris*. While a few extra species may occur in these conifer plantations the deep shade of the mature plantations virtually eliminates the natural ground flora. Ovington (1951) showed that in afforested parts of Tentsmuir the water-table was some 27 cm lower than in comparable unplanted areas, reflecting the high water abstraction power of forest trees. In addition drainage ditches have been cut recently to reduce waterlogging and thus improve the growth of the pines, and this further reduces the level of the water-table for quite a large area around. Similar effects have also been shown at Culbin (Wright, 1955) although the organic content in old woodland soils did show an enhanced water holding capacity. In addition to the changes in the water regime of the soil, Ovington (1950) showed that there was a decrease in soil nutrient level and an increase in soil acidity following afforestation. Once established, *Pinus* species seed and regenerate freely and there is often a considerable natural spread outside the area of afforestation.

The impact of rabbits

The animal that has had the greatest visible effect on sand-dunes is undoubtedly the rabbit. The loose dry soils of dunes provide the rabbits with an ideal substrate for their burrows and even in high summer the damp slacks provide some grazing. In medieval times the rabbit was deliberately cultivated on sand-dunes but control was difficult and wild populations became established permanently (Ranwell, 1972a). The impact of rabbits is considerable both through their burrowing activities and the effect of rabbit grazing on vegetation. Rabbit burrows can be the focal point for occurrence of blowouts but whether or not this will occur will depend on the overall degree of stability within the system and the extent of the area affected by the rabbits. The other effect of the burrowing activities of rabbits is the reworking of the dune soils already described. The impact of rabbits on the vegetation depends very much on the density of the rabbit population, and also the spatial distribution of their burrows. Rabbit grazing tends to be restricted to areas around active burrows and over a wide range of rabbit population densities there is a mosaic of areas with different intensities of grazing. At low densities rabbits help to maintain species diversity by keeping a check on the more vigorous growing shrubs and grasses. More open communities available for dune annuals, intolerant of competition from perennial species, occur near the burrows. As the densities of rabbits increase the grazing pressure produces a close sward of annual and perennial species together with mosses and lichens. This sward is often only 2 cm in height. However there is a delicate balance between this stage and the next, which is represented by the progressive breaking up of the sward with increasingly extensive areas of bare sand. Once the turf has been destroyed, recovery is very slow and there may be a considerable period of mobility before the grazing pressure has dropped sufficiently. The advent of myxomatosis in the 1950s caused a large decline in rabbit numbers, and in many dune areas there was a corresponding increase in the spread of scrub. Since then however, rabbit numbers have recovered to such an extent that over-grazing is again a problem in certain areas.

Use of dunes for grazing farm animals

Sheep, cattle, and ponies have been grazed on dunes, although the heavier foot pressure exerted by these larger animals tends to damage the sward with the consequent risk of erosion. Well established turf of old dune pasture is able to withstand this better than the earlier stages in dune succession, and thus the former areas can provide useful grazing without any significant risk of erosion. Areas with a long history of grazing are often very species rich. A good example of this is the Westduinen on Goeree in the Delta area of the Netherlands. An area of 163 ha provides some 100 cows and 10 horses with summer grazing. The species richness of the area is attributed to a combination of intensive grazing with a high regularly fluctuating water-table (Blom and Willems, 1971). In the north of the Netherlands (Wadden Islands) and in Scotland there is extensive sheep grazing which produces a characteristic close-cropped sward. In Britain in recent years however, there has been a decrease in the utilization of relatively marginal grazing areas as part of a general movement away from extensive stock farming, in part at least the result of increased labour costs making the supervision of low-return grazing uneconomic.

The effect of Hippophaë plantings

Hippophaë is a native dune shrub that has been extensively planted in Britain as an aid to sand stabilization. It can grow very vigorously on calcareous soils and spreads both by suckers and by seed. Vegetative spread is particularly effective with a maximum recorded horizontal rate of spread of 4 m in two years (Rogers, 1961) and it can also withstand soil accretion of up to 0·5 m per year. As already described it has root nodules that can fix atmospheric nitrogen and therefore it is largely independent of soil nitrogen levels. *Hippophaë* forms a dense cover and effectively blankets the existing herb layer although in areas where it is not at its most vigorous it does form a mixed community with other shrub and herb species (Westhoff, 1952). The problem of *Hippophaë* lies in the way it has spread over large areas in recent years. This was undoubtedly encouraged by the great decrease in rabbit numbers following myxomatosis. Clearance of *Hippophaë* is a problem because of its vigorous suckering growth and the rapid regrowth of cut stems; also, because of the high residual nitrogen levels in the soil, the reinstatement of dune grassland is difficult. Nevertheless it is a valuable species in dune stabilization and also in the control of public access. Because of the problems associated with it in Britain, it has been the subject of a special report (Ranwell, 1972b).

Direct human impact on dunes

The direct impact of people on dunes can be divided into the two rather different categories of recreation and building development. Sand-dunes form a very attractive natural target for recreation and the effects of recreation have been the subject of several studies recently (e.g. Liddle, 1975a; van der Werf, 1970; Leney, 1974; Trew, 1973). The effect of public pressure on dunes is primarily through trampling but there are side-effects e.g. through the deposition of litter (Teagle, 1966). In many cases trampling occurs because of the juxtaposition of the dunes between a car park or an access point and the beach; but an increasing number of people seek to explore the dunes, and the effects of their feet on the vegetation pose an increasing problem. Van der Werf (1970) estimated that in the Meijendel, near The Hague, a dune valley system of 104 ha, some 33 ha had lost their natural character entirely (made-up roads and tracks 7 ha, parking places 2 ha, paths 6 ha, and very heavily trampled areas 18 ha), 47 ha were moderately to strongly affected, leaving only 24 ha more or less unaffected. It might be mentioned that this was in an area where there was active management towards dune protection and conservation. A dune sward is particularly vulnerable to damage by trampling and the younger stages in the dune succession are the most vulnerable as well as being more attractive to tourists. The problem is made worse by the fact that once destruction of the sward has occurred even light trampling prevents re-establishment. At low levels however trampling can increase species and habitat diversity, and may thus be considered to be beneficial (Westhoff, 1967; van der Maarel, 1971; van Leeuwen, 1966a). In areas necessarily subject to heavy trampling pressure artificial walkways have been used to limit the damage, and fences and shrubs have been used to control access. Pony-trekking is a popular holiday recreation and where dune areas are crossed the environmental impact is considerable, as large numbers of horses follow more or less the same track completely destroying the dune sward. The only alternatives to large scale damage are either to exclude horses from the dunes completely; or to provide specially reinforced trackways such as has been done at Oxwich in the Gower Peninsula.

There are quite a number of coastal golf courses on sand-dunes. The greens themselves are intensively managed and species poor but the rough areas which are intermittently mown support a relatively rich flora (Wallace, 1953; Beeftink, 1966). Ranwell (1975a), in a study of the flora of a Jersey golf course, showed that intensively managed fairways had 5–10 species 25 m^{-2} compared with 30–40 species 25 m^{-2} in undisturbed (unmanaged) areas. He also pointed out that these undisturbed areas were always liable to be disturbed in the interests of the golfers.

The use of dunes for housing or industrial development not only destroys the dunes that are built upon, but there are also side-effects over a wider area. It is, for example, no longer acceptable to have any degree of sand mobility in dune areas adjoining industrial or residential development thus necessitating a check on natural processes. Residential development in dune areas inevitably leads to a marked increase in public pressure on the surrounding dunes. Industrial development frequently causes pollution which can affect nearby semi-natural areas. Following the industrial development at Europort a decrease in epiphytic mosses and lichens in the dune woodland on Voorne has been observed (van der Maarel, 1966b). Another aspect of the impact of human development is the extraction of groundwater from the dunes. The Wassenaar dunes near The Hague have been used for this purpose for the past hundred years (Boerboom, 1960). Boerboom's studies showed that the extraction of water particularly affects the dune slack vegetation and the rate and magnitude of changes in water level are so great that the natural plant communities are unable to compensate. The water-table has been so lowered in some areas that even agricultural crops have been affected and to such an extent that compensation payments have been made on a large scale (Colenbrander, 1975). The natural water reserves in many dune areas have been exhausted and the artificial recharging of aquifers is extensively practised (Roebert, 1975). However it is expedient to use natural riverwater for this, particularly as there is a considerable degree of cleaning of the water by natural processes while it is in the dunes. This cleaning process cannot be repeated indefinitely as the accumulation of particulate matter leads to a decrease in porosity of the sand and a corresponding loss in efficiency of the area as a rechargeable aquifer. In addition, increased levels of dissolved chemicals cause marked changes in the vegetation. Londo (1975b) showed that infiltration of the dunes near Zandvoort with polluted Rhine water caused a decrease in the mean number of species per quadrat (6 m × 0·5 m) from 8 in 1963 to 3 in 1971 with the nitrophilous species *Urtica dioica* and *Cirsium arvense* becoming dominant.

SURVEY AND ASSESSMENT OF DUNE HABITATS

With almost any habitat, but particularly with a habitat as complex as the dune habitat, thorough survey and assessment is a prerequisite for planning effective management. The design of research for a particular site depends on the information that is available but in most cases a basic survey will be required to assess the species, amount, distribution and habitats of the flora, and fauna. This will then allow specific impact studies to be made to assess the effects of particular threats on the natural fauna and flora.

Basic survey techniques

The aim of the basic survey is to determine what species are present, in what quantities and how distributed. As far as plants are concerned there are three main approaches, the classical 'Tansley' approach, the computer based survey, and the phytosociological approach based on the recognition of distinctive plant associations. These three approaches all have their own particular advantages and disadvantages.

The classical approach

The classical approach, that upon which Prof. A. G. Tansley based his description of *The British Islands and their Vegetation* (1949), is essentially a simple descriptive method based on the observed cover–abundance of the main species in each successive layer of vegetation. Taken with simple habitat information such as slope, aspect, probable depth of water-table, biotic factors, and percentage ground covered by vegetation it can give a very useful overall picture of an area. A recent example of this approach is the survey of the dunes at St. Ouen's Bay, Jersey (Ranwell, 1975b). The disadvantages of this approach is in the handling of the large amounts of data involved in the description of larger sites, and also in making comparisons between sites. It is a subjective method and thus is influenced by the observer and it means that comparisons of results obtained by different observers on different sites must be made with extreme caution. The use of standardized methods, perhaps with duplicated recording sheets, can minimize this, but a degree of subjectivity will always be present. The value of this approach can be increased considerably by the production of even a simple vegetation map showing the distribution of the main vegetation types (e.g. Saito *et al.*, 1965). The production of this sort of map is facilitated if good quality aerial photography can be obtained (Fuller, 1972, 1973), but dune plant communities do not show up on aerial photos nearly as well as salt-marsh plant communities, consequently extensive ground verification will be necessary.

The phytosociological approach

The phytosociological approach favoured on the continent is similar in outline to the classical approach with the important difference that some of the subjectivity involved in the definition of vegetation types has been removed by the use of the predefined groupings that make up this highly formalized classification of vegetation. It is not so easily applied and operators have to learn to recognize particular associations in the same way that a plant species is recognized. Whether the fundamental principles of this system are valid is perhaps less important here than the ease of application and ease of presentation of the results. A comparison of the phytosociologically-based vegetation map of Goeree (Blom and Blom-Steinbusch, 1974) with that of Sarugamori (Saito *et al.*, 1965) shows the greater complexity that is possible with the former approach. Blom and Blom-Steinbusch recognize 58 distinct plant communities based on 22 main associations. Saito *et al.* recognize 15 plant communities but these community definitions are only defined for the survey of that particular area whereas the main associations are as defined for the whole of the Netherlands (Westhoff and den Held, 1969) and are probably valid for the whole of north-west Europe. It must be admitted however that it is easier for a layman to understand a brief community description than the more precise phytosociological terms.

A new approach to vegetation mapping that seems to combine the advantages of both systems has been outlined by Londo (1974). He classifies the different plant communities on the basis of the apparent form of the vegetation. The detailed description of these vegetation units gives characteristic species as well as general ecological information. Each of the 70 vegetation types is indicated by a two-letter code, the first letter giving the general nature of the vegetation and the second the specific type. Further variants can be indicated by a third letter and a notation is given for describing simple types of patterns in the vegetation. His approach seems to offer a simple standarized method, and it merits further attention.

The computer approach

The advent of cheap computing facilities has made it possible to handle data from large numbers of quadrats located at random and this can virtually eliminate the subjective human element in the collection and analysis of data. An example of this approach is provided by the survey of approximately 100 ha of sand-dune vegetation of Holkham by Moore (1971). He used 394 quadrats 50 cm × 50 cm in size located at 5 or 10 m intervals along 24 transect lines running seaward across the vegetation. He used a computer program giving an information statistic (Lance and Williams, 1968) but the approach was otherwise similar to the classical association analysis and the results were presented in the form of a dendrogram. The quadrats recorded were divided into 15 groups. The descriptive nature of these groups falls somewhere between the Tansley approach and the phytosociological approach. Moore considers that they represent stages in the development or retrogression of the dune community and he presents a scheme showing their inter-relationship, identifying those phases where biotic factors such as rabbit grazing or trampling are considered to be having an effect. The dune slack vegetation at Tentsmuir, Fife, was classified by Crawford and Wishart (1971) using an association analysis technique and the resultant units were studied in relation to age and floristic development. It is likely that future large-scale surveys will mainly be based on computer methods because of the objectivity obtainable and because of the ease of handling very large amounts of data. Computer methods are also being applied to phytosociological studies (van der Maarel, 1969).

Habitat impact studies

Three types of impact studies will be considered here, environmental impact on or by a specific species; environmental impact by a particular factor that may affect many species (e.g. trampling); and large-scale overall impact studies of whole areas.

Species impact studies

These studies will either be directed at a species that is having a special impact on the dunes, e.g. rabbits or *Hippophaë*, or a species whose status in the dunes is under some threat. The methodologies used will depend on the nature of the species involved but it is important that they are suitably standarized so that valid comparisons can be made and so that particular species and areas are not considered in isolation. Much can be learnt by comparison of the behaviour of that species in other dune areas and possibly also in non-dune areas. The *Hippophaë* survey previously referred to (Ranwell, 1972b) is a good example of what can be achieved by considering a particular problem on a countrywide basis. Problems that seem to be acute and urgent on a local basis take on a different perspective when viewed on a national scale. A good start to any species impact study is to look at the situation over as wide an area as possible. This achievement of a degree of perspective is of equal importance in the consideration of problems relative to the safeguarding of particular species. In the interpretation of species impact studies it is important to consider results obtained in the light of the ecology of the area involved (basic survey) and against the background of the general processes involved in dune ecology.

Environmental impact studies

Two specific environmental impact studies will be considered here; the effect of changes in the water regime, and the effect of human trampling. While the level of the groundwater can easily be monitored the interpretation of the results obtained needs to be done with care, particularly with regard to the complexity of the different long- and short-term cycles involved. Identification of the causal factor (e.g. climatic changes, artificial water extraction, 'improved' drainage, greater water uptake by the developing dune shrub community) must be approached with great caution, particularly as it is unlikely that a single factor is solely responsible. The study of the impact of changing water levels has been greatly facilitated by the work of Londo (1975a, 1976) with his list of plant species and their degree of dependence on the groundwater level. He also indicated in his list those species that are likely to be affected by changes in the chemical composition of groundwater. His list is based on the Dutch situation and while it probably holds good for areas in Britain with comparable rainfall, it is possible that in some of the wetter western and northern areas the level of the groundwater-table may be less critical.

The effect of trampling on the dune flora and fauna has been the subject of a number of studies involving different approaches. Liddle and Greig-Smith (1975a,b) studied the changes in soils and in vegetation associated with tracks and paths in the Aberffraw dunes, Anglesey. They showed that passage of people and vehicles caused proportional increases in soil bulk density and soil penetration resistance, and that public pressure had marked effects on the vegetation. However, while greenhouse experiments showed that trampling caused detrimental damage to plant shoots, soil compaction alone could have beneficial effects through a greater moisture retention. Liddle also suggested that changes in penetration resistance could be used to monitor the intensity of trampling. Chappell *et al.* (1971) suggested the use of changes in bulk density for this purpose; while other workers (Goldsmith *et al.*, 1970; Burden and Randerson, 1972) have observed the actual distribution of people. Other methods available include the use of trampleometers (Bayfield, 1971a). Soft wires are inserted vertically into the soil; the proportion of them bent over is taken as a representation of the degree of trampling. It is particularly applicable to areas with low levels of trampling. Higher levels of trampling may be monitored by the use of pressure operated tally counters or by the use of time-lapse photography. Recent studies by the author at Winterton, Norfolk, indicate that the mapping of paths from aerial photographs can give a useful assessment of the extent and intensity of public pressure on a dune area. The changes in vegetation associated with different levels of trampling have been recorded in various ways. Transects are frequently used and such results are easily presented (Streeter, 1971; Liddle and Grieg-Smith, 1975b). Other workers have recorded plots and made comparisons with the use of regression analysis techniques (Randerson, 1969) or phytosociological analysis (Liddle and Grieg-Smith, 1975b). The response of the vegetation to trampling is a complex one depending on a number of different factors. Grime and Hunt (1975) have noted that plant species with a high growth potential are common on paths, and Liddle (1975b) has shown a relationship between the primary productivity of vegetation and its ability to withstand trampling. This is an interesting hypothesis that merits further investigation.

Another approach to the problem is to measure the effect of environmental changes associated with trampling on particular plant species. Blom (1976) has shown that the radicles of *Plantago major* and *P. coronopus* are more capable of penetrating compacted soils than are *Plantago lanceolata* or *P. media*. *Plantago major* is commonly associated with heavily trampled areas and it has been tempting to attribute this to the apparent high degree of mechanical resistance of its prostrate growth form, rather than considering aspects of germination and establishment. The value of species studies lies perhaps more in the direction of obtaining a long-term understanding of the ecological processes involved rather than in the immediate solution of specific management problems associated with high levels of trampling.

It is clear that dune swards are particularly susceptible to trampling and that this is, in part, due to the relatively low growth rates attributable to the low nutrient status of the soil. Recent experiments at Winterton have shown that the reduction in sward height associated with trampling can be minimized by the addition of fertilizers. Sports turfs that have to withstand high levels of trampling are normally well fertilized although Schothorst (1965) has shown that high levels of nitrogen could reduce tolerance to trampling. The rapid leaching of nutrients from porous dune soils is another problem but recently available slow-release fertilizers may provide the answer although there is evidence that they may not persist for as long as claimed (Woolhouse, 1974). It is clear that more research is needed into the use of fertilizers in increasing the resistance to wear of dune swards, and also into the side-effects of raised nutrient levels on the dune vegetation generally.

Area impact studies

So far consideration has been given to assessing specific kinds of impact on specific areas. The pace of industrial development is now such that consideration may have to be given to the assessment of a major development on a whole dune system. An example of this is the Maasvlakte development on the dunes of Voorne to the south and south-west of Europort. The coastal dunes of Voorne have an area of 1600 ha, 75% of which is a nature reserve; 700 species of vascular plant occur there (Adriani and van der Maarel, 1968). The need for land for further industrial development along the Rotterdam waterway led to the reclamation of a large area of sand-flats to the north of Voorne and to the industrialization of the reclaimed flats and adjoining dunes (Maasvlakte I). Further development is proposed in front of the coast of Voorne itself (Maasvlakte II). Adriani and van der Maarel (1968) have estimated changes in diversity on the basis of estimates of changes in the vascular flora. They consider that Maasvlakte II will cause a loss of 10%. However changes in the water-table (an estimated fall of 1 m) would increase this figure to 16%. Allowing for secondary works the total loss could be as much as 25%. Figures of this kind, with adequate backing, make the evaluation of environmental impact rather easier. It must however be remembered that the assessment of environmental impact was greatly facilitated in this particular case by the amount of information on the area that had been accumulated over many years. Forecasting changes is never easy and short-term intensive impact studies are no substitute for detailed records collected over a number of years and a careful monitoring of changes.

THE USES OF DUNES

Some of the various uses to which sand-dunes have been put have already been mentioned. It is intended here to provide a brief summary of the major uses and their implications. The final section will be concerned with a consideration of specific management techniques. The relative importance of the different uses of sand dunes depends very much on personal viewpoints, and the sequence in which the different uses of dunes are considered is based on the complexity of the involvement; from the simple use of the dunes as a barrier against the sea to the preservation of the whole complex of ecosystems when dunes are nature reserves.

Economic uses of sand dunes

In a low-lying country such as the Netherlands the value of the dunes for sea-defence has long been appreciated. The Rijkswaterstaat, the Dutch authority responsible for coastal protection, issues specifications of heights and widths, for different situations, to which the dunes must be maintained whenever they serve as natural sea-walls. The value of these natural sea-walls is perhaps less appreciated in Britain but incidents such as that at Sea Palling, Norfolk, when in 1953 the storm surge broke through the dunes and washed away houses drowning seven people, serve as reminders of the importance of dune conservation and maintenance. The main requirement for sea-defence is for a high continuous dune ridge with an adequate degree of stability. Any factor that could affect its stability is a threat to its value as a sea-wall. Thus there is a conflict here with recreational usage and also with nature conservation as the fixing of the dune ridge inhibits the natural mobility of the system and the development of habitat diversity.

Forestry

This has already been discussed at some length. Afforestation produces a high degree of surface stability, however if major erosion does occur there is not the automatic regeneration of vegetation that is possible when *Ammophila* dunes are eroded. In addition when dune woodland is felled care must be taken to avoid erosion of bare sand surfaces.

Grazing

This is a decreasing use of dunes; which is unfortunate because it is compatible with most other uses, especially recreational and educational–scientific, and it can be positively beneficial for the latter.

Industrial and residential

Demands for land for development put increasing pressure on all land areas particularly those of relatively low agricultural value such as dunes. There is likely to be total loss of the natural fauna and flora in the areas actually built on; and consequential losses over the surrounding area, through pollution, water extraction, increased recreational pressures, and through subsidiary works such as roads. In addition it is important to realize the implications of sand mobility on the development itself. Residential development and roads on the edge of the dunes at Southport, Lancashire, have suffered considerably from the incursions of mobile sand.

Water extraction

At present this is particularly a feature of the Dutch dunes. The environmental impact of the changes in the groundwater have already been discussed, but there is also the impact of the associated works and control of public access that is necessary.

Military uses

During the last war a number of dune systems were used as training areas and this led to a degree of disturbance only acceptable in a wartime situation. Most the damage has now been repaired despite problems caused by unexploded objects buried in the sand. Present military use of dunes is largely limited to relatively harmless (environmentally) installations such as the radio-masts on the Westduinen of Goeree, although these can provide a hazard to migrating birds.

Recreational uses

These can be divided into two categories on the basis of whether visitors come to the dunes for their own sake or whether they are merely a pleasant background against which to enjoy the sun, the sea, and the open air. The problem of damage to vegetation associated with large numbers of people has already been considered; but for the large number of people to whom the dunes are just a background, the degree of artificiality necessary to safeguard them may be acceptable. It is more difficult to reconcile the pressure of large numbers of people with the natural flora and fauna of the dunes.

Educational and scientific value

The educational and scientific value of dunes is based on their great floristic diversity. Not only to they have a considerable endemic flora but they also provide a habitat for many grassland species that are restricted elsewhere because of modern intensive farming and the ploughing up of old pastures. The floristic richness of Voorne has already been mentioned. Adriani and van der Maarel (1968) put it even more clearly into perspective by their statement that 52% of the total Dutch vascular plant flora occurs there, in an area amounting to 0·05% of the area of the Netherlands. There is a similar situation with regard to the birds—66% of Dutch breeding birds (111 species) occur at Voorne. The range of plant communities formed by the rich flora are equally significant with 63% of the plant communities (phytosociological alliances) of the Netherlands occurring there. Admittedly Voorne is an outstanding dune area but it exemplifies the potential richness of dune habitats. Dunes are also significant for the range of ecological processes that can be studied there. The natural degree of sand mobility that helps to maintain habitat diversity is likely to be unacceptable in the interests of coastal-defence and may also affect any roads or buildings nearby. If the teaching-use of dunes involves large numbers of people, damage can result from the trampling pressures, although it is easier to control and limit such damage as visits are usually supervised and along definite routes. In smaller dune areas a decision may have to be taken between conserving relatively large numbers of different habitats each represented by small areas for teaching purposes or concentrating on the conservation of more viable areas of those habitats best represented at that site or not adequately represented elsewhere.

DUNE CONSERVATION AND MANAGEMENT

Sand stabilization—materials and methods

Sand stabilization may be required to strengthen sea-defences, to prevent mobile sand blowing and affecting roads etc., and to reinstate areas badly affected by excessive public pressure. The early stages in dune succession are particularly vulnerable to trampling and generally people will have to be excluded from areas where dune stabilization works are attempted. The technical details of dune stabilization work have been presented by Adriani and Terwindt (1974) in an excellent booklet produced by the Dutch authorities responsible for coastal-defence. A number of different dune-forming plants, both grasses and shrubs, are considered. It is perhaps worth mentioning here that the species called in the U.K. *Agropyron junceiforme* (sand twitch) is called *Elytrigia juncea* on the continent. Data are presented by the authors of the report and also by van der Stege (1965) on the natural limits of this species. The results clearly show *Agropyron junceiforme* as a plant of flat wide beaches, that can build up dunes to a level where other dune species, with a greater vertical potential, can come in. The special feature of *Agropyron* lies in its greater salt tolerance (6% NaCl) compared with *Ammophila* (0·8%). *Elymus arenarius* is also a useful pioneer sand binder with a high salt tolerance (up to 12%) although it tends to die back in the winter and is thus unsuitable on its own in exposed situations. The growth habits of *Ammophila* have already been described and its rapid response to burial by drifting sand show that it is clearly a very important dune builder; and it can be propagated by seed or vegetatively. In certain localities a hybrid occurs between *Ammophila arenaria* and *Calamagrostis epigeios*; *Ammocalamagrostis baltica*. It is even more vigorous than *Ammophila* but it is completely sterile and has to be propagated vegetatively; and at present the stocks for planting are limited. It was planted extensively on the Norfolk and Suffolk coasts following the 1953 floods (Ellis, 1960). If a fertile strain could be produced it would have great potential for dune stabilization work. The main dune-stabilizing grasses are illustrated in Figure 56.

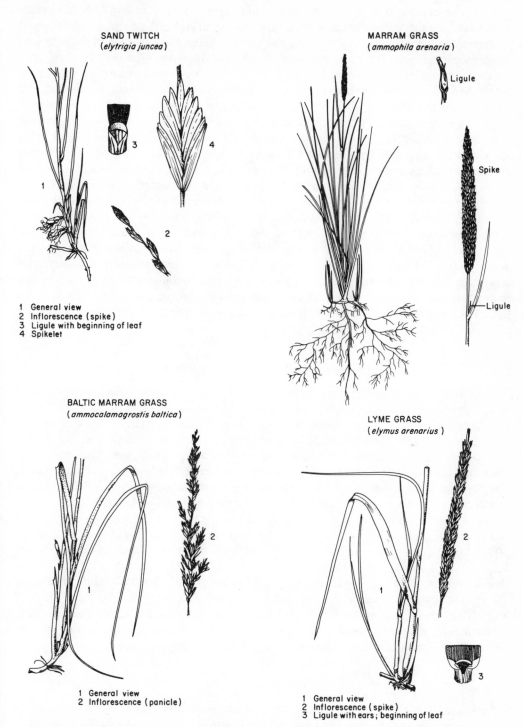

SAND TWITCH
(*elytrigia juncea*)

1 General view
2 Inflorescence (spike)
3 Ligule with beginning of leaf
4 Spikelet

MARRAM GRASS
(*ammophila arenaria*)

Ligule

Spike

Ligule

BALTIC MARRAM GRASS
(*ammocalamagrostis baltica*)

1 General view
2 Inflorescence (panicle)

LYME GRASS
(*elymus arenarius*)

1 General view
2 Inflorescence (spike)
3 Ligule with ears; beginning of leaf

Figure 56. Dune-binding grasses (Reproduced from Adriani and Terwindt, 1974, by permission of R.W.S. Communications)

It must be emphasized that *Ammophila*, and *Annocalamagrostis*, can only thrive in a situation where there is continued accretion of fresh sand and in these situations there will be active dune development. Once this has ceased *Ammophila* will lose its vigour and other species will succeed it. In the natural situation this succession may be very slow, limited by the low nutrient status of the soil. If it is necessary to maintain a good vegetation cover, then it may be advisable to utilize suitable drought resistant varieties of agricultural grasses, with the addition of fertilizers. An alternative to this is the use of *Hippophaë* which, with root nodules that can fix nitrogen from the air, is able to grow well in sands of low nutrient status. It can also stand a limited amount of burial by blown sand. If its invasive habit is acceptable it can form a useful bridge between the nutrient-poor young dune communities and the plant communities of the later stages of succession based on more mature soils. There seems to be something of a contradiction here between the situation in England where dense stands of *Hippophaë* are swamping species-rich dune grassland and in the Netherlands where *Hippophaë* scrub apparently forms a natural stage in the dune succession. It may be a question of maturity and over-maturity; certainly in the Dutch dunes there are degenerate stands of *Hippophaë* with an increasingly rich ground flora, whereas in England the oldest stands of *Hippophaë* form impenetrable thickets with a very limited ground flora. Before *Hippophaë* is used in dune stabilization work, due thought should be given to whether there are any plant communities nearby that could be threatened by the introduction of this species, and the findings of the *Hippophaë* Study Group (Ranwell, 1972b) may help here.

The use of fertilizers in sand stabilization

The mineral status of young dune soils is low and various attempts have been made to assess the benefits of artificial fertilizers in dune stabilization. Adriani and Terwindt (1974) describe a critical series of experiments on the role of fertilizers and the effects of surface binders on plant growth during the building of a sand barrier across the Brielse Gat (part of the Maasvlakte development). Various combinations of nitrogen, phosphate and potassium were used and also three binders, two based on bitumen and one on synthetic rubber. The plants used were *Agropyron*, *Ammophila*, *Ammocalamagrostis* and *Elymus arenarius* and both seed and plants were used except in the case of *Ammocalamagrostis*.

The results showed that although *Agropyron* responded to fertilizer an active sand supply caused a greater growth response. *Ammophila* responded well to the addition of 80 kg N ha^{-1} with an increase in the number of leaves, the ground cover and in vegetation height. The addition of phosphate or potassium had little effect. The growth of *Ammocalamagrostis* was also improved by the addition of 80 kg N ha^{-1} while phosphate sometimes actually impaired growth. Similar results were obtained for *Elymus*, with 80 kg N ha^{-1} being the most effective. While the binders were successful in stabilizing the sand surface they had no positive beneficial effect on the growth of either *Ammophila* or *Ammocalamagrostis* and sometimes even an inhibitory effect.

Sowing dune grasses

Adriani and Terwindt (1974) recommend that *Ammophila* and *Agropyron* should be sown from the middle of March to the end of April at a density of 100 viable seeds m^{-2}. Mechanical sowing is recommended for *Ammophila* and for both species the seed should be buried to a depth of 2–3 cm. If there is any likelihood of sand movement then the surface needs to be stabilized either by chemical binders or by harrowing a straw mulch into the surface of the soil. Both these will help to conserve moisture and improve germination.

Planting dune grasses

Adriani and Terwindt also recommend planting from mid-March to the end of April but suggest that planting is possible at any time from September to April inclusive. They recommend planting *Ammophila*, *Ammocalamagrostis* and *Agropyron* at 50 cm intervals in both directions and *Elymus* at 25 cm intervals. Cuttings should be planted as soon as possible after digging and inserted to a depth of 15–20 cm. They suggest that because of the palatability of *Elymus* to rabbits it should only be interplanted with *Ammophila* and not used on its own. The use of plants as opposed to seeds has the advantage of achieving effective plant cover more quickly, and also the young plants themselves help to stabilize the surface. This effect is helped if planting is done with staggered rows so that gaps in one row are covered by plants in the next. It is vital to avoid planting any rows in the direction of the prevailing wind. Planting is also the best way of dealing with relatively small areas of erosion. The main disadvantage is the high labour requirement when large areas are involved but the development of a suitable mechanical planter could help cut costs.

The use of sand fences

In areas where quantities of wind-blown sand are very great, and especially if it is desired to accumulate sand rapidly in a particular place, sand fences can be used. The effectiveness of a fence depends on the slowing down of the wind sufficiently for sand particles to settle out. A solid fence causes turbulence and local erosion with very little reduction in wind speeds. Suitable fences can be of wattle hurdles, reeds or split chestnut paling, or alternatively of one of the various perforated plastics now available. They should be set in lines at right angles to the prevailing wind and when gaps are necessary they must be staggered. Manohar and Bruun (1970) conclude that the optimum distance between fences is about 4 times the height. Short spur fences, set at right angles between the rows at approximately 8 m intervals will help to prevent lateral drifting of sand. The effectiveness of fences in reducing sand movement depends on their siting and it should be remembered that they are liable to produce very high rates of accretion. If it is desired to reduce sand movement into a particular area it is best to trap the sand somewhere between its source and this area, and in a position where dune growth is acceptable. It must also be remembered that it will not be possible to achieve a stable dune close to the high-water line—space should be left for a degree of embryo dune development. These transient dunes give a degree of shelter and will minimize the risk of damage from storm tides. Sometimes the problem is in the form of an isolated dune that has become unstable and mobile. In this case too, the work must first be directed at stopping sand movement at source; cutting off the major supply of sand will reduce the potential mobility of the dune and subsequently enable the whole mobile area to be stabilized. Thus stabilization measures must be applied from the windward side first.

Fixing mobile dunes

If reshaping of blown sand has to be carried out, e.g. for the removal of sand that has blown across a road, it is important to grade the surface to a natural aerodynamic shape avoiding steep slopes and sharp angles. The maximum slope should be set at 1 in 2 or less for

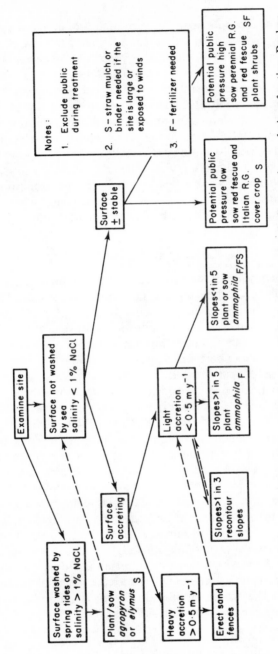

Figure 57. A key to the selection of dune stabilization techniques. Solid arrows indicate primary choice of option. Broken arrows indicate subsequent options. 'Accretion' refers to actual sand input; before stabilization this may be equalled or exceeded by erosion. It is, however, the potential that should be assessed

planting or 1 in 5 or less for mechanical sowing of *Ammophila* (Adriani and Terwindt, 1974). Steep slopes and sudden changes in slope increase air turbulence and the likelihood of further erosion.

Summary

Thus to achieve the most effective and most economic stabilization of mobile sand it is necessary to study the local situation, to select the most appropriate method and materials and to carry out the work at the appropriate time of year. There is really no substitute for an adequate understanding of the basic concepts of dune processes but as an aid in the selection of the appropriate method a key to identify the needs of the situation is given in Figure 57.

Dunes and people

Camber, East Sussex

Dune stabilization methods will not be effective if the cause of mobility remains, and the most common root cause at present is excessive trampling. The dunes at Camber, East Sussex, suffered from their use as a military training area during the last war and by the end of the war the greater part of the vegetation had been destroyed. Restoration work, consisting of the erection of sand fences and *Ammophila* planting, started in 1947 and by 1955 the state of the dunes was considered adequate for the sea-defence of the hinterland (Pizzey, 1975). Since then there has been a steady increase in the number of people using the area for recreation (Figure 58). The area of the dunes is approximately 13 ha so with a peak monthly total of nearly 60,000 visitors (August 1966–70) it is hardly surprising that severe damage had occurred, with mobile sand inundating the public roads and adjoining

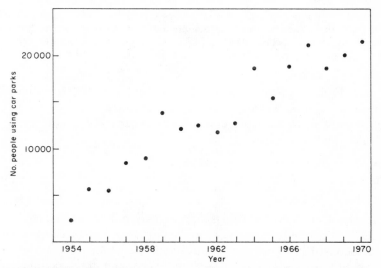

Figure 58. Annual tourist season (April–September) population visiting the dunes at Camber, East Sussex, estimated from car park returns. (Reproduced from *Biological Conservation* (Pizzey, 1975), by permission of Applied Science Publishers)

Figure 59. Vertical air photographs of Camber, E. Sussex, in October 1967 (above) and September 1971 (below). The two photographs show the increase in vegetation cover on the dunes achieved by restoration work carried out between 1968 and 1971. Bare sand can easily be seen as being very light in colour (Reproduced by permission of East Sussex County Council)

residential area. From 1967 onwards the East Sussex County Council initiated a programme of restoration aimed at fixing the moving sand and increasing the vegetation cover of the dunes so that they could be preserved as a public amenity and for sea-defence. The work was divided into two phases:

(1) The immediate stabilization of the wind-blown sand.
(2) Long-term plans to stabilize subsequent sand accumulation and increase the aesthetic value of the dunes.

The immediate work involved the removal and recontouring of blown sand to reduce wind erosion (some 85,000 m³ were removed) and the use of hydraulic seeding. Chopped straw was disc-harrowed into the sand and a mixture of seed, fertilizer and water sprayed on with a high pressure jet. This was followed by a slurry of chopped straw and bitumen emulsion thickly sprayed over the surface. Public access was prohibited and a barbed wire fence was erected, but there were five fenced pathways allowing access to the beach. Explanatory notices were put up and publicity arranged to explain the reason for the work. There was a large amount of mobile sand accumulating on the foreshore and sand fences were used here. The rate of accretion proved to be so high that the fences had to be continually raised. The long-term work included the stabilization of these new foredunes with *Ammophila* and the use of shrub species to control public access and to increase the aesthetic value. However *Prunus spinosa*, *Crataegus monogyna* and *Lycium barbarum* were used in preference to *Hippophaë* which was difficult to establish and considered to be a fire risk (Ranwell, 1972a). The changes in the vegetation during these works were recorded using existing aerial photography [Figure 59(a), (b)] and generally speaking the results were satisfactory. It was perhaps risky to use agricultural grasses in such a mobile situation and recent Dutch work suggests that the sowing of *Ammophila* might have been effective provided seed of sufficiently high viability had been used. In her survey of the results Pizzey (1975) criticizes the use of agricultural grasses as opposed to native dune grasses; however, with the addition of fertilizer that is inevitably necessary in any area subject to public pressure, agricultural grass species are likely to spread anyway and their use in the seed mixture seems likely to give better results. It is very important to use selected species and varieties rather than commercial mixtures and to use seed of proven quality. While it may seem a waste to use good seed when conditions are likely to be such that only a small proportion may germinate, it is under just such conditions that seed of high quality (germination percentage) is most needed.

Holkham, Norfolk

An example of the use of agricultural grasses to stabilize a non-accreting dune grassland worn bare by public pressure is provided by work done recently at Holkham, north Norfolk. Where a road gave access to the beach via the dunes, some 3,700 m^2 of old dune grassland had been worn bare by the passage of the occupants of *c* 12,000 cars per annum (Moore, 1971), probably about 35,000 people. The completely bare areas amounted to just under 50% of the total area and these areas were sown with the following seed mixture:

S23 perennial ryegrass (*Lolium perenne*) 60%;
S59 creeping red fescue (*Festuca rubra*) 18%;
Smooth stalked meadowgrass (*Poa pratensis*) 10%;
Westernwolth ryegrass (*Lolium perenne* x *italicum*) 10%;
Wild white clover (*Trifolium repends*) 2%.

The area was fenced off and treated with a mixture of nitrogen, phosphorus and potash. Although the work was started (October 1973) at what proved to be the beginning of a run of drier than average years, results to date have been quite promising with establishment of a grass sward over 75% of the bare area (Figure 60). The fast growing but short-lived westernwolth ryegrass was included in the mixture to provide cover for the other species in the first year. This it did, with the S23 perennial ryegrass and S59 creeping red fescue taking its place. Although there was a much higher percentage of S23 (60%) in the seed mixture than S59 (18%), the frequency of the latter increased as the experiment progressed. After twelve months the frequencies (% 10 × 10 cm quadrats occupied) were 30% S59 and 70% S23; however after two years the figures were 45% S59 and 50% S23. This was attributed to the greater drought resistance of the fescue. The smooth stalked meadowgrass had a frequency of around 10% in both years while the white clover which was never more than 1% had practically died out by the end of the second year. From these results it may be concluded that there is little to gain by including either meadowgrass or

Figure 60. Dune grassland restoration at Holkham, Norfolk. An area worn bare by excessive trampling was fenced off and sown with a mixture of agricultural grasses

clover. Thus seed mixtures for areas with little or no accretion might be based on westernwolth ryegrass and creeping red fescue mixture with perennial ryegrass to be included if high public pressure is anticipated. The length of time during which a newly sown area will have to be fenced off will vary with the climatic conditions and the local situation. It must be appreciated that it will take much longer to build up a thick sward under dune conditions than under normal agricultural–horticultural conditions.

Artificial walkways

A firm surface encourages people to use a walkway thus minimizing the damage to surrounding areas. The nature of the walkway will depend on locally available materials. For a wide path railway sleepers are very effective but expensive. Natural round timbers from forestry thinnings laid across the path and wired together are an effective substitute. When suitable quantities of shell, e.g. cockles, are available they form quite a satisfactory surface, and woody trimmings can also give some protection from erosion.

Nature conservation management

Nature conservation management may be defined as that management necessary to ensure the survival of as many as possible of the natural species, communities and habitats in an area, in other words to maintain or increase the natural diversity of an area. In many cases however there may be conflict between the preservation of the various different species or habitats and management decisions will have to be made in the light of the local situation and of overall national policy for that species or habitat. There may also be local constraints that have to be taken into account, such as the proximity of roads or buildings that might be affected if full natural sand mobility was permitted to continue. Nature conservation management is also taken to include the artificial creation of habitats (habitat building *sensu* van Leeuwen, 1969), again to preserve and so far as possible to increase the range of species and habitat types in the area. The tendencies towards uniformity can come both from within an area, e.g. by the dominance of a particularly aggressive species, and from outside an area, e.g. by reduction in the level of the water-table as a result of water extraction. This has been referred to as internal and external management (Westhoff, 1955; van Leeuwen, 1969). Historically dunes had a large degree of natural regeneration of habitats through sand mobility. This was particularly true of the dune slack communities. Natural blowouts removed sand down to the groundwater level and thus new slacks were formed as the older areas were filled in, and where the average height of the groundwater level had been altered by climatic changes or changes in coastal processes these natural blowouts produced slacks excavated to the appropriate new level. However in most areas this degree of mobility is unacceptable in the interests of coastal defence and because of the effects of mobile sand on nearby human developments. In addition changes in the groundwater level are often too rapid for natural processes to compensate (Londo, 1971). One solution to the conflict between the need for stability in the interests of sea-defence and the preservation of the natural mobility of a dune system is to adopt a sea-defence line to landward of the dunes. This approach has been adopted at the Kwade Hoek on the north side of Goeree (Figure 61).

Figure 61. Mobile dune communities at Kwade Hoek, Goeree. The sea-wall set behind the dunes to allow natural mobility can be seen in the background

Control of the water regime

Thus control of the water regime is an important facet of external management. The other most important facet is the control of public pressure. The damp slack communities (hygrosere) are critically affected by changes in the groundwater level and some of these communities have developed over many years (perhaps up to a century). Any alteration in the groundwater level would mean their end. Londo (1971) cites vegetation types on Schouwen, Voorne, and Goeree (Delta area of the Netherlands), as being in this category; and emphasizes the importance of trying to obtain a more or less stable average water-table level (seasonal fluctuations excluded). The obvious answer to the potential floristic losses that follow water extraction is to keep the groundwater replenished by infiltration. However as already mentioned this frequently leads to a marked reduction in species diversity, particularly when the infiltration water is mineral rich (Londo, 1975b). In his earlier publication, Londo (1971) gave four guidelines towards reducing the unfavourable effects of infiltration. He suggested that the water used should be as mineral-poor as possible, and in particular it should have low levels of nitrogen and phosphorus, and that it should be cleaned of fine sediments before infiltration. He suggested that the average water level should be kept as constant as possible but with regular fluctuations within each year. If the resultant vegetation is regularly mown in the autumn, and the cuttings removed, this will also help to keep the habitat mineral-poor. Mineral richness and no mowing would quickly lead to the production of species-poor rough grassland. While these guidelines were formulated to apply to the infiltration/water extraction situation they are equally valid as regards the replenishment of water lost through other causes, e.g. through field drainage in adjoining areas for agriculture or forestry.

Habitat creation

If it is not possible to restore a lowered water-table to its original level, serious consideration should be given to the recreation of the lost hygrosere by other means. This can be done semi-naturally by allowing controlled blowouts to occur, or completely artificially by excavations. There are a number of advantages in the use of blowouts: the excavation process automatically goes to the correct depth, to the groundwater level; and fluctuations in this produce, through local erosion and redeposition, natural microrelief and habitat diversity. In addition the blowout process occurs over a considerable period of time, and this again helps to increase diversity by producing habitats of different ages. Blowouts also have benefits for the plants of the dry habitats. In Braunton Burrows for example it was noted that the greatest diversity in plant species and plant communities occurred where stable areas occurred next to dynamic mobile ones (Londo, 1966). In many situations natural or artificial blowouts are not acceptable and in certain situations there would not be sufficient potential instability for them to occur; in such cases reliance must be placed on habitat creation by artificial excavations. In England dune excavations have been limited to small ones for the benefit of the Natterjack Toad, which relies on dune pools as its breeding place. The digging of new pools and the deepening of existing ones has ensured the maintenance of breeding habitats in dry years, but the overall effect on the Natterjack population has not been assessed. Excavation as a management tool has been used more widely in the Netherlands. Londo (1971) has described in detail the development of dune slack vegetation along the Grote Vogelmeer, an artificial dune lake near Haarlem (Figure 62). The lake and the surrounding damp slacks were excavated between 1951 and 1955; and a total of 4×10^5 m^3 of sand over an area of 12 ha had been removed. Botanical

Figure 62. The Grote Vogelmeer, Kennemerduinen, during a period of very low summer water levels (September 1964), nine years after the excavation had been completed (reproduced by permission of Dr. G. Londo)

development in the area was studied by van der Maarel (1959) and Segal (1960) and later by Londo (1971). The detailed records of the vegetational development in the area have enabled Londo (1971) to formulate guidelines on the creation of dune slack habitats by excavation. He considers that the greater part of the surface should lie between the groundwater level in an average summer and a level of 70–80 cm above the winter water level (normally this would amount to a range of 120–150 cm). The greater proportion of this area should be within 25 cm above and below the average high water level. He suggests that in the hygrosere there should be very gentle slopes of between 1 in 30 and 1 in 80, while steeper slopes of between 1 in 5 and 1 in 3 are appropriate in the mesosere (more than 25 cm above winter water level). A number of small slacks is considered to be more

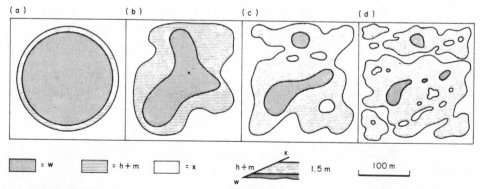

Figure 63. Creation of an artificial dune slack. Schematic survey of the botanically favourable (at the right) and unfavourable (at the left) way of habitat creation; w = water, h + m = hygro- and mesosere, x = xerosere. (reproduced by permission of Dr. G. Londo)

favourable for diversity than a few large ones, as isolation tends to produce diversity (van Leeuwen, 1966a,b). Isolation can be achieved in a large slack by the use of low dune ridges (Figure 63). This Figure shows how habitat differentiation can be encouraged. From (a) to (c) there is a progressive increase in both length and area and an increase in the degree of isolation and thus potential differentiation. Islands are undesirable as they tend to attract gulls with consequent eutrophication of the water. Gulls also disturb other nesting birds (Thomas, 1972). Londo considers that in order to avoid unfavourable changes in the vegetation the excavated sand should be disposed of in the xerosere (i.e. more than 100 cm above winter water level). This also enables *Ammophila* to be planted to stabilize the excavated sand. From what has been described it will be clear that any such work should be preceded by a careful hydrological survey and vegetation study, with particular reference to the distribution of the phreatophytes, as they indicate existing groundwater conditions.

Dune habitat management

Dune management for nature conservation can best be described in terms of specific management procedures and of these the six most important are:

(a) Turf cutting
(b) Grazing
(c) Trampling
(d) Mowing
(e) Species management
(f) *Laissez-faire*

Selection of the most appropriate management option or options will depend both on what is to be achieved and on the previous management of the area. Generally speaking, management procedures (a)–(c) act through an increase in natural dynamic activity in the ecosystem as exemplified by the rate of breakdown of organic material (van Leeuwen, 1969; Westhoff *et al.*, 1970). These authors consider that such increases occur in one of two ways; by direct influences on the physical or chemical basis of the soil (the spade effect), or by indirect factors (the cutting effect). Generally it is the spade effect that produces the most rapid habitat change and this is the effect of excavation or turf cutting. The cutting effect produces relatively less change and is exemplified by mowing. Both systems work together in the case of trampling and grazing. The different management options will be considered in order of decreasing magnitude of habitat change. According to van Leeuwen (1966a) each species and plant community requires a particular rate of habitat activity and thus the appropriate management procedure would be selected accordingly. Londo (1971) points out that the same species can be affected differently in different habitats by the same management procedure.

Turf cutting   In Britain turf cutting has not been used in dune management although it has been applied to salt-marsh to reverse the succession (Ranwell, 1972a). It has been applied in the Netherlands to regenerate certain pioneer damp slack communities, e.g. Koegel-wieck, Terschelling (Londo, 1971). The effect of turf cutting is through the lowering of the level of the soil surface and in the absence of accretion it is likely to be long lasting.

Grazing   The floristic diversity of dune areas with a long history of grazing has already been mentioned, e.g. Westduinen, Goeree (Blom and Blom-Steinbusch, 1974). Although there have been no critical studies on relative effects of different grazing animals, this seems likely to be important. At least when the grazing has been predominately by one type of animal this should be continued and at the same intensity (van Leeuwen, 1966b).

The use of grazing as a management procedure in a nature reserve is discussed by Oosterveld (1976). In the Netherlands the dunes of the Wadden Islands are grazed by sheep while the dunes of the Delta area are grazed by cattle. In Scotland the dune pastures or machairs are mainly sheep grazed while in lowland Britain it seems likely that grazing was mainly by cattle, although now the rabbit is the commonest grazing animal. Rabbit grazing is unsatisfactory as it is difficult to control the numbers of rabbits and because of their burrowing activities.

Trampling   While heavy trampling is deleterious, lower levels of trampling can contribute towards species diversity, particularly when gentle gradients from trampled to untrampled dune are maintained (Westhoff, 1957, 1967; van der Maarel, 1971; van Leeuwen, 1966a). This can happen when more or less the same track is followed through a dune slack (Figure 64). Londo (1971) reported that the trampling gradient in mown slacks on Voorne worked towards habitat enrichment.

Figure 64. A trampling gradient favourable for species diversity: grazed dune grassland in the Westduinen, Goeree

Mowing   Mowing differs from grazing because it is non-selective and is applied to all species equally. Grazing also comprises both the removal and addition of material while mowing may only be concerned with removal. Mowing is easier to use as a management tool as it avoids the necessity for fencing and stock management; it can be also applied to a much smaller area. It may be profitably applied to any area where there is a tendency for rough grassland to develop, although it has only been applied experimentally in Britain e.g. Newborough Warren, Anglesey; Saltfleetby/Theddlethorpe, Lincs; Holkham and Winterton, Norfolk. On Voorne however it has been applied to dune slack vegetation for

Figure 65. A species-rich wet slack community maintained by annual mowing, Voorne

decades (Londo, personal communication) with very favourable results (Figure 65). Round the Grote Vogelmeer small areas have been mown since 1964 and in 1970 *Gentianella amarella* appeared exclusively in areas that had been mown (Londo, 1971). The available evidence suggests that a single mowing in September or October with removal of the cuttings will be the most effective. In the nature reserve on Voorne the woody cuttings are used to give a degree of stability to the heavily used major footpaths. It may also be useful to vary the boundaries of the areas mown each year to create gradual transitions. If the tendency towards the formation of rough grass communities is low then mowing may only be required every third or fourth year or even less often; in which case there may need to be a more frequent cutting of scrub species.

Species management   This category will include those species which have a dominating influence on other species or communities and which may therefore have to be controlled. It will also include measures designed to encourage the growth of particular species. The species in the former category include both shrub species, e.g. *Hippophaë*, and tree species, e.g. *Pinus* (spontaneous seedlings from plantations). Although these are perhaps the two commonest species that cause problems many other species can prove invasive and troublesome locally, including *Alnus glutinosa* and *Salix repens* in damp slacks, *Rubus fruticosus* in dune woodland, and in one area (Winterton, Norfolk) *Rhododendron ponticum* has proved to be an invasive weed (Fuller and Boorman, unpublished). In most cases control lies in cutting or uprooting although experience with *Hippophaë* has shown that a single cutting is rarely sufficient. The lesson to be learnt is that early identification and prompt action against troublesome species is likely to save trouble later.

Measures to encourage rare species will normally be based on the conservation of suitable habitats. For example a wet slack species would be favoured by control of the water regime to keep the slack wet and by control of invading shrub species. The creation of new habitat for a particular species is always worth trying but it must be remembered that natural dune slack plant communities may have taken many years to develop and that their re-creation is not likely to be a rapid process; thus efforts at habitat reconstruction must be assessed over a suitably longer period.

Laissez-faire   In the absence of management procedures of known or probable effectiveness, the safest answer is to adopt a *laissez-faire* policy. It is also the policy to adopt when an area is not sufficiently well known to formulate proper management plans, and it is appropriate in the larger reserves when a measure of the natural processes of erosion and accretion can be allowed to occur. Whilst this procedure is as the name implies, leaving the natural situation alone, the importance of adequate monitoring of the changes that occur cannot be over emphasized. Only by this means can the appropriate time be selected for the implementation of one of the more positive management procedures already described.

Conclusion

In conclusion, effective management depends on a sound knowledge of the species of an area, their distribution and amount, the communities they form, the historic factors that led to their present status and a knowledge of the patterns and processes in the communities, the changes that are taking place and the rate of change. In addition it is desirable to have an overall policy for whole lengths of coastline with areas allocated according to the most appropriate use. This has been done for the Dutch coast between Wassenaar and Camperduin (Doing, 1974) as part of a long-term project to map all coastal dune areas of the Netherlands. Doing identifies nature reserves and infiltration areas, both of these closed to the general public; forest reserves; recreation areas, including areas where there is excessive recreational pressure and potential recreational areas; built-up areas and agricultural areas. This type of information enables an effective overall national coastal strategy to be developed. As more information becomes available it is clear that there is much that can be learnt by a study of both fundamental dune ecology and practical applied dune management on an international basis.

Chapter 10

Shingle formations

R. E. Randall

INTRODUCTION

Shingle is a frequent component of the coasts of Britain, France, and the Baltic lands. At most locations, the shingle occurs as a fringing beach or other ribbon-like phenomenon, which is continually mobile and bears a sparse but characteristic vegetation. Such areas are described in Chapter 3. However, at a few points along the coast, larger quantities of shingle build up, resulting in stable formations that are more terrestrial in nature. Where a whole series of parallel shingle ridges have been thrown up, a cuspate foreland is produced; at a smaller scale this may be just one or two apposition beaches. Large masses of shingle may also form offshore barrier islands (Figure 66). In Britain, the foreland is best represented by Dungeness, Kent, and the barrier island by Scolt Head, Norfolk. In Scotland, the largest apposition beaches occur on the Isle of Arran. The division between various features is pedantic: Orford Ness, for instance, is a spit, part of which is a series of apposition beaches; Blakeney Point is virtually an offshore barrier island but is connected to the coast at its eastern end and is thus technically a spit. Furthermore, it is often impossible to treat shingle formations as separate entities, since in many places they are intricately linked with other coastal features (Fuller, 1975). Scolt Head Island has a shingle skeleton, on parts of which dunes have formed, and the whole unit contains salt-marsh within its laterals. Thus any management programme would have to include all ecological types.

Because of the physiographic importance of the larger shingle formations, certain of them have been examined by geomorphologists, and some detail is known of their origin and development. In Britain, the earliest studies were those of Elliott (1847) on Dungeness and Coode (1853) on Chesil Beach. Considerable recent work has been carried out on Dungeness (Lewis, 1931, 1932; Lewis and Balchin, 1940; Hey, 1967), Scolt Head Island (Steers, 1960), Blakeney Point (Hill and Hanley, 1914; Oliver and Salisbury, 1913a) and Orford Ness (Steers, 1926; Carr, 1969; Carr and Baker, 1968). Other formations are considered by King (1959).

Apposition beaches are thought to form when shingle, instead of continuing its course along an existing spit or fringing beach, piles up in front of it and is then driven landwards by high waves produced by storm at sea. When this process is repeated a number of times, a series of roughly parallel ridges may be produced and an extensive area of stable shingle results.

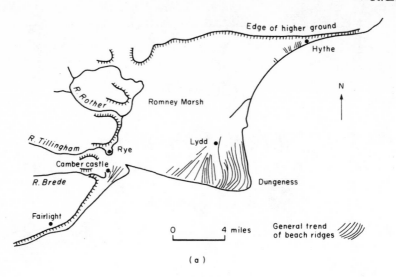

(a)

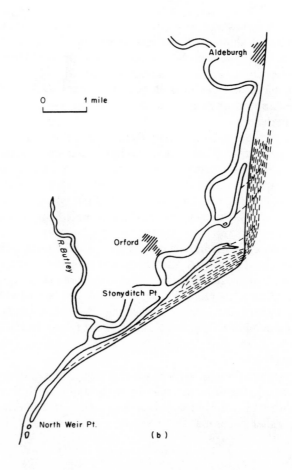

(b)

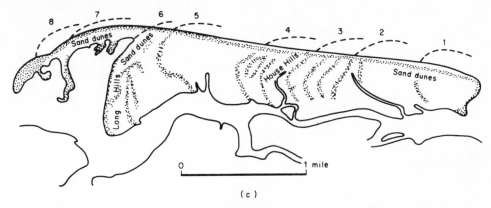

(c)

Figure 66. Shingle formations. (a) The cuspate foreland of Dungeness. (b) Apposition beaches of Orford Ness; south of Stonyditch Point, this feature becomes a simple spit. (c) An offshore barrier island, Scolt Head, showing the main shingle ridges. (Reproduced by permission of Longmans from Sparks (1972))

Beach mobility is less of an ecological consideration on the larger shingle formations than it is on shingle foreshores. Scott (1963a) recognizes two vegetational classes relating to 'terrestrial shingle':

(a) Long-lived perennial species—beach subject to occasional inundation: lichens present on shingle.
(b) Heath—beaches entirely stable.

Naturally, all shingle formations must be protected on their coastal fringes by shingle foreshores of greater mobility, such as those described in Chapter 3. The greater stability and greater expanse of apposition beaches and offshore barrier islands results in a more complex ecological development, including plant, invertebrate, and vertebrate life-forms. Thus, Dungeness is important for its succession to holly wood (Scott, 1965; Peterken and Hubbard, 1972), its entymology and its bird life.

The amount and type of fine fraction of the substrate (that material under 2 mm diameter—see Chapter 3) remains an extremely important factor right through the succession on stable shingle formations, and it is probably the primary control over the establishment of vegetation. Fine fraction is of greatest importance at the time of seed germination, because, without it, enough moisture may not be present for growth of most plant species to begin. The major difference in fine fraction between shingle foreshores and shingle formations is that the latter do not have the advantage of external inputs of organic matter through drift. A certain amount of nutritive material will be blown on to the formations by wind, but the bulk of organic material will be produced over time *in situ*, by the plants themselves. This hypothesis was tested by Scott (1963a), who later suggested the form of successional sequence taking place at Dungeness (Scott, 1965). This succession was updated by Peterken and Hubbard (1972). The tentative conclusion is that natural succession on Dungeness is doubly cyclic (Figure 67). In the first cycle, which begins when the substrate is stable, increasing quantities of humus are added to the shingle as heath vegetation develops under prostrate *Sarothamnus* bushes. This cycle develops gradually into a second involving a number of shrub species.

Shingle formations suffer from the same water-supply and water-retention problems as shingle foreshores. Pendulant water from precipitation, including fog and mist, must be the major source of water for most plants, but the quantities retained are extremely dependent

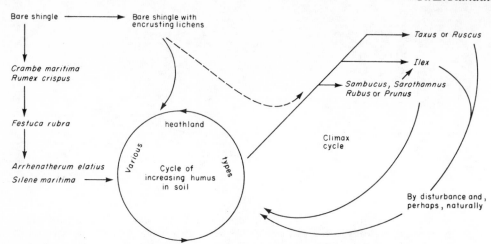

Figure 67. Summary of presumed successional relationships of communities on the shingle at Holmstone. Heavy lines indicate the changes which have been most important in recent decades. (After Peterken and Hubbard, 1972 reproduced by permission of Blackwell Scientific Publications Ltd.)

upon the amount of fine material and humus in the soil (Fuller, 1975). At Dungeness the beaches act as a vast reservoir of freshwater, but this is at least 3 m below the surface and is not available to most plant species. The freshwater marshes of the Open Pits are, however, related to the water-table.

A final ecological consideration on large shingle formations is wind, which blows unchecked across the wide expanses of exposed apposition beaches or offshore barrier islands, and accounts in part for the prostrate or deformed nature of the vegetation.

Although stable shingle formations have few unique species of plant or animal life, succession frequently results in an unusual assemblage within communities, and certainly many species exist locally on stable shingle which are not present elsewhere within that region.

From the ornithological viewpoint, stable shingle formations are particularly interesting because of the colonies of rare species, or common species nesting in unusual locations, which they contain. Orford Ness/Havergate National Nature Reserve has a colony of the avocet (*Recurvirostra avosetta*), and Dungeness used to be a breeding ground for the Kentish Plover (*Leucopolius alexandrinus alexandrinus*).

Special features of the plant ecology of stable shingle formations are the successions on stony substrates and, where lakes occur, the successions of aquatic vegetation. Examples of the latter can be seen at Shingle Street, Suffolk, behind the 1893 ridge (Lagoons 6a, 6b in Cobb, 1956) and in the Open Pits at Dungeness (see Rose, 1953; Harden, 1963).

Open shingle provides an inhospitable habitat because of the basic ecological factors discussed above and only a few higher plants are suited to this environment; e.g. drought-resistant forms or winter annuals. A few species, such as *Arrhenatherum elatius*, appear to be able to colonize bare shingle lacking in fine fraction, but the development of heathland vegetation, as on Denge Beach, Dungeness, which occurs as discrete patches surrounded by bare shingle, is entirely dependent for its survival on the continuing accumulation of organic matter.

At Dungeness, any one of six shrub species can invade heath but cannot invade bare shingle. These are *Rubus fruticosus* agg, *Ulex europaeus*, *Sambucus nigra*, *Ruscus aculeatus, Ilex aquifolium*, and *Taxus baccata*. Once established all these species suppress the heath vegetation, but they produce large amounts of litter so that when the scrub degenerates (see Figure 67) the heath rapidly returns.

Some shrubs can invade established thickets of other species. For example, *Ulex, Rubus*, and *Sambucus* can be invaded by *Ilex*. The latter species is very long-lived—in the order of 200–300 years—and may, in fact, be the climax vegetation of south-eastern shingle formations.

Elsewhere, there is nothing comparable with the succession at Dungeness. The nearest to it is the development of woody cover on the raised shingle/cobble beaches of western Scotland.

TYPES OF PRESSURE

The pressures affecting the stable shingle habitat are considerable, including both direct and indirect intervention by man, and by animals, especially the rabbit (*Oryctolagus cuniculus*). For centuries, cattle grazing has been common on shingle areas, especially during periods of high tide when cattle were moved off adjacent marshes. Grazing is not practised today on any of the shingle formations in south-east England, but apposition banks on Arran are grazed by sheep. Rabbits were particularly common on the shingle formations of the Channel and north Norfolk coasts prior to the introduction of myxomatosis when the populations were virtually eliminated. Rabbits have now returned in quantity to several areas, especially along the landward fringes of shingle masses. In places on Dungeness, *Ulex* bushes are so heavily grazed that they become compacted and will stand the weight of man. The changes in vegetation resulting from changes in rabbit pressure have been reported from Scolt and Blakeney (White, 1961), and Dungeness (Thomas, 1960, 1963). The general effect of rabbit removal is an increase in cover of grasses and *Sarothamnus*. Rabbits, and hares (*Lepus capensis*), locally wipe out all plants except the unpalatable (*Teucrium scorodonia, Silene nutans*, and *Prunus spinosa* at Dungeness), allowing the removal of surface organic matter and the degradation of the heath.

Surface disturbance can also result from treading or vehicles. One vehicle passing over the shingle may leave tracks that will last for years. Continued vehicular use, as at The Forelands and Jury's Gap, Dungeness (an army range since 1883), can destroy the whole ridge system. The thin carpet of acidic heath is easily torn, especially where *Cladonia impexa* is an important component. At Blakeney Point, tyre tracks to the observation post can be distinctly seen as lines of different vegetation. Rose (1953) described the compressed shingle of an old track, which contained *Vicia lutea, Trifolium ornithopodioides, Trisetum flavascens, Anthoxanthum odoratum, Echium vulgare*, and *Digitalis purpurea*. In some areas of heavy use, heathland vegetation has completely disappeared.

Vegetation change also results where mortar bombs and landmines have pitted the surface of Dungeness. Such craters fill with *Ligustrum vulgare, Silene nutans, Digitalis purpurea*, and *Teucrium scorodonia*. The distribution of *Ulex* on shingle formations is, as elsewhere, primarily related to areas of human disturbance. It is found as a fringe to agricultural land, along roads and railways, near old buildings and in disused shingle workings. Many of these plants of disturbed areas were introduced in association with the building of roads and railways across shingle formations, such as at Dungeness or Shingle Street.

The final and perhaps greatest pressure on shingle formations is excavation. Removal of shingle not only destroys the nature of the various habitats, but also creates new ones dependent for their species composition on age and on the relationship between the water-table and the new surface. At Shingle Street, large areas that were once clothed have been bare shingle since the wartime excavations until colonization by *Lathyrus japonicus* over the last five years. At Dungeness, excavation began in 1883 for the building of the South-Eastern Railway and has increased dramatically over the last two decades. This has resulted in *Sarothamnus* or *Rubus* scrub, *Salix* scrub, *Phragmites* marsh or aquatic vegetation according to depth.

METHODS

Significant ecological information is difficult to obtain for shingle formations because of the limited number of areas where such features exist. Therefore it is extremely difficult to generalize on such factors as shingle succession, and most data obtained will only be relevant to the particular area under study.

Where data are available, archival information is often extremely useful. Peterken and Hubbard (1972) were able to improve considerably on Scott's (1965) idea on shingle success to *Ilex* at Dungeness, by examination of the first 6-inch edition of the Ordnance Survey map, and later aerial photographs. Detailed examination was also made of the structure and growth of *Ilex*.

Aerial and ground-based stereophotography are becoming extremely important methods in the conservation and management of coastal areas (Hubbard and Grimes, 1972). A common problem of large shingle formations is the lack of vantage points from which to gain a general impression of diversity. Stereophotography can be used for a wide range of purposes from aerial description to the monitoring of detrital breakdown.

Vertical aerial photography is best left to one of several firms specializing in this service and equipped with the necessary type of aircraft, cameras and processing equipment. The task of the applied ecologist is to survey in the ground control. The success of the whole operation depends upon the accuracy with which this is completed. On shingle, control points are best located with 1 m lengths of scaffolding tube driven into the substrate. Squares of yellow polythene (60 × 60 cm) or boards painted in black and yellow triangle attached to the top of the scaffold tubes make easily visible ground-to-air markers. Interpretation of vertical aerial photographs can be simple and inexpensive using a mirror stereoscope, with detail transferred to plastic overlays. More complicated methodology is described in Howard (1970), and the use of various film types is discussed by Grimes and Hubbard (1971). Scale of photography varies according to use, but the most useful is within the range 1 : 10,000 (basic types such as heath or bare shingle can be recognized) to 1 : 2,500 (individual bushes such as *Sarothamnus* or *Ulex* can be defined).

A major problem with many shingle formations is that vegetated areas are small and fragmented in relation to the whole. Ground-based vertical stereophotography is especially useful for detailed examination of such small areas. Grimes and Hubbard (1969) describe work on shingle formations to study regeneration of vegetation following rabbit grazing. A camera is suspended above a plot of known size by means of a rectangular frame. It slides into three predetermined positions along a horizontal bar to produce stereo-overlap. The exact position of each corner of the frame is marked by a stake driven into the shingle. Photographs are taken regularly three times each year. Using 20 × 20 cm prints beneath a mirror stereoscope, it is quite possible to identify lichens, mosses, and grasses and thus obtain accurate estimates of changing vegetation cover, or the breakdown of rabbit droppings. A related method of obtaining ecological information is the use of the traditional permanent quadrat as described in Chapter 3.

Because apposition beaches and offshore barrier islands are relatively large shingle formations, there will be considerable environmental change with distance from the open sea. This change and others that result from human use of the area will be reflected in vegetational changes, which are best shown graphically by means of a transect which traverses a line across the formation at right angles to the shore. Several kinds of transects have been devised to illustrate the character of vegetation. The simplest of these is a line transect (Figure 68) which consists of a record of the plants occurring along a surveyed line across the research area. In fact, it is not possible to record or reproduce a complete record. Therefore the most common practice is to record those plants which occur at fixed distances along the line—every 10 cm or 26 cm, for example. This is facilitated by running a tape along the survey line. The major objection to this method is that it introduces personal error, since several plants may be found equally close to the point where a record is to be made, but selection will inevitably be biased by plant size or ease of recognition, so that inconspicuous or difficult taxa may well not appear in final species lists.

A much better but more time-consuming method is to produce a belt transect—for this a 1 m wide strip rather than a line is used. This strip or belt is divided into metre squares, each of which is examined separately. Many methods have been employed for this examination, from the highly quantitative to the artistic. The most instructive method which involves the least time is to work out a cover-abundance value for each species in each quadrat. Percentage cover is extremely difficult to assess by eye and thus cover-abundance classes such as those of Domin or Braun-Blanquet (Table 14) are recommended (Bannister, 1966). It is wise to prepare duplicated sheets of likely species in advance to save field time. Environmental information (soil samples, pH measurements, damage etc.) should be collected concurrently.

Table 14   The Domin Scale and its relation to the Braun-Blanquet Scale

Cover–abundance Rating	Domin Scale	Braun-Blanquet Scale
Cover ≈ 100%	10	5
Cover > 75%	9	5
Cover 50–75%	8	4
Cover 33–50%	7	3
Cover 23–33%	6	3
Abundant, cover ≈ 20%	5	2
Abundant, cover ≈ 50%	4	2
Scattered, cover small	3	1
Very scattered, cover small	2	1
Scarce, cover small	1	1
Isolated, cover small	X	X

The results of a belt transect survey should be drawn up in histogram form as shown in Figure 69. A separate histogram is compiled for each species, the graphs being arranged under one another. It is important to include in the same figure the distance from the shore and levels of the transect, using a fairly large vertical scale, and any other environmental detail that can be portrayed. A great deal of useful information on the correlation between the environment and species cover–abundance can be obtained by the study of such belt transects.

In theory, the transect is merely one part of a systematic grid system, which if extended could cover the whole area of the formation. Obviously aerial photography gives the most

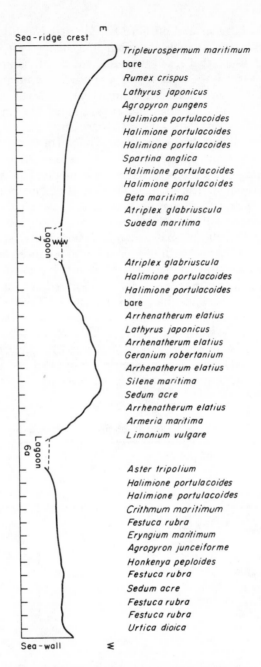

Figure 68. Line transect across Shingle Street
North End. Scale 1:600; v.e. 4:1

complete descriptive coverage, but detail can only be obtained via some form of quadrat sampling. The mesh of the grid will depend very much upon the time and money available for recording, but naturally the finer the mesh, the greater the detail. Strictly random sampling may also be used, but this does not give such good areal coverage. Other sampling designs are discussed in detail by Kershaw (1964) and Greig-Smith (1957) and an

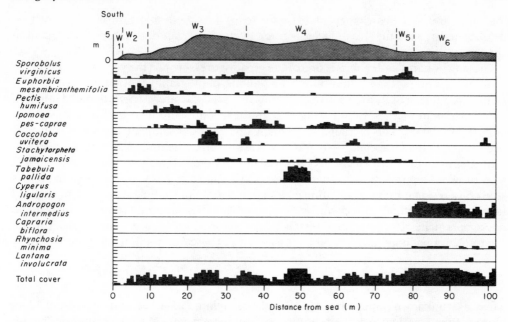

Figure 69. A belt-transect histogram showing the results of a survey through the coastal vegetation of Barbados. Each small square of the histograms vertically represents 20% of a m² covered by the species concerned, successive metres being represented by small squares of the histogram horizontally. The actual squares have been omitted from the figure for clarity. The slope of the ground is shown at the top of the figure; estimation of total cover at the bottom. Six different communities are recognizable within this complex (W1—W6). (After Randall, *J. Ecol.*, **58,** 161. Reproduced by permission of Blackwell Scientific Publications Ltd. and the British Ecological Society)

excellent survey has been made by Haggett (1964). Most of the data concerning the collection of environmental information are covered in Chapter 3.

Whereas the pressures on shingle foreshores are the physical ones associated with the action of the sea, shingle formations suffer much more from human pressures. The nature and degree of these pressures can be assessed in a number of different ways. Land ownership, access routes and potential uses are the major factors upon which pressure depends. On large features, such as Dungeness, there may be several owners or users of the land. Maps should be compiled showing multiple ownership and multiple use, and surveys should be made at access points of numbers of people and vehicles entering the shingle area. Similar work has been carried out by Moore (1971) on the dunes at Holkham, Norfolk. He has shown that leisure use is dependent upon access and that little use is made of areas distant from access points. This does not apply, however, to commercial use, and it is wisest to prepare some type of classification grading of the area for further use. In this way, if pressures must continue, advice can be given concerning the priorities within each area. Such mapping and survey work should be regularly updated as conditions of pressure change, so that really important wildlife resources are not endangered by other uses.

As with ecological information, some of the most useful assessments of pressure on shingle formations can be undertaken by means of photographic survey. Large-scale aerial photographs taken from aircraft are of use for monitoring physiographic change. Hubbard (personal communication, 1973) has used 1 : 2,500 air photography to map Dungeness at 1 : 1,000 with ridge contrours at intervals of 25 cm. Comparison of old and new work over the same area enables changes and trends to be recorded as pressures alter. Surface disturbance by man and his vehicles is often evident and can be recorded by vegetation change.

Grazing by hares and rabbits is one of the greatest pressures on the vegetation of shingle formations. Ground-based stereo- or even oblique-photography is often the best method to monitor this activity by repeated filming of particular areas (Grimes and Hubbard, 1972). Another method of assessing changes in this pressure is to take a census of the animal population. This is fraught with many difficulties of sampling error and statistical bias, but a systematic capture–mark–recapture is often helpful in obtaining changing estimates of abundance. The ratio of marked to unmarked animals in subsequent trapping runs provides a population estimate known as the Lincoln Index. If, for example, one is studying a resident rabbit population on a shingle formation and succeeds in capturing and tagging 100 individuals, all of which are released, and if on a second trapping programme one again catches 100 individuals of which 20 were previously tagged, one would estimate the total original population at 500 animals. This is based on the Lincoln Index ratio:

$$\frac{P}{M_1} = \frac{T_2}{M_2}$$

where P = unknown population; M_1 = number caught and marked at first capture; M_2 = number found marked at second capture; T_2 = number caught at second capture. The use of this method involves several major assumptions, especially relating to the extent to which marked individuals disperse evenly through the parent population after release, and further details should be read in Adams (1959).

USES

The uses of shingle formations frequently overlap with those of shingle foreshores and other coastal types, and reference can usefully be made to Part IV of Ranwell (1972a). From an educational and scientific viewpoint, shingle formations are of considerable use in the study of both physiography and ecology.

Physiographic interest in shingle foreshores is concerned with current processes, but shingle formations give added depth to this study since, within their depositional activity, they combine a history of major shoreline changes with records of other related changes such as river estuaries, ports, and settlements (viz. Dungeness or Orford Ness). Studies can involve investigation of the shingle composition of ridges of different ages, in an attempt to understand the cause of growth of the formation. At present, there is a wide divergence of views among scientists as to how and why massive deposition of shingle takes place. Dungeness, particularly, is important in this respect because of its documentation and age, but much useful work could be carried out on other smaller shingle formations.

Botanically such formations are useful because they can be used for teaching and research on the succession from stony vegetation to more advanced communities. At Dungeness, there is a great deal of teaching material within the build-up, spread, and die-back of the vegetation associated with *Sarothamnus*. Ecologically, this is the finest example of naturally developed plant and animal communities on coastal shingle in Britain, but other areas are also of great botanical and ecological interest, as repeated field parties from Flatford Mill Field Study Centre to Shingle Street will testify. Much of the basic detail required for understanding succession in vegetation within such simple communities can be gathered within educational programmes and should involve local students wherever possible.

Entomologically, shingle formations play an important part in overseas insect migrations and are useful study areas for such movements because they are usually undisturbed and isolated compared with other coastal types. Ornithologically, too, shingle formations are extremely useful because of their open, undisturbed nature, away from human habitation. Ornithological work continues at Dungeness, Scolt Head, Blakeney Point and Orford Ness/Havergate, and could well be extended with profit to other shingle formations. Thus shingle formations make good educational centres with a wealth of scientific interest, and a great deal remains to be done to initiate field study in many areas of this country.

Shingle formations have never been greatly used as recreational areas and they remain one of the most unused parts of our coastline. In many places, especially at Dungeness and Shingle Street, fishing causes considerable modification via summer residences and boat-dragging equipment although this is confined to the coastal fringe. Difficulty of access to formations such as Scolt Head, Blakeney Point or Orford Ness also limits recreational use, and only ardent dog-walkers or nature-lovers will venture on to stable formations, where walking is uncomfortable and vistas are desolate.

Economically, shingle formations have been a relatively unimportant resource used only for grazing cattle and sheet at very high acreages per head. In the past they were mainly used within a farming method which used adjacent marsh for the bulk of grazing and stock were moved on to the stable shingle during high tides. Nowadays stock is not reared for this type of farming since it is not profitable, and grazing is no longer practised. An exception to this is hill sheep-farming in Scotland, where stock include shingle within their travel for food.

During the second half of this century, as the commerical value of undisturbed shingle has been seen to be negligible, so its exploited value as a mined commodity has risen greatly. In fact this is the main danger to the survival of wildlife on shingle. If suitable management controls are not enforced, many of the most accessible shingle formations will be mined out of existence for the construction of roads, houses, and other buildings. Dungeness has suffered dramatically in this way in the past and continues to suffer at the present. Not only are large areas excavated, but many tracks are formed and buildings erected on site. As building materials decrease in quantity and increase in value, exploitation will increase commensurately, to the detriment of the wildlife environment.

Another economic use of shingle formations is as large underground reservoirs of freshwater, partly surrounded by yet insulated from the sea. The effect of water-extraction on most of the flora will be negligible, but where freshwater lagoons exist, such as the Open Pits at Dungeness, the maintenance of purity and water-level are of the highest importance for scientific study and maintenance of the habitat. Van der Voo's (1964) work on the Wassenaar dune-slacks shows clearly the results of excess extraction.

A final use of shingle formations is as military training areas. As with inland areas of low farming importance such as Westleton Heath, Suffolk, parts of Salisbury Plain, or the Breckland, so shingle formations are thought suitable for army use: Orford Ness was until recently Ministry of Defence property, as are large parts of Dungeness. The offshore barrier islands do not suffer in this respect. Military use if unsupervised is, of course, highly detrimental to the wildlife of shingle formations. Damage caused by tracked vehicles and bombs are obvious problems, but so are extension of communications networks, building and other levelling or grading operations. On the other hand, this damage is partially offset by prohibition of the access of other users which improves wildlife conditions. If military use of certain shingle formations is overseen by ecologists and conservationists its dangers may be minimized.

CONSERVATION AND MANAGEMENT

In the consideration of conservation and management policies for shingle formations, the rarity of such features must play an important part. In Britain, large areas of stable shingle are relatively uncommon, and unfortunately few of these areas are within nature reserves. In fact, Dungeness is a good case study of the failure of the then Nature Conservancy to obtain National Nature Reserve (N.N.R.) status.

The importance of Dungeness from a conservation point of view was first shown when the Royal Society for the Protection of Birds (R.S.P.B.) appointed a warden in 1907. In 1915, the Society for the Promotion of Nature Reserves listed Dungeness as a category A area of primary importance in their Report to the Board of Agriculture. It was starred as being of especial interest. Reserve land was not owned in the area until 1931 when the R.S.P.B., with a Mr. R. B. Burrows and other subscribers, bought the 102 ha known as Walkers' Outland. In 1931, a further 110 ha were conveyed to the R.S.P.B., and an additional 283 ha were added in 1936.

However, during the Second World War the whole area was requisitioned by the War Department. In 1944, the local regional subcommittee of the Nature Reserve Investigation Committee (N.R.I.C.) again recommended the area for conservation because of its geomorphic, botanical, and ornithological interests, and this was endorsed by the main committee of the N.R.I.C. in 1945, when it was included in the main list of conservation areas. In 1947, the Wild Life Conservation Special Committee accepted the N.R.I.C. recommendations on Dungeness and listed it as a Conservation Area.

From its inception, the Nature Conservancy supported the reserve status of Dungeness. In 1949, a party of natural history experts visited the area and submitted reports. In 1951, part of Dungeness was notified to the Kent County Council as an S.S.S.I. (Site of Special Scientific Interest) and in 1952, Dungeness Bird Observatory was established, with the Nature Conservancy providing a grant for a full-time warden. By 1954, the Conservancy had added the area to their list of Proposed National Nature Reserves, and over succeeding years negotiations were opened with the R.S.P.B., the War Department and other owners, for a Nature Reserve agreement. In 1958, the Kent Development Plan, including the reserve, was approved by the Minister of Housing and Local Government.

Under the National Parks and Access to the Countryside Act, 1949, the Nature Conservancy could have compulsorily acquired Dungeness as Crown Land. However, less drastic means were used and negotiations were still in progress when, in 1958–9, the Central Electricity Generating Board saw the site as suitable for a nuclear power station. A public inquiry was held in December 1958 at which the Nature Conservancy put forward all the evidence against the proposed nuclear power station (Nature Conservancy, 1958); but the Conservancy lost. Unfortunately, it seems that the Nature Conservancy felt that in the battle for Dungeness it was 'all or nothing'. The R.S.P.B. area and the Observatory have been maintained on the S.S.S.I. with a warden financed by the C.E.G.B., but the Conservancy have expressed little further interest. This is a great pity; although the power station is damaging both physiographically and ecologically, N.N.R. status would prevent further deterioration of this unique shingle formation.

Happier histories can be reported for the shingle formations along the north Norfolk coast, which is designated as an Area of Outstanding Natural Beauty (Steers, 1971). Blakeney Point was the first nature reserve to be owned by the Norfolk Naturalists' Trust [itself the first county Naturalists' Trust to be formed (editor)] and has been a reserve since 1912. Scolt Head Island has been a nature reserve since 1924 and is owned jointly by the National Trust and the Norfolk Naturalists' Trust. In 1953, it was leased to the Nature

Conservancy and is now managed by a joint committee of the representatives of the three organizations and local residents.

It is important to point out, however, that the offshore barrier islands of Norfolk (Thornham, Scolt Head, Holkham Meals before reclamation, and Blakeney Point), although basically shingle formations, are ecologically important because of their marsh–sand–shingle complexes.

When Blakeney Point was established as a Nature Reserve in 1912, Professor Oliver declared his pleasure 'that so wonderful a collection of natural habitats should have been secured by the wise generosity of the donors against the possibility of any interference with the operation of natural factors'. The Blakeney Point Reports in the Transactions of the Norfolk and Norwich Naturalists' Society through the years (see Oliver, 1929) show the successful research that has been carried out into all aspects of its biology and geography. However, although almost an island, Blakeney Point is too convenient for public access, and vandalistic visitors did not permit unhindered conservation and scientific study at Blakeney. Therefore an alternative and less accessible system was sought where research scientists should continue their work. In 1923, an appeal was launched for money to buy Scolt Head Island for £500 and within 12 weeks the island became National Trust property.

Wardening of Scolt was initiated in April 1924 for protection of breeding birds. This activity began in the Farne Islands in the 1880s, but the appointment of a woman to the lonely and unusual post at Scolt made a great impact on public opinion and established the general principle of wardening for successful conservation. Despite the early successes and continued research at Scolt (Steers, 1960), the whole public relations and educational aspect of management is still incomplete. The terns have been carefully preserved, but no bird observatory has ever been set up. Difficulty of access has kept Scolt Head unimpaired, but has also kept away the interested public. Despite the great deal of work already carried out, many groups of plants and animals remain to be adequately studied there.

Our lack of knowledge concerning shingle foreshores and the more terrestrial formations is obvious from the literature. There is no book equivalent to those by Salisbury (1952), Chapman (1960b) or Ranwell (1972a) on sand-dunes and salt-marshes. Even the general reviews are limited to a very few pages (Chapman, 1964; Gimingham, 1964; Hepburn, 1952; Tansley, 1939).

The value of a site depends very much upon the length of time during which it has been studied. After a time, it is seen to need protection and the result comes as conservation and management. Chance plays a very large part in which areas of the country are conserved and protected, since the most important shingle formations are generally those where specialists have worked. Dungeness, without a road until 1930, tended to be a neglected region, whereas Blakeney Point and Scolt Head Island, being relatively near to Cambridge University, were studied by Professor Oliver and Dr. Long respectively during the early part of this century. (Nicholson, 1960) and still receive considerable attention from, amongst others, the Department of Zoology.

Shingle systems, both foreshores and larger formations, should be recognized as a valuable and very limited resource, not just for plant and animal life but also as an overflow area from adjacent recreational land. Every country that possesses shingle formations should develop a national plan for their protection and use, particularly because of their economic importance when exploited as a mineral resource. This national plan should record the size and distribution of shingle areas and designate the importance of each area in terms of coastal protection, recreational use, plant and animal protection, military use, and gravel exploitation. In this way, the pressures that have been created on shingle formations by access could be controlled. What there is could be safeguarded as a whole, yet released bit by bit as certain developments become of over-riding importance. Destruction of shingle formations is so simple that this sort of planning is vital at the present time.

Any shingle formations which are managed for coastal protection, recreation or for military use will also contain considerable wildlife resources. The protection of these resources, such as *Suaeda fruticosa*, is obviously important to the stability and maintenance of the formation for all these uses. However, because of the prime uses of the areas, initiative for protection of local or rare species must be with local naturalists of Naturalists' Trusts. They can go a long way towards avoiding unnecessary damage by compiling maps and showing owners and planners the location of especially important sites. At Dungeness, the army has been a willing owner in this respect. In certain situations disturbance must occur. This, of course, is especially true of areas designated for mineral exploitation. Here it is often actually desirable to transplant species to nearby safe areas: such transplanting should be recorded and reported to the Biological Records Centre, Institute of Terrestrial Ecology, Monks Wood, Abbots Ripton, Huntingdon PE17 2LS. Another method of reducing damage is to restrict the number of trackways made to the mining area.

Where coastal protection is the main aim, any gravel mining should be stopped and a regular maintenance commitment be accepted. In the West Country, much of the shingle facing the Atlantic waves must be pushed back into place after every large storm. In such locations, aerial photography should be carried out every year, if possible, to record the success of the management programme and to further physiographical understanding of the shingle in relation to adjacent coastal types. Continued research on various types of vegetational stabilization is particularly important in these areas. The major concentration should be placed on plants like *Suaeda fruticosa*, with binding root-systems, and *Lathyrus japonicus* or *Silene maritima*, with mat-like aerial parts which stabilize the surface. Success is most likely to come, if anywhere, with joint use of both types.

Recreational management depends mainly for its success upon access. It is essential to make convenient access to the shore, which is where most people want to be, whether for fishing or merely for relaxation. In this way, pressure is taken off other areas where stabilization or specific populations of wildlife are important. Where visitors are few, natural vegetated pathways are sufficient, but in most cases an artificially surfaced path such as the sleeper walkway at Walmer, Kent, and ample parking facilities must be provided. Car parking on the shingle is often one of the greatest menaces to management and protection, and uncontrolled parking leads to expensive control measures and restabilization of the surface.

Recreational use of shingle areas is increasing at the present time. Much can be done to relieve the pressure on adjacent sand-dunes by judicious siting of shingle nature trails, etc. The fulls and lows of shingle ridges can provide similar privacy to that enjoyed by people on sand-dunes, and an informed public will use such areas to avoid crowds elsewhere. Much more research on recreational–educational use of shingle is required.

Management for wildlife protection is much more complex and is, as Ranwell (1972a, p. 226) has stated, 'to maintain the plant and animal communities for which they were originally selected, to utilize them for educational and research purposes and, where there is scope for this, to increase the diversity of habitats within them by controlled disturbance'.

One of the most interesting problems of shingle formations is the role which rabbits play in their ecology. On offshore barrier islands, it might be useful to eradicate rabbits on some islands and build-up populations on others: rabbit populations would not be acceptable on mainland apposition beaches. In areas such as Dungeness, where a large number of alien plants have entered along the access roads, removal is a useful and practical policy, as invasion may affect the local flora. Potential water-level problems, such as those at Dungeness where the Open Pits are vulnerable, must be tackled by active management of the water-table. The deliberate creation of lagoons on shingle by excavation can help to recreate late-stage hydroseres if these are lost by water-level changes. A prime effort must be made to include examples of each shingle type within nature reserves and special conservation areas should be set aside for rare or declining species.

Perhaps the most useful means of management of shingle is geographical zonation of activities, so that no area gets over-used and destroyed. Temporary fencing can frequently be used on shingle formations to rest them from educational or recreational over-use. Foreshore floras and terneries are particularly susceptible to over-use and can be improved by resting. Educational use can be separated from remoter research sites to eliminate disturbance in the latter.

As Ranwell (1972a) has described in connection with sand-dunes, there is thus a pattern in shingle management and shingle use. Some areas close to large urban complexes will be lost to gravel exploitation, and others are best managed for recreational and intensive educational use. Those more distant from towns, but with ready access to research stations, are more appropriate for research use; while small, remote areas should be left untouched.

Chapter 11

Earth cliffs

V. J. MAY

INTRODUCTION

The management of earth cliffs has fallen mainly to engineers concerned with the prevention or control of erosion, and it is upon this aspect that most of the available literature has concentrated. The ecology of marine cliffs has been neglected at least partly because of the difficulties of access. The author has as a result relied heavily on his own field observations so that the examples discussed will be limited both in number and location.

BACKGROUND INFORMATION

Few geomorphological studies provide evidence for any subdivision of cliffs into 'rock' or 'earth', 'weak' or 'strong'. Most geomorphological texts (see, e.g., King, 1972; Zenkovich, 1967) devote no more than a few pages to cliff form and process. The most important characteristic of so-called 'earth cliffs' is their tendency to instability and rapid change, unlike 'rock' cliffs where change is infrequent and particularly localized. In western Europe, frequent change and instability leading to mass-movements are most common in sand and clay cliffs in south and east England (Bird and May, 1976), north France, Denmark, and the low Baltic coast (Figure 70). Because of their rapid retreat, some chalk cliffs are also included in the earth cliffs category (Table 15).

Table 15   Mean annual cliff-top retreat in certain cliffs facing the English Channel

	Number in Sample	Mean Retreat (m)	Standard deviation
All cliffs	499	0·29	0·36
All chalk cliffs	160	0·21	0·20
Sussex chalk	38	0·38	0·19
Thanet chalk	31	0·23	0·10
North Downs chalk	37	0·11	0·15
Poole and Christchurch Bay Tertiaries	30	0·59	0·38
North Kent Tertiaries	42	0·81	0·49
North-west Isle of Wight Tertiaries	33	0·26	0·32
South-west Isle of Wight Wealden	38	0·41	0·27

Table 15 (continued)

Particular locations	Rock type	Mean retreat (m)	Period	Source (other than authors)
Cranmore, Isle of Wight	Hamsted Beds	0·61	1868–1963	
Christchurch Bay	Barton Beds	0·58	1896–1962	
Brightstone Bay, Isle of Wight	Wealden	0·51	1872–1962	
Seven Sisters, Sussex	chalk	0·51	1873–1962	
Ault, Somme	chalk	0·42	1825–1912	Briquet
Ringstead Bay, Dorset	Kimmeridge	0·41	1888–1963	
Folkestone, Kent	Gault clay	0·28	1872–1962	
Ault, Somme	chalk	0·25	1825–1960	Precheur
Ballard Down, Dorset	chalk	0·23	1882–1962	
North Isle of Thanet	chalk	0·23	1877–1962	
Ault, Somme	chalk	0·20	1825–1960	Precheur
West of Dover, Kent	chalk	0·09	1872–1962	
Ault, Somme	chalk	0·09	1825–1960	Precheur
Isle of Purbeck, Dorset	Portlandian	0·00	1870–1964	
North of St-Jouin, Seine Maritime	chalk	0·00	1834–1894	Precheur

Figure 70. Locations referred to in the text: 1. Antrim; 2. South-central England; 3. Seven Sisters, Sussex; 4. Folkestone; 5. North Kent; 6. Denmark; 7. Gotland, Sweden; 8. North France; A. West Dorset landslips; B. Ringstead Bay; C. Ballard Down; D. Arne and Shipstal Point; E. Bournemouth; F. Hengistbury Head; G. Highcliffe; H. Barton; I. North-west Isle of Wight; J. Brighstone Bay. Areas marked in solid line are coastlines dominated by earth cliffs

The form of earth cliffs and the processes by which they are changed depend primarily upon:

(a) rock cohesion;
(b) groundwater conditions;
(c) the effectiveness of marine erosion of the foreshore, the cliff foot, and the debris derived from the cliff;
(d) the amount and effectiveness of cliff-foot protection, both natural and man-made;
(e) offshore relief;
(f) sea level change.

Marine cliffs are usually considered to be a special case of slope development in which the removal of weathered material from the base is especially efficient (Young, 1972). Carson and Kirkby (1972) have suggested that slopes range between those which are *transport-limited*, i.e. where the rate at which weathered material is removed from the slope is slower than the rate of supply by weathering processes, and *weathering-limited*, where transport processes are sufficiently effective to remove all debris supplied by weathering. In the slope process–response system, the conditions operating at the foot of the slope are of primary importance in determining the extent to which slopes are transport- or weathering-limited. Marine cliffs are thus mainly weathering-limited slopes on which threshold angles, i.e. angles at which mass-movement is no longer significant, are only rarely attained.

The type and structure of the rock in which the cliffs are cut affect to a considerable extent not only the cliff form but also the dominant processes modifying it. For example, in eastern Dorset, the angle and form of the cliff profile (Figure 71) depend to a substantial extent on rock cohesion. Despite vigorous marine erosion, chalk stands at high angles, whereas clay slumps to lower angles.

Although many writers find evidence for important seasonal differences in the efficiency of cliff-retreat processes, much depends on the frequency of storm waves and on fluctuations of groundwater levels. Large cliff-falls or landslides, although more common in the winter half of the year, do bring about significant change in summer. Thus, although Hutchinson (1971) showed that falls from the chalk cliffs of Kent are associated with high average values of effective monthly precipitation and with high values of average number of days with air frost, May (1971a) demonstrated that for the chalk cliffs of Sussex many large falls occur during the summer half of the year. Strong marine erosion associated with winter storms at spring tide periods brings about deep-seated rotational slips in the London Clay cliffs of North Kent (Hutchinson, 1967; So, 1967), and Cambers (1976) confirms the role of storm-surges in cliff-retreat in eastern England. Where considerable mass-movement is taking place, however, much of the storm wave action is concentrated on cliff-foot debris and not on the cliff-foot itself (Hutchinson, 1976).

The efficiency of this wave attack on the cliff-foot and on the accumulations of debris depends not only on the frequency of storm waves, but also on the extent of shelter from such waves and their efficiency in lowering the foreshore. Such foreshore erosion may be comparatively rapid. For example, chalk platforms around the Isle of Thanet are lowered by about 25 mm annually (So, 1965), and bedrock at Byabuaura, Japan, at a rate of 20 mm per annum (Horikawa and Sunamura, 1970).

When the rapid removal of the cliff-foot and cliff-foot debris by wave action ceases, slope processes continue (Figure 72), which in time brings about a threshold slope. According to Fisher (1866), as the upper cliff degrades, a lower aggrading slope is formed by debris which protects the foot of the cliff and thus lowers its angle of slope. Mass-movements, albeit on an increasingly small scale, continue to modify the former marine

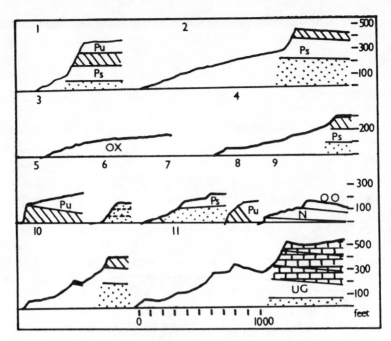

Figure 71. Relation of cliff profile to rock type and structure in Dorset.
Key to locations: 1. Isle of Portland (West); 2. Isle of Portland (East); 3.
Furzy Cliff; 4. Isle of Purbeck (East); 5. Langton Matravers; 6. Kim-
meridge; 7. Osmington; 8. East Lulworth; 9. Jordans Cliff; 10. Emmits
Hill; 11. White Nothe. Key to rock types: 'bricks' = Chalk; UG = Upper
Greensand; Pu = Purbeckian; shading = Portland Stone; Ps = Portland
Sand; dots = Kimmeridgian; OO = Osmington Oolite; N = Nothe Grit;
Ox = Oxford Clay

cliff, which is likely to be affected by substantial environmental change as variations in
climate, vegetation, or sea levels occur. Changes thus occur in the limiting conditions for
the maintenance of threshold angles. Hutchinson (1967) shows that slopes in the London
Clay ultimately reach stability at angles of about eight degrees, but many clay cliffs
maintain angles well in excess of this figure. As environmental changes occur, former
marine cliffs at angles close to threshold values may become reactivated. Between
Folkestone and Lympne, Kent, former marine cliffs show considerable variation in slope
form and angle (Figure 73). Slopes at Lympne have probably not been affected by marine
erosion for three or four thousand years, yet there is evidence in the slopes around Stutfall
Castle that slope movements have occurred since Roman times (Hutchinson, 1968).

Many recent landslips appear to have been initiated on cliff-lines abandoned earlier as
sea levels rose during recent post-glacial times to reach their present levels about 1600 B.C.
or even more recently (Jelgersma, 1966). Prior, Stephens and Archer (1968) consider that
late Pleistocene slumping along the coast of north-east Ireland is one of several factors
contributing to the formation of composite mudflows, and Hutchinson (1969) suggests that
the landslips at Folkestone Warren were initiated during the period when sea levels were at
or close to present levels.

Cliffs, then, are affected by two groups of processes:

(a) weathering and slope transport processes which produce deposits of rock debris at
    the foot of the cliff and reduce its upper slope;

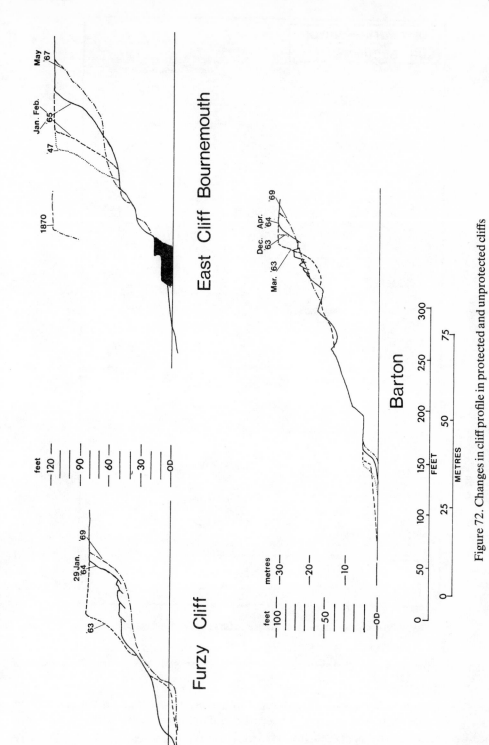

Figure 72. Changes in cliff profile in protected and unprotected cliffs

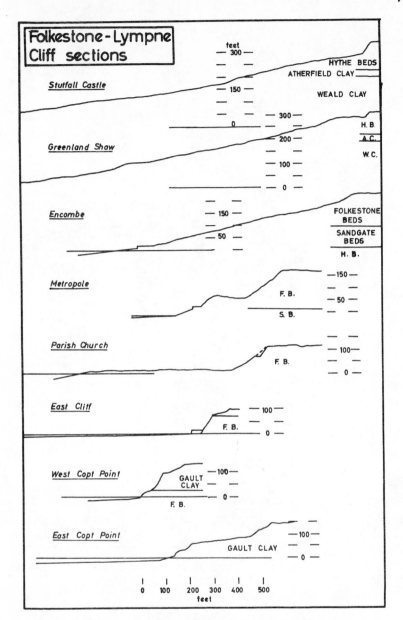

Figure 73. Development of former sea-cliffs towards threshold angles

(b) marine processes which are responsible for the removal of this debris, the erosion of the cliff-foot, and the lowering of the foreshore (Figure 74).

If the second group of processes are inhibited by the growth of stable beaches or by the construction of effective protection works, the cliffs become dominated by the slope processes and ecological management may become a realistic proposition. It is possible, however, for marine processes to become important once more if beaches or protective walls are removed, and the slope processes may pass through phases of greater or lesser activity as environmental conditions change.

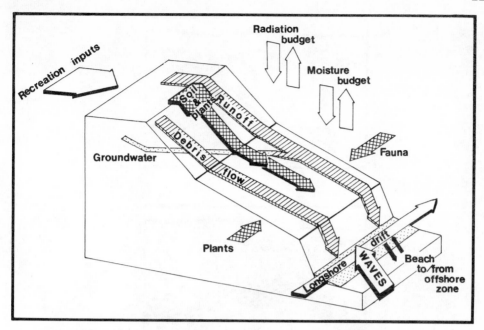

Figure 74. Mass and energy movements within earth cliff systems

With such instability in the cliff environment and frequent cliff-top collapse, most plant life on earth cliffs is derived from cliff-tops or from comparatively stable areas of cliff face. The extent to which any area is colonized depends on:

(a) the rapidity and nature of cliff changes;
(b) the ability of the available plants to colonize bare sand or clay;
(c) the nature of the cliff-top and cliff-face vegetation.

Thus whereas the rapidly retreating chalk cliffs of the Seven Sisters in Sussex are normally bare of vegetation, many equally rapidly changing clay cliffs sustain considerable communities, because slopes are less steep and slips or mudflows more localized in area. For example, on the clay cliffs and slips of the Isle of Wight, where cliff-top retreat rates in excess of 1 m annually have been recorded (May, 1964, 1966), areas of recent movement are quickly colonized by coltsfoot (*Tussilago farfara*) and great horsetail (*Equisetum telmateia*). More stable areas are colonized by the orchids *Anacamptis pyramidalis* and *Gymnadenia densiflora*, yellowwort (*Blackstonia perfoliata*), marsh helleborine (*Epipactis palustris*), pale flax (*Linum bienne*), slender birdsfoot trefoil (*Lotus tenuis*), and golden melilot (*Melilotus altissima*) (Westrup, 1964).

If cliffs are protected from marine erosion, colonization takes place on the talus slopes developed by the normal processes of slope degradation. For example, low cliffs at Hengistbury Head, Bournemouth, which were eroded by the sea until 1938 (May, 1971b), show a greater vegetation cover (a) the longer the period of protection from marine erosion, (b) the larger the talus fans, and (c) the more sand that is blown on to the foot of the cliff. Areas standing behind mature dunes have as much as a 36% cover, whereas more recently protected areas have only 17% and 1% cover (Figure 75).

The climate of coastal cliffs is one in which wind and salt-spray play an important role (Chapman, 1964). Nevertheless the role of aspect in considerably modifying the radiation budget, and thus the moisture budget, of coastal cliffs must not be underestimated (Geiger,

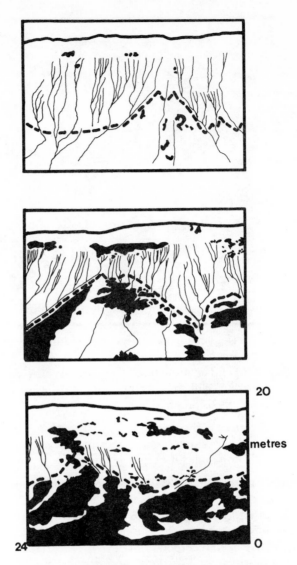

Figure 75. Colonization of cliff-foot talus at Hengistbury Head, Bournemouth; first protected in 1938.
Key: black = areas of vegetation; dashed line = top of
talus slope; fine lines = cliff-face gullies

1965). North facing slopes will receive less solar radiation and will lose less moisture by
evapotranspiration. The geological and groundwater conditions playing a dominant role in
the slope process–response system also affect the availability of soil moisture for plant
growth. Thus some temporarily stable cliffs support patches of bog vegetation, and many
areas possess *Juncus*.

The earth cliff system is, then, one in which much change, both in geomorphological and
biological conditions, may take place within a short time scale. Over-riding factors are the
geological character of the cliffs, the groundwater regime, the microclimate, and the
colonizing plants close to the more stable parts of the cliffs.

SPECIAL FEATURES

No studies have been made which detail the requirements of vegetation on earth cliffs. In rapidly and uniformly retreating cliffs most plants are derived from the cliff-top, whereas in cliffs where retreat is spatially more intermittent, colonization may take place from adjacent areas of cliff-face. The cliff-face environment is frequently very different from that of the cliff-top and thus colonizing plants must be particularly adaptable to changing soil and moisture conditions.

In broad terms, the particular problems of management of earth cliffs arise from:

(a) their intrinsic instability;
(b) the occurrence of cliff-top dunes and blown sand, which are particularly susceptible to trampling and burrowing damage;
(c) their attractiveness for recreation, and in particular as means of access to beaches;
(d) the use of some cliff-top areas for grazing (Crofts, 1973, refers to damage to sand cliffs resulting from prolonged and intensive grazing on crofters' common grazings in west Scotland);
(e) the occurrence of small colonies of important plants in more stable landslips;
(f) the use of the same comparatively stable areas for camping or caravan sites.

It is recreational trampling, and to a lesser extent grazing and rabbit damage, which cause most problems at the stage when marine processes have been reduced or controlled, and both human and animal activity become more concentrated. The problems arising from recreational damage can best be illustrated by reference to three sample areas in the Bournemouth area.

The three areas are Hengistbury Head, Shipstal Point, and west Arne—all within Sites of Special Scientific Interest. The cliffs are cut in Eocene sands and clays commonly capped by gravels, and at all sites marine erosion and gullying combine to modify the cliff forms. The main contrast, however, lies in their recreational usage. Hengistbury Head is easily accessible both by public transport and car from Bournemouth. At periods of peak usage, several thousand visitors use the area, with most concentrated on cliff-top grassland (46%) and beaches (37%). Shipstal Point lies on the southern shoreline of Poole Harbour at a distance of 1·5 km from the nearest road. Comparatively small numbers of visitors walk down a track to this area. Many follow a nature trail prepared by the Royal Society for the Protection of Birds and thus remain on the beach. The third site, at west Arne, also lies within an R.S.P.B. reserve but no access to this area is allowed except for the occasional survey party. Some visitors reach the shoreline by water but their numbers are minimal (Table 16).

Table 16   Site summary

	Site length	Public access	Visitors daily (max)
Hengistbury Head	2,500 m	main road, bus, footpaths	3,000
Shipstal Point	1,000 m	road within 1·5 km; track and nature trail	300
West Arne	2,000 m	none	15 (once a year)

At all three sites, trampling occurs when visitors climb up or down the cliff-face. At Hengistbury and Shipstal, many visitors walk along the cliff-top and at Hengistbury this is the most severely trampled area (Brooke, 1967), as most visitors walk here for views of sea and beach which cannot be obtained elsewhere. Moreover the cliff-top is capped by blown sand and many of the slopes have developed directly downslope. Gullies have developed which are over 1 m in depth and drain towards the cliff-top. In 1972 one such gully had sufficiently high discharge to destroy the beach at the foot of the cliff, thus exposing the cliff to marine erosion. Some trampling occurs when visitors to the beaches climb the cliffs as a means of access to the cliff-top. On the cliff-top, as gullies develop along paths, new paths are soon trampled out and the area of bare ground is extended (Figure 76).

At Shipstal, the cliffs are lower and stand at a gentler angle than those at Hengistbury. They suffer considerable trampling, thereby accelerating the downslope transport of debris to the beach (May, 1976). Furthermore, heathland paths slope towards the cliff-top where accelerated gullying is destroying both trampled and undamaged areas. Areas of gorse are not trampled! At west Arne there is no cliff-top trampling.

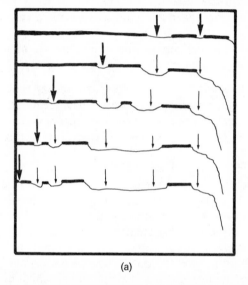

(a)

(b)

Figure 76. (a) Effects of path-shifting on a sandy cliff top. Key: heavy arrows = existing paths; light arrows = former paths, now gullied; thick black = vegetation; thin black = bare ground. (b) Path shifting

In summary, then, the primary effects of recreation are:

(a) the development of cliff-top paths which assist or accelerate gullying of both the cliff-top and cliff-face;
(b) downslope transport of slope debris during trampling of bare ground;
(c) destruction of cliff-top and cliff-face vegetation by trampling.

In all cases, the result is a lowering of the cliff angle of slope, thus making the cliff more suitable for climbing, playing, and sitting upon, and more readily colonized by gorse, *Armeria* spp, sea couch (*Agropyron pungens*), ling, marram (*Ammophila arenaria*), and similar plants.

At Chale on the south-west coast of the Isle of Wight, similar damage to the cliffs has occurred as a result of the wind and gully erosion associated with a large rabbit warren in blown sand, and in Folkestone Warren, gullying of low chalk cliffs has increased as recreational demands on the area have grown, with the consequent development of camp and caravan facilities.

From an ecological point of view, the extreme case of pressure occurs when cliffs are graded to improve their stability characteristics. Bedrock, soil, and plants are completely removed, to be replaced later by artificial means.

METHODS OF STUDY

To examine effectively the ecological characteristics of earth cliffs, analysis may be needed of:

(a) the nature, frequency and causes of change in the cliffs—the *geomorphological* and *geotechnical* characteristics;
(b) the nature of the flora and fauna inhabiting the cliffs, patterns of colonization, and their particular habitats—the *biological* characteristics;
(c) the nature of human use and management of the cliffs—the *human* characteristics.

Each of these groups of characteristics requires different methods of study and will have greater or lesser importance at different stages of cliff development and management. Each is therefore dealt with in turn. It cannot be stressed enough, however, that a thorough local search of available literature should be made, for such references as are quoted here do not include the wealth of local authority reports or the many geomorphological and geological studies which have been made of restricted locations.

When Wood (1950) suggested the following procedure for studying coastal erosion, he considered that the approach was one which would provide sufficiently complete and detailed information to be used for planning proposals. First, there must be a systematic collection of all data concerning rates of erosion and shoreline characteristics, and these data are then plotted on a diagram which shows the following features:

(a) the height and slope of the cliffs;
(b) geological cross-sections of the cliffs, including simple mechanical analyses of the materials forming them;
(c) miscellaneous shoreline characteristics such as vegetation, sea-walls and groynes;
(d) the average rate of erosion and accretion, obtained by the comparison of old maps with up-to-date aerial photographs;
(e) the severity of wave attack, classified by wind direction.

Such an approach provides a full reconnaissance survey of the physical characteristics of the cliffs, but does not consider the processes by which changes occur. With two reservations, the approach is to be recommended, for it sets any particular site in its context both regionally and temporally. The reservations concern points (d) and (e). Problems with the use of old maps and surveys, and their comparison with aerial photographs and more recent surveys, arise from such problems as inaccurate surveying and cartography, instability of map paper, lack of revision, and difficulties of cliff-top definition (De Boer and Carr, 1969; Horikawa and Sunamura, 1967; May, 1964, 1966). Such comparisons usually provide only long-term values, whereas many management problems arise from short-term pressures. Cliff-top retreat is usually assessed by measurement from known points or markers. Two problems can arise from this:

(a) markers may be removed—Rudberg (1967) comments ruefully that all his prepared sites were completely destroyed with one exception;
(b) without some reasonable knowledge of the form of short-term changes, the spacing of markers may be too sparse to allow adequate sampling.

Furthermore, important changes may occur on the cliff-face and at the cliff-foot which cannot be accurately assessed by comparison of maps at lengthy time intervals. The author knows of one site where in a particular year no change would be measured in either the foot or the top of the cliff, yet because of a mid-cliff fall, actual cliff-foot erosion of over 1 m is masked by debris. Figure 77 shows that during one twelve month period at Hengistbury Head most falls occurred on the middle and lower cliff face.

The second reservation arises from the fact that much cliff-top retreat cannot be correlated directly with marine erosion. For example, on the Isle of Wight, parts of the

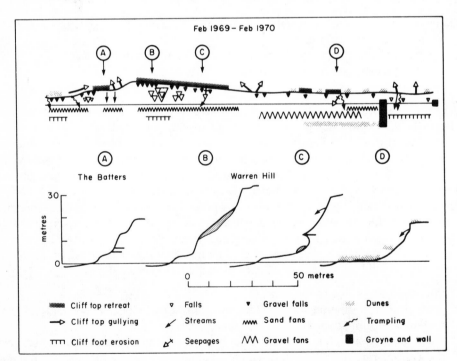

Figure 77. Cliff-face process and change, Hengistbury Head, Bournemouth (from *Field studies in South Hampshire*, 1971, p. 114. Reproduced by permission of the Southampton Branch Geographical Association)

north-west cliffs exhibit rapid cliff-top retreat and minimal cliff-foot retreat. In contrast, in the south-west both the cliff-top and cliff-foot retreat rapidly (May, 1966). Wave efficiency, as measured by wind frequency and fetch (Bruun, 1960; May, 1964), shows little relation to cliff-top retreat rates when the cliff-foot is particularly resistant compared with the cliff-top, or the cliffs are protected at their foot.

The importance of Wood's approach is that it allows particular problems to be seen in a wider context. Too often the coast protection or management problem has been seen as a localized, even unique phenomenon, thus disregarding its regional context and the great geomorphological similarity of many coastal locations. Cliffs cannot be considered in stability terms without knowing something of the past, present, and future behaviour of both the cliff-top and the cliff-foot. The behaviour of the cliff-foot depends to a great extent on the nature and stability of the protection afforded to it by beach (Sunamura and Horikawa, 1971), debris or walls. Examination of active cliffs must therefore include some understanding of changes in the beaches and platforms fronting the cliffs, especially as worldwide retreat of shorelines has characterized the past century (Bird, 1976).

Geotechnical studies are primarily concerned with stability of the cliff and prediction of slope failure. Such measures as the shear strength and liquid limit of the materials forming the cliffs give important indications of the cohesiveness of the rocks, and can be obtained by standard soil engineering tests (Terzaghi, 1950; Terzaghi and Peck, 1948; Zaruba and Mencl, 1969). Groundwater conditions are particularly important as they affect the pressures exerted within the rocks of the cliffs and thus its stability. Monitoring is usually carried out by the use of piezometers. Simple wedge-type stability analysis in terms of effective stresses has been used by Hutchinson (1970) for chalk cliffs, and the methods proposed by Morgenstern and Price (1965, 1967) have been applied by Hutchinson to non-circular landslips (1969).

Examination of the biological characteristics of an easily eroded cliff poses particular problems, for as has already been shown, trampling damage to both the cliff-face and the cliff-top can bring about much change in the vegetation and can initiate gullying. Furthermore, some parts of the cliffs are not accessible. With few available reference points in cliff sites, marks usually have to be placed for future reference. Unfortunately, the position of markers placed on cliffs cannot be relied upon to remain unchanged. Prior, Stephens and Archer (1968) established base pegs outside existing unstable areas and used a combination of peg lines, selected surveyed profiles and comparative photography to investigate mudflow movements. In Hutchinson's study of a mudflow at Beltinge, north Kent (1970), base pegs moved negligible amounts with one exception.

A first step must be the identification of species on the cliffs. This is particularly important in this habitat as there is a serious lack of studies on cliff flora (Chapman, 1964). Random sampling using quadrats may be of only very limited value, as plants are often isolated or confined to limited areas of the cliff face. The cliff-top may be treated as a normal terrestrial habitat, and normal ecological methods applied. To assess the pressure of trampling on cliff-top areas, sample quadrats should be used within which proportions of bare ground and damaged plants are counted and vegetation height measured. The number of visitors passing through these quadrats must be counted or sampled. To observe changes in plant cover and colonization patterns, comparative photography may be particularly effective. The use of time-lapse photography on areas of mass-movements was used by Hutchinson (1970), and comparative photography employing overlapping photographs for plotting purposes has been used by this author. Problems arise from the fixing of base points and tilt distortions, which must be overcome by standard photogrammetric methods (Howard, 1970).

Examination of the human characteristics of cliffs requires the application of methods used for the assessment of land utilization, particularly recreation. Most recreation on cliffs is informal; for example, walking, picnicking, and nature study. Few specific facilities are required, although in some areas beach huts and chalets, and even camp sites, may be located on relatively stable parts of cliffs and landslips. To assess the effects of recreation (or agriculture) it is necessary to know:

(a) the type of activity;
(b) its timing;
(c) its detailed location;
(d) the broader regional conditions affecting the demands made upon a particular site.

Whereas the number of visitors to a particular area can be counted comparatively easily by direct and automatic means (Hammond, 1967), their behaviour within the site is more difficult to assess. Direct observations may be made from vantage points or time-lapse photography can be used. The latter method depends for its effective use however on careful selection of the time-lapse period to suit the frequency of visitors. Whether all users of a particulr site can be plotted on large-scale maps depends on the manpower available or on the use of special aerial photographic surveys. To assess the catchment area, type of visitor and length of stay, questionnaire surveys are usually employed. None of these methods, however, can assess the infrequent users, the occasional very damaging vehicle, or the shingle damaging observer.

In summary, the methods to be used should assess the geomorphological situation and characteristics of the cliff, its geotechnical properties, the nature of its vegetation, and the uses to which it is put. It is necessary, therefore, to use the methodology of several inter-related disciplines and so team investigation of earth cliffs would appear most fruitful. Studies should be carried out as follows:

(a) examination of geological maps, and comparisons of maps, surveys and aerial photographs of various dates;
(b) reconnaissance surveys along the lines proposed by Wood (1950) and Carlson (1972);
(c) identification of species on the cliffs;
(d) examination of colonization patterns, using, where practicable, sample quadrats, linked to precisely surveyed base and reference points;
(e) examination of the area of bare ground, plant height, and plant damage in relation to usage of sample quadrats;
(f) geotechnical studies of cliff processes;
(g) direct observation of usage and user behaviour;
(h) questionnaire surveys of users to establish regional demand for use of sites and likely future pressures.

Clearly the data derived from such surveys must be subjected to rigorous mathematical and statistical analysis and evaluation, wherever possible or practicable.

USAGE

Cliffs and the immediate cliff-top are used mainly for the establishment of towns (particularly resorts), all forms of agriculture, education, and recreation. The cliff-face itself and a narrow strip of land very close to the cliff-top are rarely used for any activity. The cliff-face may however provide access to the beach, if slopes are sufficiently low in angle. It is this access which causes most damage to the cliff-face. Examination of cliffs at most reasonably accessible points shows that most users concentrate close to the cliff-top and along the beach at its foot. Narrow strips of cliff-face are used for access to and from the beaches (Figure 78).

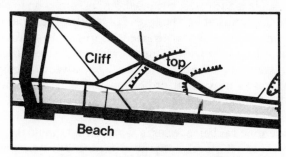

Figure 78. Characteristic user-flows on part of Hengistbury Head (shaded area = cliff face)

One form of use which cannot be overlooked is the increasing educational use of cliffs. Since many educational visits occur outside the main holiday season, often in inclement weather, severe damage to cliff-face paths may occur. Moreover, many parties visit cliffs for their particular value as geological sites, and some particularly noteworthy or fossiliferous horizons suffer considerable 'hammer erosion' at their hands. It is alarming to find large parties of small children hammering the lower parts of steep and even overhanging cliffs without any very clear purpose.

It is, however, the urban use of cliff-top land which poses most problems for the management of cliffs. As former small fishing villages have expanded to become resorts or dormitory towns, building has taken place on adjacent cliff-top land. Once established, the settlement may be threatened by cliff retreat, and measures are required to prevent or reduce losses of land and property to the sea. The presence of such cliff-top settlement severely limits the possibilities available for management of the cliffs.

MANAGEMENT PROCEDURES AND PRACTICE

Management of rapidly changing cliffs has been the concern of coast protection authorities in Britain since the Coast Protection Act of 1949: elsewhere it is mainly the concern of the landowner. The use of coastal protection measures is mainly determined by the expected losses of cliff-top land and property which might ensue if cliff retreat were not controlled. Management and conservation policies are often quite different for the three main groups of cliff situations:

(a) cliffs with land of low value at the top (usually non-urban agricultural or wilderness) and thus unlikely to attract support from rates or national funds for schemes of protection against erosion;
(b) cliffs with urban or transport uses where protection from erosion is needed but not implemented;
(c) cliffs which have already been protected against marine erosion, and are thus only affected by subaerial processes of slope development.

Each group is found on the coasts of west Europe, but it is at the English resorts constructed on cliffs of Tertiary sands and clays that the most extensive management policies have been undertaken. It was the rapid retreat of clay cliffs at Herne Bay in north Kent and around several east-coast towns which stimulated much public and official pressure for national action during the last decade of the nineteenth century and brought about the Royal Commission on Coast Erosion which reported finally in 1911. The coastal erosion problem arose, in many cases, because housing spread on to cliff-tops which were already or were potentially unstable. The spread of housing itself may have contributed to the instability of some cliffs, as water drained into them from roads (So, 1967). The arguments about the value of land lost to the sea were extensively aired during the course of the Royal Commission's hearings of evidence. The Royal Commission did not accept the strongly argued case that coastal protection was a national duty, and the Coast Protection Act of 1949 makes it clear that, although grants may be available from the national Government, coast protection is the responsibility of the local authority.

Areas which are identified as being zones of rapid retreat should therefore be avoided by high-value land uses, unless the local authority in its capacity as coast protection authority, is prepared to carry out protection works in the future. Planning authorities, however, cannot control buildings in areas of potential difficulty without information about the nature of cliffs, their ecology and the present and likely future changes in them. In this respect, former landslips present a particular problem, as although they may appear to be stable, they can readily revert to their former instability. At Sandgate, a former fishing village between Folkestone and Hythe in Kent, houses, some built since 1962, have been affected by movements in reactivated landslips on several occasions (May, 1964; Hutchinson, 1968). Areas of rapid retreat and former landslips which have not been affected by urbanization offer considerable opportunities to the engineer, geomorphologist, and ecologist to examine natural processes without the political and social pressures which develop when valuable cliff-top land is endangered. Where areas of particular scientific value are identified, they should be protected by the use of such designations as Site of Special Scientific Interest. Such a designation has been given to some landslips; for example those of west Dorset first examined in detail by Arber (1940, 1941) and still the subject of detailed investigation.

The first group of cliffs, then, should be avoided by urban or high-value land uses. They should be the subject of detailed scientific investigation, and, where investigations show it to be necessary, protected from severe educational and recreational pressures.

Where cliffs are retreating rapidly into areas of residential land, public calls for protection against erosion are common and frequently effective. Since coast protection authorities must have regard to the cost and the benefits of any coastal protection works, account must be taken of the value of the land which will be lost before the cliff achieves a natural equilibrium. It is questionable whether protection works in purely residential areas are always cost-effective (Clark *et al.*, 1976). Phillips (1972) suggests that coastal protection works at Barton, Hampshire, have cost more than the monetary benefits derived from them. It must be emphasized, of course, that social and recreational benefits are only partially quantifiable and that the over-all gain to the community probably more than outweighs the monetary losses.

If cliffs are to be protected, the primary concern must be the prevention or reduction of marine erosion at the cliff-foot. In many cliffs, however, although the mass-movements resulting from the instability caused by cliff-foot erosion can also be reduced, much instability exists merely because the cliffs are oversteepened slopes. Cliff protection schemes must therefore control the retreat of both the cliff-foot and the cliff-top (Thorn, 1960; Mitchell, 1968). The most effective long-term means of cliff protection is a stable beach, as this protects both the foot of the cliff and the platform. Being permeable and mobile, the beach absorbs wave energy more efficiently than impermeable man-made structures. Some of the material forming the beach is derived from erosion of the cliffs, but some is also supplied by longshore drift and by seabed movements of beach material. In some circumstances, the continuing stability of the beach may depend on beach processes themselves, in others upon the continuing erosion of the cliffs. Zeigler *et al.* (1964) show that at Cape Cod, sea-defence works would make the maintenance of the beach dependent on longshore drift, since roughly half the material reaching the beach and offshore bars is derived from cliff erosion. The beach would, according to their calculations, disappear within 86 years. A similar problem is evident at Bournemouth, where beach maintenance requires a continuing source of coarse sand: a source that can be supplied neither by the now protected cliffs nor by longshore drift because of extensive groyne systems.

As an alternative to maintaining a stable beach in front of the cliffs, protective walls may be constructed. Their form depends to a substantial extent on the capital available, and ranges from massive concrete walls to simple wooden walls and revetments, sometimes backfilled with stone (Thorn, 1960). The protective walls themselves may be severely damaged if the beach or platform is lowered by erosion. Much of the coastline of southern England is now protected by such defence works (May, 1966; Bird and May, 1976).

To control cliff-top retreat, different measures are required from one site to another. All are concerned to reduce the effectiveness by which processes of slope development modify the oversteepened coastal slopes. The use of filter and adit drains to remove water from landslips is common, for, with reduced porewater pressures in the clays or sands forming the cliff, increased stability can be expected to occur.

At Barton, Hampshire, the cliffs are characterized by three main morphological features (Clark, 1971): an upper cliff in gravels and Barton sand, a lower cliff of Barton clay rising directly from the beach, and a complex undercliff of former slips and seepages. The slipping has taken place as a result of seepages and higher porewater pressures along a band of silty clay which lies above the impermeable Barton clay. The clay cliff has been affected in the recent past by slips and mudflows resulting from instability associated with marine erosion (May, 1964). Cliff-top retreat reached a maximum annual value of about 1·0 m, with an average annual retreat of about 0·6 m during the first part of the present century (May, 1966). Clark reported in 1971 that stabilization and desiccation of the former slips at Barton had been generally successful, following the use of an impermeable wall of interlocking sheet-piles driven into the slip debris, and drainage of the seepage water from this area. Access to the beach had been improved, and the undercliff itself was now of considerable amenity value to residents and visitors alike. Gravel paths and steps both then and now allow access without severe trampling. The cliff-top is level and there is little surface runoff to the cliff-edge, so that there is little danger of the severe damage which occurs at Hengistbury Head. With increased stability, the undercliff was being naturally colonized by vegetation—where grass had not been planted. During the early winter of 1974, however, the sheet-piles were swept away at one point by a large movement accompanied by advance of the cliff-foot. Cliff-top property has since been demolished. What was regarded locally as effective cliff-management was revealed as ineffectual (Clark *et al.*, 1976).

The cliffs farther to the west at Highcliffe provide an interesting contrast in the effectiveness of coastal protection schemes. Formed mainly in clays with narrow bands of sand, these cliffs have not be affected by such rapid retreat as those at Barton (May, 1964; Clark, 1971). Nevertheless a serious problem exists. Although a wooden revetment has been constructed at the foot of the cliffs, and cliff-foot retreat retarded, slipping and gullying of the cliff-face continues, for the problems of marine erosion and cliff-face processes were not tackled effectively at the same time (probably for financial reasons). Cliff-face processes will continue to lower the slope towards an angle of about ten to twelve degrees where drainage works were not carried out. Even paths built to provide access to the beach are affected by movements, and in winter much of the undercliff cannot be crossed on foot. In 1963, similar conditions prevailed at Barton.

Cliffs already protected by walls or stable beaches show varying stages of stability and colonization by vegetation. At Folkestone, Kent, where Lower Greensand cliffs have been protected for at least 100 years, mature trees and shrubs, many introduced, cloak a narrow undercliff and the cliff face (Figure 73). Although small movements and falls take place occasionally, management is confined mainly to minor drainage works and maintenance of the existing vegetation. At the cliff-foot, however, considerable investment has been necessary since 1947 to maintain adequate beaches and repair storm-damaged walls. With surfaced paths and steps providing access to the lower parts of the cliff and beach, there is little recreational damage.

At Bournemouth, a positive programme of cliff protection has been carried out, involving not only the construction of walls, but also the planting of vegetation on cliff-faces to assist stabilization. Protective walls have been progressively constructed since 1907 to control erosion of the cliffs, which are capped by plateau gravel and blown sand, but are formed mainly in interbedded sands and clays of Eocene age. Many of these beds are very thin, and fine sands overlying impermeable clays are particularly susceptible to erosion. Seepages from the top of the clay bands sap the overlying sands, and gully those lower down the cliff-face. As a result, the cliffs are often very complex in profile. Although most of the mass movements here are shallow and affect limited area of the cliff-face, occasional larger slips may affect the whole cliff (Figure 72).

Management of these cliffs has taken three main forms:

(a) the construction of paths and steps to provide access from cliff-top to beaches (cliff-top paths also reduce damage to the easily eroded sand capping);
(b) terracing and planting on the cliff-face to prevent surface erosion;
(c) grading and seeding of cliff-faces to reduce the possibility of both large and small movements.

Terracing was accompanied by the planting of privet hedges at right angles to the trimmed cliff-face. The areas on the terraces were then filled with soil and seeded with grasses. Shrubs such as *Euonymus japonicus,* sallow (*Salix* spp) and evergreen oak (*Quercus ilex*) were also used to provide stability to the cliff-face. In parts, ice plant (*Mesembryan-themum*) was used as an alternative to grasses, and marram (*Ammophila arenaria*) was used to stabilize the embryo dunes on the cliff-top. Such measures, however, do not prevent deep slips. The likelihood of deeper movements can be reduced by the use of filter drains and by grading the cliff-face so that it is closer to its threshold angle. This is the policy now used at Bournemouth. For example, following a slip on the East Cliff in January 1965, when cliff-top paths and a road were undermined, the cliff-face and debris were graded and seeded (Figure 79). This cliff is now covered by a mixture of grasses and gorse. To overcome the difficulties of access and of providing quick germination conditions, the original seeding was carried out by hydromatic techniques. Figure 80 shows clearly the result of the older privet and grass method contrasted with more recent grading and seeding techniques. The former leaves steeper sections of the cliff bare, but involves little loss of cliff-top land. The latter method makes full use of modern earth-moving equipment, but produces a slope of sufficiently low angle for soil-erosion control to be undertaken efficiently.

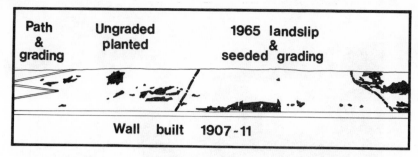

Figure 79. Cliff-face management, East Cliff, Bournemouth, March 1973
(black areas = bare ground)

Figure 80. Cliff-face management, Bournemouth. Although not at threshold angles, management of these cliffs must take account of the proximity of cliff-top buildings and roads

The measures for erosion control used at Bournemouth represent one of the most
extensive programmes for the ecological management of rapidly-eroding sand cliffs in
Europe. It could not have been undertaken without considerable investment by the former
County Borough in schemes which have been seen not merely as cliff protection works but
also as a considerable amenity to a major resort. The older cliffs support sufficient small
fauna to have supported in the recent past an occasional kestrel. The bird population is
varied, mainly as a result of extensive planting of shrubs and the lack of public access to the
cliff-face. Fences inhibit access and individuals convicted of trespassing on the cliffs may be
fined up to £20. During the grading process, damage to plant and animal life is complete,
but by planting a variety of shrubs, colonization can take place from adjacent areas. The
cliffs have never been regarded as a reserve, but they serve such a purpose nonetheless.
The attractiveness of the cliff-top at one point has brought about probably the worst case of
recreational damage to cliffs anywhere in southern England. The problem at Hengistbury
Head (Figure 81) has already been described; the measures which have been undertaken
so far go some way towards reduction of the problem.
    While the obvious 'solution' is to prevent any public access to such sensitive areas, this
has never been the policy of the local authority. Some degree of control over the
movements of visitors has been brought about by the introduction of a nature trail, which
avoids the worst areas of trampling damage, and by the construction of gravel- or

Figure 81. Cliff-top gullying at Hengistbury Head. This gully was over
one metre in depth in 1974 but has now been filled in

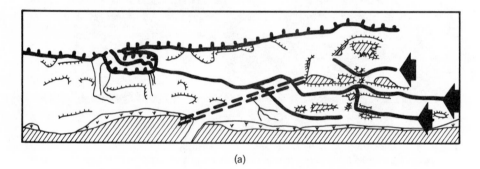

(a)

(b)

Figure 82. Cliff-top gullying and bare ground at Hengistbury Head
in 1974. Key: arrows = former paths; shading = areas of vegetation fronted by mobile sand

tarmac-surfaced paths. On Easter Monday 1973, however, only 26% of the visitors were on constructed paths. Moreover, one path constructed directly downslope had been gullied to a depth of 0·2 m. Severe cliff-face erosion by the main cliff-top gully has been reduced by diversion of its flow by a low earth embankment (Figure 82), and by the dumping of brushwood and topsoil. Some cliff-face erosion has been reduced by the construction of steps on formerly steep and deeply-gullied paths, but on the main cliff-face trampling damage continues. Such damage will continue until access from cliff-top to beach is rigorously controlled, as it was at Bournemouth itself. The dilemma between the use of fences and paths, and leaving the area as a wilderness is a considerable one, as pressure is increasing and damage by trampling can be considerable.

SUMMARY

The management practices which appear most effective are as follows:

(a) any individual site should be seen in its regional context, both in terms of its geomorphological and ecological characteristics and its recreational or usage value;

(b) control of erosion both of the cliff-top and the cliff-foot must be approached in terms of the complete coastal system, for individual points are but part of coastlines which are adjusting their form to reach some degree of equilibrium with oceanic and climatic conditions (Silvester, 1970; King, 1972);

(c) where possible, control of change should use natural methods of protection—for example, stable beaches should be encouraged;

(d) in areas of heavy recreational use, surfaced paths along cliff-tops, and paths or steps down cliff-faces, are the only effective means of preventing severe gullying;

(e) in areas of moderate use, paths can be constructed using local materials, but must avoid positions directly downslope (Food and Agriculture Organization, 1965);

(f) the use of nature trails and well-sited maps and indicators can do much to channel visitors to areas of particular interest and away from particularly sensitive zones of cliff-top;

(g) in certain areas, some control, in the first instance voluntary, of educational visits may become necessary.

The management of earth cliffs therefore requires measures to control marine erosion, soil erosion, and damage to vegetation.

ACKNOWLEDGEMENTS

My thanks are due to the former County Borough of Bournemouth Engineer's and Parks Departments which have allowed access to documents and gave permission for surveys, and to Bryan Pickess, of the R.S.P.B. at Arne, for access to the reserve.

Chapter 12

Rocky cliffs

F. B. GOLDSMITH

INTRODUCTION

Any estimate of the length of coastline of the British Isles depends not only on its definition (are estuaries for example included?) but also on the scale of the investigation. The Countryside Commission have produced a figure of 4,413 km for England and Wales, and Flinn (1974), for Shetland alone, an estimate of 1,610 km. The proportion of this coastline which is rocky varies, being low for England and high for Scotland, Ireland, and Wales. Generally speaking, metamorphic and igneous rocks are hardest, but there are also many tough sedimentary rocks in Scotland, Ireland, Wales, and south-west England. Young sediments occur most commonly in England, especially in the south-east, and are softer and form lower cliffs. Angle of dip, rock depth or thickness, and coastal erosion, however, generate conditions in which even these rocks can form spectacular, steep cliffs, as at Dover and elsewhere on the south coast.

The distinction between earth cliffs (Chapter 11) and rocky cliffs is artificial and totally arbitrary, but is nevertheless a useful one recognized by layman and specialist alike. Earth, in this context, means a depth of soft material or soil which can itself be the product of biological as well as physical and chemical processes. Earth cliffs are low, unstable and, unless eroding rapidly, have a gradual slope and a high cover of vegetation. Rocky cliffs by comparison are high, steep, rock dominated, stable, wave or spray beaten and with little vegetation.

The principal value of rocky cliffs does not lie in agriculture, forestry, or any other commercial development, but in their exhilarating scenery and wildlife interest.

BACKGROUND INFORMATION

Rocky cliffs are usually composed of hard material such as sandstone, limestone, slate, basalt, serpentine, or granite. They are rarely smooth: the action of wind and sand, waves and spray picks out the softer materials to form a complex pattern of ledges, flats, cracks, and gullies. This provides some flatter surfaces on which vegetation can establish with lichens nearest the sea, salt-tolerant flowering plants higher up and more normal, inland vegetation towards the cliff-tops (Figure 83). Mosses, liverworts and ferns, with one or two notable exceptions, are salt-sensitive and not usually represented. On ledges and in cracks soil develops as a result of the processes of rock weathering and the accumulation and incorporation of dead plant material by the activity of micro-organisms and the soil

(a)

(b)

Figure 83. Contrasting sea-cliff scenery: (a) Chalk cliffs in Dorset
between Lulworth Cove and Weymouth; (b) Gneiss on Yell,
Shetland, showing ungrazed cliffs in the foreground

microfauna. Food-webs develop as they would on bare ground inland although the
number of species involved is considerably fewer, and those in the zone of salt-splash and
spray have anatomical, morphological, and physiological modifications in order to survive
in an extreme environment. Many species are restricted to this habitat being either
sensitive to competition as for example many of the characteristic flowering plants
(sea-pink or thrift, *Armeria maritima*, and samphire, *Crithmum maritimum*) or perhaps
escaping predation as does the large isopod, *Ligia*.

   Rocky cliffs vary in height to a maximum of 274 m in England at Countisbury, North
Devon, to 426 m on St. Kilda, Scotland, and 668 m on Achill Island, County Mayo,
Ireland. In an area such as the British Isles with a range of latitude from about 50 °N on the
south coast to 60 °N in Shetland the range of climate is marked, although the extremes of
temperatures are moderated by the local effect of the sea. Wind velocities are characteristically high on the coast and humidities higher than would be encountered inland. Climatic
variation may be responsible for the appearance of some salt-marsh species occurring on

sea-cliffs in the west and north of Britain, for example sea-aster (*Aster tripolium*), sea-milkwort (*Glaux maritima*) and sea-poa (*Puccinellia* spp).

The relationship between geology and cliff-form is naturally close and both affect the flora and fauna. Sandstones, shales, and slates which may have inclined bedding-planes sometimes show smooth rock surfaces on one side of a headland but a series of steps on the other (Figure 84). Soil only develops on the face with the steps or ledges and depending on whether this face is north-facing or south-facing, a very different association of species appears (Figure 85). Aspect affects the temperature of the foliage and therefore the rate of photosynthesis and also the rate at which soils become desiccated.

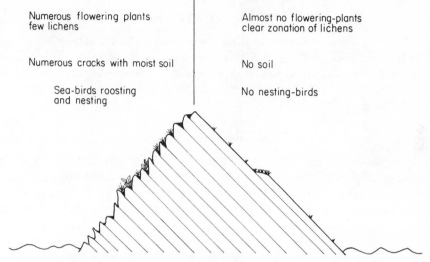

Figure 84. Diagrammatic representation of the effects of angle of dip on cliff appearance and ecology. These effects may be further modified by aspect

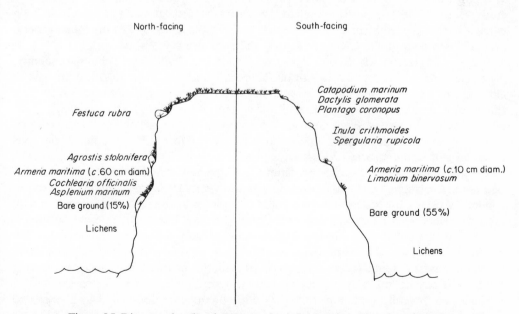

Figure 85. Diagram showing the contrast between north and south facing cliffs

In Britain the prevailing wind is from the west or south-west and as wind is the vehicle that transports the salt-spray the quantities that are incident on south to west facing cliffs are usually very much greater than on cliffs with a northerly to easterly aspect. This is well demonstrated on islands and often on headlands too. Salt-spray both affects the foliage of plants, burning the leaves of sensitive species, and raises the osmotic potential of the soil solution. After periods of desiccation soil salinities can exceed that of seawater so rendering the soil inhospitable to most plants and animals. After heavy rain however the salinities may drop very rapidly and in this respect exposed cliffs may be comparable to salt-marshes because of the environmental extremes experienced.

SPECIAL FEATURES

Rocky cliffs are distinctive in terms of their habitat, vegetation, and their internationally important sea-bird colonies. These are the principal resources of this facet of the coastline and whilst the rocky foundations and vegetation are not generally subject to human pressures, the sea-birds are or have been the victims of cropping, oil-spillage, and toxic chemicals in the sea.

Resource

All biological resources are capable of self-replenishment through the processes of reproduction and growth and therefore if properly managed can withstand continuous exploitation. In some places, however, excessive trampling by tourists has denuded cliff-tops of vegetation and caused the soil to erode, and the cropping of sea-birds and their eggs has reduced their numbers.

Vegetation

The vegetation of sea-cliffs is probably the least man-modified habitat in western Europe. It is not usually cut, grazed or burnt and is therefore particularly suitable for scientific study. But it is rather surprising that so few serious ecological investigations have been carried out; (e.g., Gillham, 1953; Malloch, 1971; Goldsmith, 1973a,b). More general accounts may be found in Chapman (1964) and Hepburn (1952).

The principal factors which control the distribution and abundance of plants are salinity, aspect, grazing, geology, and competition. There is a gradient of species from cliff-base to cliff-top which parallels soil salinity. Not far above the high-water mark the rocks are often thickly coated with maritime lichens, such as *Verrucaria maura* and *Lichina confinis*, which appear as a blackish band. Higher up there is usually an orange belt of *Xanthoria parietina*, sometimes mixed with the white *Ochroleuca parella*. Higher still, large fluffy, glaucous green tufts of *Ramilina* species appear.

The first flowering plants to be encountered and the most strictly maritime are *Armeria maritima*, *Crithmum maritimum*, and *Spergularia rupicola*. *Festuca rubra* (red fescue) commonly occurs low on the cliff-faces and in moist situations may be associated with *Agrostis stolonifera*. There are numerous other species, each with its own ecological requirements, so that a simple ranking in terms of only salinity is not possible. The maritime fern, *Asplenium marinum*, occurs in moist or shaded cracks, and *Inula crith-moides* on warm, south-facing ledges. Nearly all the characteristic species are very sensitive to grazing, many are succulent, and most are slow-growing (Figure 86).

Higher on the cliff-faces there is often a fairly broad zone dominated by grasses and above this vegetation dominated by species typical of more inland situations. Only here

(a)

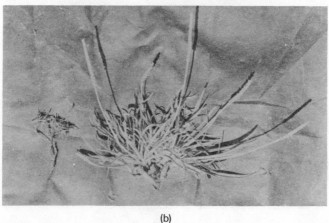

(b)

Figure 86. (a) Sea-pink (*Armeria maritima*), one of the common-est and most characteristic sea-cliff species (South Stack, Anglesey); (b) Sea-plantain (*Plantago maritima*), showing differ-ences due to grazing (on the left, grazed; on the right, ungrazed) (Inishark, Co. Galway)

does one see the effects of geology acting through the soil conditions and dictating whether heather, gorse, cocksfoot, or false-brome will be dominant. On the lower, spray-washed faces the influence of geology and soil acidity are not detectable.

Most perennial, slow-growing maritime species occur on sea-cliffs not because they have a requirement for salt or any other physical or chemical characteristic of this habitat but because they are sensitive to competition from faster-growing inland species. Salinity

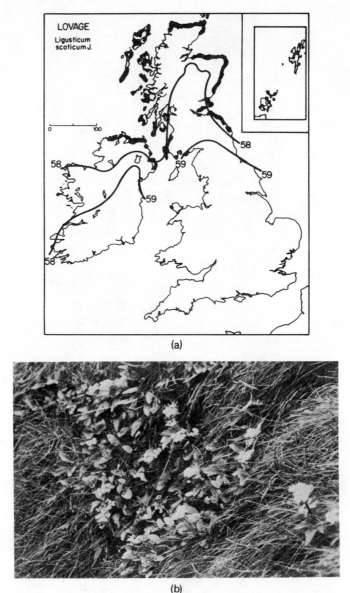

Figure 87. (a) The interesting phytogeographical distribution of
lovage (*Ligusticum scoticum*); (b) The umbelliferous species
growing on ungrazed cliffs on Unst, Shetland

reduces the vigour or eliminates inland species and so creates an environment with a low
intensity of competition.

Aspect has a marked effect on sea-cliff vegetation, determining the amount of insola-
tion, temperature, humidity, and salt-spray deposition. There is often a better correlation
between salinity and aspect than either salinity and height above or distance from the sea
(Goldsmith, 1973a).

In biogeographical terms there are noticeable differences between the northern and
southern extremes of the British Isles (Table 17), but the change is a gradual one (Figure

87). The richest zone in the continuum appears to be around Anglesey and the Isle of Man where a mixing of some northern and southern extremes occur (Goldsmith, 1975a). It is more difficult to identify an east–west gradient, partly because cliff vegetation is not well developed in eastern England although *Glaux maritima*, *Sedum rosea*, and *Aster tripolium* can be considered western elements.

Table 17   Species occurring on sea-cliffs in the north of Scotland compared with southern England as well as a selection of those with much broader ecological tolerance.

Northern (60 °N)	Both	Southern (50 °N)
Silene acaulis	*Armeria maritima*	*Inula crithmoides*
Sedum rosea	*Crithmum maritimum*	*Catapodium marinum*
Saxifraga oppositifolia	*Silene maritima*	*Spergularia rupicola*
Ligusticum scoticum	*Festuca rubra*	*Daucus carota*
Cochlearia scotica	*Agrostis stolonifera*	*Euphorbia portlandica*
Puccinellia maritima	*Plantago coronopus*	*Lavatera arborea*

Locally there are some unusual cliff floras, the best known being on the Lizard Peninsula where the combination of a very southerly location and serpentine soils results in the appearance of lusitanian species such as *Erica vagans*, *Schoenus nigricans*, and some very rare clovers. Other species, such as ox-eye daisy, *Chrysanthemum leucanthemum*, occur as distinct, often dwarfed, ecotypes.

The marked altitudinal range over which some so-called 'sea-cliff' species occur in the British Isles has been referred to by Matthews (1955) and Turrill (1959). *Silene maritima* has been recorded from 969 m, *Armeria maritima* from the summit of Ben Nevis (1,268 m) and *Plantago maritima* from 792 m on Snowdon. However the distribution is disjunct with an abundance of plants at sea level and small populations on mountain tops and cliffs. It has been suggested that these species were widespread in post-glacial times and that their present habitats are relics of their earlier, wider distribution. This suggestion is supported by the occurrence of pollen remains in the Lea Valley and in Norfolk (Matthews, 1955). The populations that survive today occur in environments with a low intensity of competition and it is probable that some ecotypic differentiation has occurred. A series of transplant experiments might indicate that the mountain-top and coastal populations were adapted to their own habitat whilst retaining their sensitivity to competition.

Sea-birds

There are 24 sea-bird species in the British Isles, of which all except the petrels, shags, terns, and skuas normally nest on cliff-faces. There are the four species of auk (guillemot, puffin, razorbill, and black guillemot), seven species of gull, the gannet, the fulmar, and shearwater. It is possible to generalize about the combination of site characteristics that make an area suitable for breeding purposes. Areas should have:

(a) freedom from human disturbance;
(b) abundant planktonic food for fish offshore, usually caused by the mixing of two or more currents offshore;
(c) vertically-bedded rocks such as sandstones which provide suitable nesting ledges;
(d) inaccessibility to predators such as cats, rats and foxes.

This combination of factors is found principally in north and west Scotland (Hermaness, Noss, and Foula on Shetland; Westray and Hoy on Orkney, North Rona, and St. Kilda). These are the most famous sites each containing over a dozen species with often over 100,000 breeding pairs. Elsewhere in the north and west of the British Isles and the east of Scotland there are numerous other important sites such as Rathlin Island, Clare Island, and Great Blasket Island in Ireland, as well as the Pembrokeshire Islands, Isles of Scilly, and Farne Islands. These sites provide the bulk of sea-birds that are to be seen around the coasts of western Europe and much of the north Atlantic. Their importance in food-chains is considerable and to the large numbers of ornithologists in Britain, beyond fiscal evaluation (membership of the Royal Society for the Protection of Birds is about 250,000).

The populations of these sea-birds are far from static. In places, herring gulls have become so numerous that their numbers have had to be controlled, and others like the little tern have at times become so scarce as to cause concern. So the Seabird Group was established in 1965, in order to record such changes and to try to distinguish between those which are 'natural' and those which are the result of exploitation or environmental crises. Its primary task was a census of breeding birds in Great Britain and the results of this mammoth undertaking are published by Cramp, Bourne, and Saunders (1974). This excellent book contains accounts of the biology, movements, and population changes as well as distribution maps of each species. From this account species can be ranked from the most numerous with about 500,000 pairs in Britain and Ireland (guillemot, puffin, and kittiwake) to the least numerous with less than 2,000 (little tern and arctic skua).

The survey provides baseline data which will permit ornithologists to monitor future changes in each species and it will be possible to see if the fulmar, for example, continues to

Figure 88. Part of a large, cliff-top gannet colony. (Photograph by Carl Gibson-Hill; kindly provided by Robert Campbell)

increase in numbers as a result of either some genetic change or its successful utilization of the surpluses of fishing fleets. Similarly some auks such as the puffin are declining in numbers although it is still not completely clear whether this is because it is physiologically more sensitive to the toxic wastes of man or because its behaviour is different from other birds. With respect to oil-slicks for example auks will often dive and re-emerge more contaminated, whereas a herring gull is more likely to fly away. The inter-relationships of behaviour, physiological sensitivity, feeding habits, and distribution mean that there is much still to be learnt, but reliable baseline data is the first and most important step towards being able to maintain 'normal' populations of these interesting and beautiful species.

Although large sea-bird colonies are the most spectacular ornithological features of rocky cliffs (Figure 88), it would be wrong for conservationists or planners to overlook the land-birds which also occur in the habitat. Rock doves, rock pipits, starlings, and jackdaws have been recorded nesting on cliff-faces, whereas corn buntings, meadow pipits, skylarks, and linnets nest on the cliff-tops. Wheatears and whinchats can be quite common about the fields and fences during spring and autumn, along with goldcrests and willow-warblers in the hedges and bushes.

Pressures

The principal pressures on rocky cliffs affect the sea-birds. Traditionally human populations around the coast have supplemented their diet and often risked their own lives collecting eggs and birds. On Flamborough Head, for example, the sea-birds were pillaged for over 250 years by men who descended the cliffs on ropes to collect their eggs. It is estimated that this practice, known as 'climming', yielded about 130,000 eggs per annum, chiefly those of guillemots. The adult birds, especially kittiwake, were shot in large numbers until the passing in 1869 of the Sea-birds Preservation Act. 'Climming' was apparently resumed as a result of food shortages in the Second World War but was stopped completely by the Protection of Birds Act of 1954.

Jewell (1974), however, suggests that many people who exploited these animal populations were conscious of their actions and deliberately adopted a conservationist attitude and regulated their activities. He quotes the sea-bird fowlers of St. Kilda who up to the end of the nineteenth century lived in almost total isolation 65 km west of the Outer Hebrides. The population of about 100 persons subsisted primarily on the great colonies of sea-birds which remained stable for a long period at about 20,000 pairs of fulmars, 40,000 pairs of gannets, and about 1,000,000 pairs of puffins. Each species was exploited with regard for its biology so that the eggs of the fulmars (which only lay one) were not collected. Instead, about half the population (10,000 young birds) were taken just before they left the nests. However, as the gannet responds to the theft of its egg by laying another the St. Kildans collected large numbers of these eggs as well as plump young birds and adults. This can be justified by the occurrence of large numbers of spare or non-breeding birds in gannetries. Puffins were also eaten and taken for the feathers which were traded and in one year nearly 90,000 were cropped. In spite of this at no stage was there any indication of a decline in their numbers on St. Kilda.

Another pressure on sea-birds comes from oil-spillages. When these occur at terminals they are fairly speedily dealt with, but where they involve accidents at sea the equipment and chemicals for dealing with oil are less readily available: The *Torrey Canyon* which went aground in 1967 and released 30,000 tonnes of crude oil is the most notorious example and is believed to have caused the deaths of thousands of guillemots, kittiwakes, and puffins. It was impossible to count the number on that occasion but it is known that only 150 tonnes of fuel-oil killed about 40,000 birds in Holland in 1969.

In addition, the effects of the detergent are still detectable not only on the fauna and flora of the Cornish intertidal zone but also on the flowering plants of the cliff-tops from which it was distributed. We have not yet seen the consequences of spillages from drilling-rigs or pipelines but many conservationists are extremely concerned about the effects of the accidents which will sooner or later occur.

The organochlorines which are used as agricultural pesticides have been found in the eggs of all sea-bird species examined and presumably cause some mortalities. Related compounds used in industry called polychlorinated biphenols (PCBs) are believed to be an even more insidious pollutant. They probably caused the Irish Sea disaster of 1969 when over 15,000 sea-birds were found dead and a further 35,000 probably died. The official report (Holdgate, 1971) showed that the majority were guillemots and suggested that PCBs combined with starvation was the most likely cause.

Other pollutants known to occur in marine ecosystems and recorded from sea-birds include mercury, copper, lead, zinc, cadmium, and arsenic but these are not believed to cause widespread mortality, though much less is known about their sublethal effects.

Another pressure on the rocky cliff habitat is from the visiting public. Many come simply for the spectacular coastal scenery, often to admire sandy beaches from above or to gain access to them. Few people value the biological resources of the cliffs; more often they come to witness seascapes or, for example, to have been to England's most southerly point (The Lizard) or Land's End. It is hard to see how the complex of buildings, vast car parks and eroded ground at Land's End is attractive to hundreds of thousands of visitors annually.

At Kynance Cove on the Lizard Peninsula where the pressures are considerably less, 500 m^2 of Cornish heath vegetation and 150 tonnes of soil have been lost from a Cornwall Naturalists' Trust Nature Reserve on National Trust land. Similar problems may be encountered on the Isle of Purbeck and Hengistbury Head (Dorset), and Beachy Head (Sussex). High use-rates have led to the breakdown of the naturally fragile and slow-growing vegetation and, in turn, to sheet and gully erosion of the underlying soils. Comparable headlands in France, for example Point du Raz in Brittany, experience the same problems although regrettably it appears that the average visitor is in no way discouraged by commercial exploitation or extensive damage to the resource.

METHODS

Vegetation

The techniques used by the botanist depend on the objectives of the investigation. There are major differences between those used in studies of a single species or of a whole community for example and it is impractical here to discuss the relative advantages of the wide variety currently in use. The interested reader is referred to Willis (1973) or Goldsmith and Harrison (1975).

Sea-cliff vegetation can be classified in a traditional continental manner as by Malloch (1971) for the Lizard Peninsula and Birks (1973) for Skye. On the other hand Goldsmith (1973a) used ordination as a means of representing the complicated multidimensional inter-relationships of different groupings of species.

Birds

Operation Seafarer in 1969–70 was a survey of coastal sea-bird colonies over a large area which involved over 1,000 observers (Cramp, Bourne, and Saunders, 1974). It developed from the Sea-bird Group which involves representatives of the British Trust for Ornithology (B.T.O.) and the Royal Society for the Protection of Birds (R.S.P.B.).

The survey methods involve counting birds during the breeding season in their colonies. This is relatively easy for some species such as gannets but more difficult for species living partly in burrows such as puffins and for nocturnal species such as petrels. Interested persons who would like to know more about the group or obtain its reports should contact its secretary in the Zoology Department, University of Aberdeen.

Human impacts

The measurement of human impacts on vegetation has developed principally from concern about the effects of trampling (Goldsmith, 1974). Studies have involved either experimental plots which are trampled at various intensities and frequencies, or the correlation of different vegetation types with numbers of people trampling them, often by examining transects across paths. Over larger areas it is possible to determine visitor distribution by using questionnaire maps and then comparing this data with the extent and distribution of vegetation damage or soil erosion (Goldsmith, Munton, and Warren, 1970). These studies are however complicated by the fact that there may be a lag between the activity and the resultant ecological effect.

Human impacts on sea-birds are determined by regular monitoring of the numbers of breeding pairs of each species at as many colonies as possible. Even so it is difficult to know whether changes in density are due to natural causes, to human activities or to some combination of the two. By the time a change has been detected it is usually too late to prevent the cause. This is also true for the effects of oil-spillages and many toxic chemicals, but it is hoped that in the future, as a result of better monitoring of environmental pollutants and the activity of the Sea-bird Group, the situation will be more closely controlled.

USES

Economic

The economic uses of rocky cliffs are extremely limited. The rock itself is sometimes used for the extraction of valuable minerals, e.g. chromium from serpentine, or as a roadstone as with granite and basalt. These activities usually produce serious scars but are fortunately localized.

A few industrial plants and chemical works, such as bromide plants, and nuclear power stations, e.g. Wylfa, Anglesey, are located on cliff-tops and whilst these are generally considered visual intrusions in the landscape their effects are again fairly local.

The cropping of sea-birds, discussed above with reference to St. Kilda, used to be widely practised in bird colonies on the west and north coasts of Britain. There is no doubt that up to the end of the last century it was vital to many coastal people and produced one of the few marketable commodities to the St. Kildan people.

Educational and scientific

Natural history is a hobby of increasing numbers of people and as sea-cliffs are undisturbed habitats they are suitable for many biological activities, although their steepness and height might deter some visitors. Geologists and geomorphologists also use cliffs for teaching at all levels from school to postgraduate university classes. The study of the vegetation has been somewhat neglected, but this is more than counter-balanced by the thousands of ornithologists interested in sea-birds. Their studies constitute both a recreation for birdwatchers and scientific research of the highest standards.

Recreation

Many thousands of people walk long-distance coastal footpaths each year, and even more drive to a convenient point of access and walk relatively short distances, often less than a kilometre. Cliffs are also crossed to gain access to beaches and in Britain and other parts of Europe there is a seasonal, almost lemming-like migration to places like Land's End.

At Kynance Cove 600–700 cars and up to 20 coaches arrive daily during August and disgorge 1,500–1,800 visitors (or 175,000 per annum). Of these, 80% cross an area of extremely interesting cliff-top vegetation in order to reach the beach. Only 2% have any interest in the biological resource they are crossing and many are unaware that it is National Trust property. The pattern of use, visitor attitude and damage recorded at Kynance Cove is repeated on many other coastal areas in Britain and western Europe.

At Point du Raz, a westerly headland in Brittany, a car park frequently holds 400 cars and visitors engage guides at seven francs a time to conduct them around the spectacular granite cliffs. On the cliff top a square surrounded by souvenir shops and cafés caters for the material needs of the tourists but hardly improves the scenery.

CONSERVATION POLICIES

More than 70% of all holidays taken by British people in this country are spent on the coast. The majority of these are based in seaside towns where the principal attractions are a sandy beach and the entertainments of a resort, for example at Bournemouth or Blackpool. However, increasing numbers of holidaymakers are car-borne and travel on several days to relatively natural coastal areas such as rocky cliffs and these areas are sometimes over-used to such an extent that the resource itself is damaged.

Thus conservation policies need to protect rocky cliffs from:

(a) development such as urbanization or industrialization;
(b) excessive use of the area for recreation;
(c) excessive use of the area for scientific study or natural history activities.

These must all be negative policies and can be achieved by protective ownership (for example purchase by the National Trust, local authorities, or Nature Conservancy Council) or appropriate legislation such as designation of special areas, as for example Areas of Outstanding Natural Beauty, or protection of a particular feature of the resource as with the Protection of Birds Act.

Policies of this type are essential but are even more effective if accompanied by more positive action such as the encouragement of people to inform themselves of the sensitivity of the resource and their own impacts. The mass media can be very effective in this respect and, in the case of many B.B.C. and Anglia programmes, make for good entertainment.

Other positive action includes information in the form of adult education courses (for example the University of London Certificate in Ecology and Conservation), interpretation centres and nature trails. In some places long-distance coastal footpaths and restoration works on damaged areas as at Kynance Cove, Cornwall, are other ways of more positively providing for the visitor.

In the past it has often been the policy of landowners or managers to let semi-natural areas look after themselves whereas more recently it has been realized that areas accommodating tens or even hundreds of thousands of visitors per annum need to be actively managed. Management costs money and in future there will be heated debate as to who should pay for the various categories of coastal land.

Protective ownership

The coasts of Britain are protected by a wide variety of state, municipal and independent forms of protective ownership. Probably the most effective is ownership by the National Trust. Their first acquisition was $4\frac{1}{2}$ acres ($1\cdot8$ ha) of cliff land near Barmouth, Wales in 1895, but this length of less than 2 km had grown to 280 km by 1965. Nevertheless, this was considered insufficient and an appeal, called Enterprise Neptune, was launched. The National Trust's coastal holdings grew to 480 km in 1971 and are still growing especially in Wales and south-west England (Patmore, 1970).

National Parks were first established in 1949 and serve to protect land for its scenic and recreational potential by restriction of activities rather than ownership. The Pembrokeshire Coast is the only park where the coastline is the major feature whereas Exmoor, the Lake District, Snowdonia, and the North York Moors Park include important areas of coastline.

Areas of Outstanding Natural Beauty (A.O.N.B.) were established as a second tier of special areas to supplement the National Parks. They include several important coastal areas with biologically important and scenically attractive rocky cliffs such as the Isle of Purbeck in Dorset and the Gower Peninsula in South Wales.

The Nature Conservancy Council (N.C.C.) own or lease over 130 National Nature Reserves in Great Britain. There are eleven with important sea-cliffs: Hermaness, Noss, North Rona, St. Kilda, Monach Isles, Rhum, Isle of May, Lindisfarne, Skomer, Gower Coast, and Axmouth–Lyme Regis. The N.C.C. are also responsible for the designation of areas as Sites of Special Scientific Interest of which there are many thousands. This is not a particularly effective method of protection but ensures that local authorities or owners notify N.C.C. before dramatic land-use changes are put into effect.

The Royal Society for the Protection of Birds is responsible for many of the rocky cliffs which have large sea-bird colonies. These include some owned by N.C.C., for example Noss on Shetland, and they assist with monitoring and advice on National Trust properties with other colonies, e.g. Fair Isle.

The Ministry of Defence owns several areas of cliff-top land and the customary exclusion of the public is seen by many conservationists as a mixed blessing. The tank and gunnery ranges around Tyneham on the Isle of Purbeck are the centre of almost continuous controversy, although recently the coastal footpath has been extended. The Castlemartin ranges in Pembrokeshire also include important stretches of sea-cliffs.

There are other designations which serve to protect areas of coastal cliff such as Local Nature Reserves (usually owned by local authorities) and special trusts such as that for Steep Holme in the Bristol Channel. Ownership however is not necessarily the end of the conservationists' problems. Visitors' activities may have to be redirected, predators on sea-bird eggs and chicks may have to be controlled and the vegetation managed. On one area of Cornish cliff, scrub has had to be controlled to encourage wild-thyme, which is important in the life-cycle of the very rare Large Blue butterfly. This type of important management is often carried out by members of the British Trust for Conservation Volunteers (often known as the Conservation Corps).

Positive provision

As a result of increasing concern about the use, development and protection of the coast the National Parks Commission commenced in 1966 a comprehensive study of the Coastline of England and Wales. The study was based on nine regional conferences with maritime local planning authorities and other interested bodies, each of which published reports. They emphasized that coastal areas were under increasing pressure from a wide variety of demands—commercial, industrial, recreational, etc.—while the existing planning protection policies were in many cases vague and their implementation of varying and sometimes questionable efficiency.

In 1970 the Countryside Commission (which had replaced the National Parks Commission) published two reports—*The Planning of the Coastline* and *The Coastal Heritage* in which an attempt was made to safeguard those parts of our undeveloped coast with the highest quality scenery, by designating them as *Heritage Coast* (Countryside Commission, 1970). In England and Wales, 34 areas covering 1,170 km of coast, were identified as being worthy of this status and as being areas which demand special measures to ensure their conservation. However, this activity in the late 1960s has not led to any over-all policy about coastal conservation. Although the Nature Conservancy Council embarked on a massive review of the premier nature conservation areas in Great Britain, the two substantial volumes have remained secret and only the criteria of selection used have been published (Ratcliffe, 1971).

Probably the most important development for those seeking active recreation around our coasts has been the provision of long-distance coastal footpaths of which that around the south-western peninsula (Devon and Cornwall) is the longest and best known. The Countryside Commission has grant-aided the local authorities and National Trust to warden these paths for they are an important recreational asset. Other coastal footpaths are to be found in Pembrokeshire and the Isle of Purbeck in Dorset.

Management of sites affected by recreation

Because of the sensitivity of coastal vegetation, areas which are subject to intensive visitor pressure are vulnerable to damage. A Countryside Commission restoration project on National Trust land at Kynance Cove, Cornwall, found that the maximum use rate during the summer period for the survival of vegetation cover of any kind was about 10 persons per metre per hour (O'Connor, Goldsmith, and Macrae, 1974). If areas are to exceed this capacity the visitors should be channelled onto paths which might need artificial surfaces. At Kynance Cove serpentine chippings were recommended because it was the natural rock of the area. Revegetation experiments showed that use of local cliff-turf was the quickest and most effective way of revegetating bare ground and recommended the initial scarification of the surface. Grazed areas from which turf had been taken developed a 70% vegetation cover within twelve months and had more species than the surrounding turf.

At Kynance Cove a new return route attracted 57% of visitors returning from the beach or 28% of all traffic. The latter figure increased to 50% after the surfacing of the path. Thus the strategy recommended was to concentrate visitors on paths, including a separation of upward and downward flows, and then to revegetate the adjacent bare ground.

Rocky cliffs are characterized by a hard substratum but their vegetation and fauna are more sensitive. Fortunately human impacts are usually slight and it is only in a few, heavily visited areas that the kinds of damage referred to about Kynance Cove occur. The high salinity of the soil solution suppresses the growth of the vegetation so that recovery from damage by trampling or pollution, such as by detergents, is slow. However, the sensitivity of this vegetation type is less than that of sand-dunes where the substratum is also highly sensitive. Rocky cliff vegetation requires careful monitoring to ensure that the species composition is not changing.

If it is, and if the change is considered undesirable, the organizations such as the Nature Conservancy Council, Countryside Commission or the local trust for nature conservation should be asked to give advice about management. It is difficult to generalize about the type of management which would be appropriate, as it depends on the characteristics of the area, its ecological value (however that should be measured: Goldsmith, 1975b) and the causes and degree of the damage. Usually drainage, protection by rerouting, or path improvement, are needed and occasionally reseeding, returfing or fertilizer application are useful, although some areas could be even more seriously damaged by some of these treatments.

Chapter 13

Reclaimed land

A. J. GRAY

INTRODUCTION

Since the reclamation of coastal land is a human activity the habitats it creates are somehow 'derived' or 'artificial'. They range from those in which there has been little obvious human interference such as partly-ditched, high-level salt-marshes, to those which are intensively managed at the present time, most often for agriculture. However, this chapter deals with a category of land somewhere in between these two extremes which may be defined as resulting from a distinct reclamation bank and containing, in addition to the land surface behind the bank, subhabitats such as drainage ditches, earth embankments, and small bodies of open water as well as the enclosing bank itself.

Excluded by this definition are areas to seaward of the sea-bank including those where land improvement is being carried out by artificial drainage, by grazing, or by accelerating natural accretion rates by planting selected species, ditching, or constructing brushwood groynes. Whilst such activities are often referred to as 'land-reclamation', and are frequently followed by the building of enclosing banks, they create what are essentially modified intertidal habitats different in character from the reclaimed land defined above (see, for example, accounts of such methods in the Wadden Zee in the Netherlands (Kamps, 1962) and Denmark (Jakobsen and Jensen, 1956). At the other extreme is land reclaimed in historical times but which has been developed for arable farming or industry and has entered a phase of land-use which is either relatively uninteresting to the ecologist, or produces an ecosystem differing little from an inland one in character and management requirements.

Reclaimed land may be produced either directly for agriculture or industry, or indirectly from the construction of banks for other purposes. In the past, most land has probably originated from the need for agricultural and building land as, for example, in The Wash, Lincolnshire, and Norfolk, where almost all of the 32,000 ha reclaimed since the beginning of the seventeenth century were enclosed to produce high quality arable farmland. However, considerable areas of land have been enclosed as a result of the construction of sea-defence banks, road and railway crossings, harbour improvements, and so on. For example, less than a third of the 1,300 ha reclaimed in the northern part of Morecambe Bay, Lancashire, since the thirteenth century was deliberately reclaimed for agriculture and about 450 ha of the total resulted from a single operation, the building in 1857 of the Ulverston–Lancaster railway crossing (Gray and Adam, 1974). Similarly, the enclosure in 1969 of the Lauwerzee, a former inlet of the Dutch Wadden Zee, was principally for hydrological and sea-defence reasons and has created an area of 9,100 ha, approximately a third of which is open water and only a part of which will be used eventually for agriculture.

The many current proposals for port development, coastal recreation areas, road crossing, tidal barriers, estuarine water storage schemes and so on are likely to increase the proportion of land reclaimed in the future for purposes other than agriculture. If a third London airport had been built at Maplin Sands in Essex, for example, it would have created in the space of a few years a surface area of reclaimed land (roughly estimated to be in excess of 7,000 ha) approximately twice that reclaimed for agriculture in The Wash this century.

Two basic reclamation methods may be recognized. The first consists simply of enclosing an area, the former intertidal land surface, be it vegetated salt-marsh or, more rarely, unvegetated tidal flat, forming the land surface within the enclosure. In the second method, the land is enclosed and then either pumped dry or infilled, often with material dredged up from the nearby sea-bed and pumped in a slurry into the enclosure (although in some cases artificial materials such as domestic and industrial refuse may be used). Reclamation by the first method is widely practised and there are many such areas, for example, in south-east England. The IJsselmeerpolders provide an example of reclaimed land which has been pumped dry, whilst infilled reclamations include Calshot, Hampshire (slurry), and Benfleet, Essex (refuse). The two methods produce land with striking differences in terms of the physicochemical properties of their soils, particularly their structure, maturity, and organic content, and in terms of the drainage patterns, which are generally well developed in salt-marsh soils prior to enclosure. Land enclosed after artificially accelerated accretion, either by planting cordgrass (*Spartina anglica*) or by methods such as those used in the sedimentation fields of the Wadden Zee, may be regarded as intermediate between these two types. The soils produced are generally less mature and less well consolidated and, particularly in former *Spartina* marsh, contain more areas of locally impeded drainage.

A further subdivision may be made into (i) reclaimed land which is more or less ephemeral, being a transitional stage between the enclosing of the land and its eventual use for industry or arable farming, and (ii) that which is more permanent, having a land-use which retains a high wildlife interest or is specifically designed to conserve this interest. The two types have different ecological features and may be managed in different ways. Hence the distinction between 'temporary' and 'permanent' areas referred to below.

SPECIAL FEATURES

The principal ecological feature of reclaimed land is its intermediacy, both temporal and spatial. First, it provides an opportunity to study the changes which occur in time as the new land is colonized by plants and animals. Second, it consists of a series of habitats physically interposed between those which are under predominantly inland and freshwater influences on one side and predominantly maritime and saltwater influences on the other.

Temporal intermediacy—colonization studies following reclamation

Most studies of the colonization of newly-enclosed bare land surfaces by plant and animals have been undertaken in the Netherlands, where extensive reclamations of this type have taken place. In particular should be mentioned the research on the IJsselmeerpolders created by the reclamation of Lake IJssel (the former Zuider Zee) which was isolated from the sea by a dam completed in 1932. Although the land was reclaimed by pumping dry enclosed sections of a large shallow lake, the subsequent changes are similar in general to those which occur in infilled areas or on former intertidal mud-flats where the water-table is lowered following enclosure. For example, the former subaqueous soils of the IJssel-meerpolders are subject to processes such as dehydration and shrinkage (de Glopper, 1969; Pons and Zonneveld, 1965), penetration of oxygen into the surface layers, desalinization (Zuur, 1938), and organic nitrogen transformations (van Schreven, 1963a,b) including the utilization of available nitrogen by pioneer vegetation (Bakker, 1958; van Schreven, 1963c): all are processes known to occur in some other reclaimed sediments and are thought to be closely related and almost universal features of the ripening of reclaimed soils.

Similarly the plant migration studies in the IJsselmeerpolders (Feekes and Bakker, 1954; Bakker and van der Zweep, 1957; van der Toorn *et al.*, 1969) have elucidated a number of general principles of plant dispersal and migration into bare areas and have included classic studies on the reproductive and dispersal biology of individual species (e.g. coltsfoot, *Tussilago farfara*, creeping thistle, *Cirsium arvense* (Bakker, 1960a), and marsh fleawort, *Senecio palustris* (*S. congestus*) (Bakker, 1960b). Studies on animal migration into the new polders include those of phytophagous insects, especially the cabbage white butterfly, *Pieris brassicae* (Mook and Haeck, 1965), the chloropid fly, *Lipara lucens*, which produces a reed gall (Mook, 1972), several species of carabid beetles (Haeck, 1969a), earthworms (Meijer, 1972), and number of mammals (Cavé, 1960; Mook, 1969), particularly the mole, *Talpa europaea* (Haeck, 1969b).

There are relatively few published accounts of the colonization of former bare land surfaces from areas outside the Netherlands. Notable exceptions include those of Barnes and Jones (1974) on the flora and fauna of infilled reclamations near Calshot in Southampton Water, and of Glue (1971) specifically on the birds in the same area.

The colonization of surfaces which had a cover of vegetation prior to enclosure has been less well studied. One reason is that this is often the most ephemeral type of reclaimed land, destined to become arable farmland within a few years of enclosure. Another reason may be that such land is more diverse and often well differentiated into many niche types, and thus offers less opportunity for the monitoring of natural immigration rates, seral development, or the growth and expansion of populations from a few founders under competition-free conditions. It is therefore less attractive for study from a theoretical point of view.

The margins of the Veerse Meer, enclosed at the seaward end in 1961 as part of the Delta Plan in the south-west Netherlands, contain both former salt-marsh and tidal flat areas, and the plant colonization of both types has been studied by the monitoring of permanent quadrats (Beeftink *et al.*, 1971). The sequence of appearance of birds on the nature reserves within the Veerse Meer is described by Lebret (1972). The work of Heydemann and his colleagues (e.g. Heydemann, 1960) on the invertebrates, particularly spiders and beetles, in the newly reclaimed polders of the German North Sea coast has produced extensive documentation of the population changes in these animals with time. Von Brehm's studies on the birds of the Hauke–Haien–Koog reclamation in Schleswig-Holstein are an important contribution in this field (von Brehm, 1971). From England, the results have recently been published of an interesting study which has followed the changes in vegetation since enclosure in 1964 of a former salt-marsh area at Egypt Bay on the north Kent coast (Side, 1973). Two other accounts of the vegetation of reclaimed marsh in south-east England which refer to the sequence of changes in time are those of Petch (1945) on the west coast of The Wash, and Myers (1954) on the inned salt-marshes at Leigh-on-Sea, Essex.

Thus the reclamation of coastal land has provided ecologists with a chance to examine the ways in which organisms are dispersed into and then exploit newly available sites and the changes which occur in the interval between enclosure and the eventual exploitation of the land for agriculture. The studies referred to above emphasize that there are many differences in detail between different areas. These differences result from a large number of local variables such as type of sediment, rate of water-table lowering, rate of leaching and desalinization, availability of local seed parents, remoteness of the site, and so on.

However, four very broad generalizations emerge. First, there is a tendency for the production on bare land surfaces of large monospecific stands of a comparatively few individual plant species. This is reinforced by the early chance arrival of some species excluding equally suitable competitors which happen to arrive at a later stage. Similarly with spiders, for example, the communities of newly-reclaimed land are characterized by large numbers of individuals of relatively few species (Heydemann, 1960). Second, there appear to be measurable but overlapping 'phases' or development as one species (or group of species with similar dispersal mechanisms) succeeds another. There is frequently a decline in vigour among early colonists of open habitats in the year following their arrival, particularly among the nitrophilous plant species which invade former sea-bed areas. It is clear that the word 'succession' as understood by ecologists is an inappropriate one to apply to the sequence of very often rapid and disorderly changes in species composition of many reclaimed areas. This sequence is frequently controlled by changes of a high amplitude in the physical and chemical soil environment associated with the conversion of the land to its eventual economic use. A third point is that the persistence of salt-marsh species in formerly vegetated areas is related to the management of the area once it has been reclaimed. (This subject will be returned to later.) Finally, the studies indicate that the early immigrants into newly reclaimed land may come from both maritime and inland habitats; the features which they share in common being efficient dispersal mechanisms and tolerance of open habitats. Thus, depending on the local variation in soil factors, the successful pioneer plant species may be those generally found on salt-marshes (e.g. the wind-dispersed sea aster, *Aster tripolium*), on drying lakeside mud (e.g. the water-dispersed celery-leaved crowfoot, *Ranunculus sceleratus*), or as farmland weeds (e.g. wind-dispersed coltsfoot, *Tussilago farfara*). Barnes and Jones (1974) noted that 86% of the invertebrate aquatic fauna of the pools on the Calshot reclamation were freshwater species capable of penetrating varying degrees of brackishwater. In general efficiently dispersed groups such as insects and birds provide a high proportion of the animals to be found on newly reclaimed land.

Spatial intermediacy—the paramaritime habitat

The type of reclaimed land referred to earlier as 'permanent' land has a unique character, defined, at least floristically, by a number of characteristic species and combinations of species. Despite this, there are very few published studies of such habitats, and the need for more research, particularly of a management-orientated type, is apparent. For this reason I propose to illustrate the special character of reclaimed land with a specific example, confined to plants and to south-east England, based on personal experience of botanical surveys of such areas from Humberstone, Lincolnshire, to the Isle of Sheppey, Kent. The example has a wider application, many of the species and the habitats occurring elsewhere in Britain and Europe.

The major subhabitats which may occur in this area are illustrated in Figure 89. Of these, sea-walls (frequently earth embankments), disturbed and vehicle-rutted ground, grassland

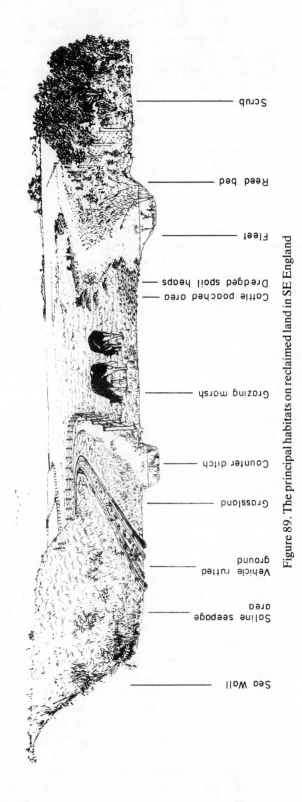

Figure 89. The principal habitats on reclaimed land in SE England

and counter ditches or 'soak-dikes' (which run parallel to the sea-wall collecting the drainage of the hinterland) are the most constantly found. Conversion to arable land and lowering of the water-tables have reduced the areas of grazing marsh, shallow fleets, small ponds (often former duck decoys) and ditches (often former salt-marsh creeks), whilst extensive reedbeds and scrub (generally of hawthorn, *Crataegus monogyna*, in dry situations and of willows, *Salix* spp, and alder, *Alnus glutinosa*, in wetter ones) are becoming rare outside nature reserves and similar areas.

In spite of the fact that the habitat is man-made it is very interesting botanically, containing not only the largest populations of a number of relatively common coastal species but also some local and rare species which are more or less confined to reclaimed land. In addition there are populations of species the habitats of which are increasingly being destroyed in inland Britain. Some examples of these three categories of species are given in Table 18. To these might be added a number of species from the higher levels of salt-marsh and shingle, such as shrubby sea-blite, *Suaeda vera* (= *fruticosa*), and golden samphire, *Inula crithmoides*, for example, which often occur on sea-walls in areas where the upper marsh habitat is being eroded or destroyed (as in marina development). Similarly the largest populations of the rayed forms of sea aster, *Aster tripolium*, which is being replaced in the lower zones of salt-marshes in south-east England by the rayless morph (var. *discoideus*) (Gray, 1974) now occur on the edges of shallow fleets and ditches to landward of the sea-walls.

An interesting contrast is provided by the high diversity of flowering-plant species (the large number of species per unit area) in areas such as the landward base of sea-walls and the large almost pure stands of species such as sea-clubrush, *Scirpus maritimus*, which occur in the fleets. In a recent survey, 36 different species of higher plants were recorded from one 2×10 m quadrat at the landward base of and running parallel to a sea-wall on reclaimed land on the Isle of Sheppey, Kent (Institute of Terrestrial Ecology, unpublished report). Such a figure would rarely, if ever, be obtained from even the most species-rich area of salt-marsh, and must be considered very high by the standards of many 'semi-natural' vegetation types. Reclaimed land is further enriched ecologically by the fact that a number of plant species which occur in large, almost pure, stands are important food plants. For example, the seeds of *Scirpus maritimus* are popular food for the dabbling ducks, whilst the rayed *Aster tripolium* attracts a wide variety of insects, including butterflies.

However, there is still much to learn about the basic ecology of the plant communities of reclaimed land and in particular the conditions necessary for their maintenance. Some species, such as the so-called 'salt-marsh grasses', *Puccinellia distans*, *P. fasciculata*, and *P. rupestris*, are plants of bare and disturbed ground, occurring in vehicle tracks, on the spoil dredged from ditches, in cattle-poached mud and on the drawn-down edges of fleets and drains, and rapidly disappear when these subhabitats are invaded by other grasses. Newton (1965) has suggested that such species were probably even more restricted in distribution prior to human activity on the coast and de Visser has recorded the spread of *P. fasciculata* in the Netherlands following the 1953 floods and the construction of new dikes (de Visser, 1971). To conserve these species a regular programme of dredging or controlled 'disturbance' is apparently necessary. On the other hand some marshes consist of old and relatively stable pastures with a balance of species maintained over the many years by traditional grazing practice under high water-table conditions. These areas are especially vulnerable to disturbance by turf-cutting and drainage.

Thus, despite its 'artificial' nature, reclaimed land is often highly diverse and has become an important habitat for those paramaritime species the native habitat of which may have

Table 18   Evaluation of the floristic importance of reclaimed land in SE. England

Status of species in Britain	Main subhabitat within reclaimed land			
	Sea-walls and banks	Wall bases, cart tracks, and disturbed ground	Grassland (wet and dry)	Aquatic habitats: ditches and fleets
Relatively common species with a coastal distribution and for which reclaimed land is the most important habitat in which the species occurs	*Agropyron pungens* *Beta vulgaris* spp *maritima*	*Spergularia marina* *Puccinellia distans* *Parapholis strigosa* *Hordeum marinum*		*Scirpus maritimus* *Ruppia maritima*
Local or rare species for which reclaimed land is the major habitat	*Bupleurum tenuissimum* *Trifolium squamosum* *Peucedanum officinale* *Lactuca saligna*	*Chenopodium botryodes* *Carex divisa* *Puccinellia rupestris* *Puccinellia fasciculata* *Polypogon monspeliensis*	*Alopecurus bulbosus* *Ruppia spiralis*	*Ranunculus baudotii* *Ruppia spiralis*
Species with a partially inland distribution but for which reclaimed land is becoming an increasingly important regufia (varying in rarity)	*Carduus tenuiflorus* *Tragopogon porrifolius*	*Eleocharis uniglumis*	*Carex distans* *Althaea officinalis* *Hordeum secalinum*	*Scirpus tabernaemontani*

been destroyed by general reclamation. These features, coupled with the presence of large populations of food plants and in many cases the remoteness from human settlement, and therefore lack of disturbance, combine to produce a coastal habitat of extreme interest and potential importance. An indication of the richness of such areas for wildfowl, for example, is given by Harrison's study of the north Kent marshes (Harrison, 1973).

LAND-USE AND PRESSURES ON THE HABITAT

Agriculture, both arable and pastoral, remains the major use of reclaimed land and changes in agricultural practice constitute the greatest source of change in the habitat. The biggest single change comes from the improvement of drainage and the conversion of grazing marsh to arable.

Even in those areas where reclamation is undertaken specifically to produce arable land the rate at which the enclosed land is converted to arable has accelerated. For example, in The Wash the practice in the 1930s was to plough only after as many as ten years of grazing (Stamp, 1939), a practice recommended as recently as 1957 (Dalby, 1957), whereas at the present day newly-reclaimed land may, depending on the soil type and the amount of rainfall, be cultivated as soon as 18 months after enclosure and is often cultivated in under three years. In addition, the development of improved earth-moving and ditching machinery, together with the availability of relatively cheap fertilizers, has encouraged the early levelling and drainage of new land and the sowing of temporary leys. Similarly, within the IJsselmeerpolders experience of the weed problems in earlier reclamations has led to faster and improved soil drainage and ripening techniques involving the aerial sowing of reed, *Phragmites australia* (= *communis*). For example, in 1968 over 40,000 ha of the southern Flevoland polder were sown at a rate of 500 g of seed ha^{-1} in only 14 days (*Flevoland Facts and Figures*, 1972).

Whilst 'temporary' reclaimed land is becoming even more temporary, the pressures to convert 'permanent' reclaimed land to arable are increasing. These pressures which, as Gramolt's detailed study (Gramolt, 1960) reveals, are part of a long-term trend, are almost entirely economic, being generated by the rising cost of land and the greater profitability of arable agriculture compared with traditional grazing of marginal pasture. They are being fostered by improved farming techniques, cheap fertilizers, the incentives of Government grants for improvement schemes, and, in many cases, the added security afforded by the reconstruction of sea-walls following the disastrous floods of 1953. A discussion of these and other factors and their effects on the grazing marshes of north Kent is included in a recent working party report (Green, 1971).

Harrison (1973), quoting figures which throw doubt on the relative profitability of cereal production on the north Kent marshes in comparison with stock rearing and fattening, has called for a re-evaluation of the agricultural usage of reclaimed marsh. The need for a controlled experiment monitored by agricultural economists is clear.

Where grazing is still continued on reclaimed land the wildlife interest may be diminished by changes in management including the general lowering of water-tables to reduce areas of shallow fleet and marshland, the use of herbicides, and the sowing of agronomic strains of perennial grasses. The availability of cheap fertilizers has also led to the decline of 'warping', the controlled periodic flooding of estuarine meadows, formerly practised in large estuaries such as the Elbe, the Ems, and the Humber.

Widespread use of fertilizers reduces the species richness of grazing marshes (by encouraging the growth of the more vigorous agricultural grasses at the expense of other species) and enriches the areas of open water and ditches. The counter dikes for example carry the drainage water of not only the adjacent reclamation but often of a very large area of hinterland and the production in them of extremely eutrophic conditions by the runoff water from agricultural land is a high potential hazard to their wildlife (and, judging from their appearance in many places a phenomenon which is possibly already widespread). The water bodies in reclaimed land are slow-moving and even stagnant, often being removed only by pumps or by flap-valves in tidal sluices, and nutrients may accumulate in them over long periods without regular flushing.

Contrasting with the intensification of agriculture is the increase in many areas of the other forms of land-use referred to earlier. For example, in the IJsselmeerpolders only 50% of the most recently reclaimed polder (S. Flevoland) will be brought under agricultural use compared with 87% in the first polder to be reclaimed (Wieringermeer in 1930). At the same time, 25% of S. Flevoland is being set aside for woodland and nature reserves compared with 1% in the Wieringermeer polder, and the amount of land devoted to residential areas has increased from 1% in the Wieringermeer to 18% in S. Flevoland (*Flevoland Facts and Figures*, 1972).

These other forms of land use bring their own management problems. The effects of urbanization and industrial development are well known and widespread in other habitats. Particularly relevant to reclaimed land are the problems of pollution of water (cf. the earlier point about eutrophication) and the dumping of domestic and industrial waste. Odd corners of apparently unproductive land are major targets for infilling with refuse which together with that dumped onto the adjacent salt-marshes often forms the land for the expansion of building.

Similarly the increased recreational use of reclaimed land brings to it a set of problems which are familiar in other types of land. Some recreational activities (e.g. birdwatching, sailing, fishing, and wildfowling) are more likely to be compatible with the management of reclaimed land for wildlife than are others (e.g. motorboating, water-skiing, and caravanning). Wildfowling, for example, and particularly the use of duck decoys, has been a widespread activity on reclaimed land in Britain which has done much to create and preserve its unique landscape and character.

The military use of reclaimed land comes in a similar category. Favoured for its remoteness, reclaimed land provides both the security and safety required for military training and, most commonly, weapons testing. These activities in turn ensure that the land remains inaccessible to large numbers of people and unsuitable for development.

METHODS OF ASSESSMENT

Set out in this section are a series of 'guidelines' which might be followed by a potential manager or a student of reclaimed land. Hopefully, whilst not exhaustive, they should provide him with a means of measuring the major factors affecting the ecology of the land, of assessing its value in ecological terms, and of assessing its potential for management. Listed below under the headings, *History*, *Survey*, *Hydrology* and *Populations* are those investigations which need to be made and questions which require some answers before a realistic programme of management can be drawn up. The questions are listed in what is thought to be their logical order of asking.

History

When was the land reclaimed?

It is helpful to know not only the reclamation date of the particular piece of land in which one is interested, but also that of adjacent reclamations. The age of the land and the time interval between it and both previous and subsequent reclamations provide an indication of the pattern and pace of coastal reclamation in the general area. In particular, the rate of development of reclaimable salt-marsh to seaward of the present sea-wall can often be gauged and the speed with which a recently reclaimed area may become incorporated into a larger unit of land or further removed from saline and seepage influence can be estimated.

Detailed historical studies of reclamation are unfortunately rare and those that exist tend to be published in fairly local journals. Examples of this type of study are those on Poole Harbour (May, 1969), on parts of Essex (Gramolt, 1960), on The Wash (Hydraulics Research Station, 1975; Gray, 1976) and on Morecambe Bay (Gray and Adam, 1974). However, there is usually a good deal of local knowledge about reclamation and the local water authorities or drainage boards are generally most helpful to the serious student. It is expected that where the potential manager is neither owner nor tenant he would naturally look to that person for information when seeking permission to work in the area.

For what purpose was the land reclaimed?

The answer to this question is often blindingly obvious, particularly if one is confronted by a cluster of cooling towers! However, it is worth remembering that a proportion of reclaimed land is created incidentally as a result of sea-defence works or rail crossings and is often unsuitable for agriculture or development. A knowledge of the short- and long-term intentions of the owners of the land, and of those of adjacent areas, is vital in order to judge the potential success of an attempt to manage the area for wildlife. In short, the land must be set in the context of its surroundings and likely development (bearing in mind the increased pressures on all land types due to rising land values).

How has the land been managed in the past?

The history of management of reclaimed land contains many of the clues to its present ecology. In the case of grassland or marsh, the extent to which it has been grazed, turf-cut, drained, and treated with fertilizers or herbicides is largely responsible for its present botanical composition. Knowledge of these factors helps to explain not only existing features of the topography, soil-type and vegetation but also may account for notable absentees in the flora and fauna (absentees which might be encouraged to return under the appropriate management regime). Management in the past includes the use of the salt-marsh or intertidal area before as well as after its enclosure.

Survey

What are the major subhabitats within the reclamation?

A preliminary, albeit fairly crude, classification of the habitat types is a useful starting point for a survey. Something on the lines of that described earlier for reclaimed land in S. England and indicated in Figure 89 may be satisfactory although refinements are obvious. For example, sea-wall habitats may be divided into landward and seaward, grazed and ungrazed, stone-faced and earth-faced, north-facing and south-facing and so on. Once classified, the amount and areas of each subhabitat can be calculated either by ground survey or from maps or aerial photographs (see below).

What are the major vegetation types?

The description and classification of vegetation types is a complex subject, full of pitfalls for the unwary. The reader who requires discussion of the various techniques (of random versus non-random sampling, of quantitative versus qualitative methods, of quadrats, transects and communities etc.), is referred to the many textbooks on the subject (e.g. Kershaw, 1964; Greig-Smith, 1957; Shimwell, 1971). What should be remembered is that the choice of method should be determined by the purpose of the classification, the level and detail of information required, and the time available.

With these considerations in mind it may be convenient to use the subhabitats as strata for a sampling programme (in say a stratified random sampling procedure). The sub-habitats have the virtue of being easily recognizable in the field and, in many cases, on aerial photographs.

It is often sufficient, especially in the early stages of a management programme, to have a generalized vegetation map. For this, aerial photographic methods may be the answer. As with vegetation analysis, aerial survey and air photo interpretation are complex and often highly technical subjects. However, much can be achieved using black and white aerial photographs and a quick and relatively easy method for assessment of the major subhabitat and vegetation types. Such a method has been described for salt-marsh vegetation by Fuller who stresses that it can be applied by people untrained in sophisticated cartographic techniques (Fuller, 1972, 1973). Local councils and other authorities in Britain often hold sets of 9×9 in black and white aerial photographs of the coast and, where obtainable, the making of maps of reclaimed land from these using Fuller's method is possible with acceptable accuracy due to the flat terrain involved.

Hydrology

Where is the water within the reclamation coming from?

The next stage in assessing the land for management potential is understanding the pattern of drainage. This involves discovering not only the direction of flow of ditches within the reclamation but also of those outside the reclamation which drain into it.

Where is the water going to?

Information is vital on the flow of water out of the reclamation and indeed within the whole catchment of which the reclamation forms a part. Reclaimed land managed entirely or partly as a wetland habitat is often an island surrounded by well drained farmland. The chances of maintaining high water-table conditions at the appropriate seasons in the face of continued improvement of the surrounding farmland drainage must be assessed realistically at the outset. Sometimes the drainage pattern within the area and the large number of sluices which would have to be built to control the water level will make the creation of locally high water-tables too costly to contemplate. On other occasions, the natural conditions enable high water levels to be maintained very easily by, say, the construction of a pair of sluices and the use of a small pump. Occasionally the natural conditions are such that wetland will develop without any artificial control of water levels, but the use of improved drainage machinery and techniques is rapidly reducing the number of such cases.

What variations in water level occur within the reclamation?

Clearly a complete answer to this question can only be obtained over a number of years. However, a rough idea of the amplitude of fluctuations in water level in any one year can be obtained by measurement of summer draw-down in open water bodies and in ditches, and by measurement of groundwater-table fluctuations in sample cores. These may then be related to the general rainfall level of the particular year to give an over-all picture of the potential variation in water level.

Populations

What are the limits of the plant populations of special interest?

Initially this can be simply a question of mapping the physical limits of those species or groups of species of special interest. These may include common species which are important food plants (e.g. *Scirpus maritimus*) or species of very restricted distribution (e.g. *Puccinellia rupestris*). In both cases a distribution map is a first step in the process of understanding the factors which affect their survival and spread, and ultimately of taking measures to ensure their conservation or to extend their distribution by positive management.

Any attempt to assess the ecological limits of these species requires detailed study of their biology, much of it experimental. However, a great deal can be learned from careful observation of the prevailing conditions within the species' present distribution and a successful extension of this distribution can often be achieved, quite empirically, by the provision of these conditions. For example, the creation of bare ground areas adjacent to existing populations of species such as *Puccinellia distans*, *P. fasciculata*, *P. rupestris*, and *Spergularia marina* is an obvious and simple way of increasing their populations which has a high probability of success.

What are the sizes of the animal populations of special interest?

As with the classification of vegetation there is not space here even to enumerate the various approaches and methods which might be used to survey the animal populations of reclaimed land. Again, the reader must be referred to standard textbooks for details of census sampling and marking techniques, of sampling and trapping equipment, and so on. However, it is worth repeating the maxim that methods should be determined by the objectives and by the time available. For general extensive surveys of a particular group of animals, for example insects, simple methods which again use the subhabitats as sampling strata, may be appropriate (Morris, 1960, gives a general account of insect sampling methods, and a detailed discussion of these is contained in Southwood, 1966). For more particular studies, there are often special methods available—for example, in the case of breeding birds a census method standardized by the International Bird Census Committee (1969).

CONSERVATION AND MANAGEMENT

Defining management objectives

It is perhaps pertinent at this point to reassert a central piece of dogma; namely that all biological management should be directed towards clearly defined objectives. This is of particular relevance to reclaimed land, the management of which can have a very wide range of objectives. The fact that the land has been created artificially widens the range of possible management options in at least two ways. First, it frequently creates a set of physical and topographical conditions, in particular drainage and hydrological regimes, which are more easily controlled and varied than is possible on 'natural' areas. Second, its 'artificial' origin removes many of the purist objections to those forms of management which do not adhere to the concept of the 'natural' condition. This latter may seem an academic point, but it is one which in the past has constrained the activities of those managers who have sought to introduce new species or create new habitats within a reserve consisting of natural or semi-natural plant and animal communities. Of course, maintenance of the *status quo* may still be a perfectly legitimate management objective, as for example on certain types of reclaimed grassland.

Extending the categories drawn by Harper in a more restricted context (Harper, 1971) it is possible to recognize at least four broad management objectives, viz:

(1) Maintenance or promotion of high species diversity (i.e. large numbers of species per unit area).
(2) Maintenance or imitation of some specific floral or faunal *status quo*.
(3) Preservation, increase or introduction of specific desirable species (because they are rare, beautiful, interesting biologically, or an important part of the food-web of other species with these characteristics),
(4) Preservation of natural assemblages of plants and animals as a 'museum of types'.

To which might be added a fifth:

(5) Experimental studies on the immigration of species into newly created areas and their spread within them.

The final objective, whilst it involves predominantly a *laissez-faire* approach, also includes positive management actions. As in recent studies in the Netherlands (e.g. the Lauwerzee polders, Joenje, personal communication), these may include attempts to restrict, control or monitor access of the principal agent of dispersal, man.

Before going on to discuss some examples of management two further general points should be made. The first is that specific management objectives will frequently conflict and be unattainable within the same parcel of land. An obvious example of conflicting objectives is the preservation of say an arthropod species confined to tall flowering grassland and the maintenance of a plant species restricted to the short turf of grazed areas. Objectives do not always conflict as in the case of, say, the maintenance of the grazed-sward plant species and the preservation of wild geese populations. However, the existence of conflicts means that management will involve a system of priorities. Occasionally compromise is possible and if the area is sufficiently large it may be possible to divide it into zones and to operate several management programmes side by side.

The second point is that the management of land, particularly reclaimed land, for wildlife conservation is an infant science. It is still very much a hit-and-miss affair. We know enough about ecosystems to appreciate the high interdependence of their component parts but too little to predict with any degree of certainty the exact consequences of interfering with them by imposing a particular management regime. Practically nothing is known about the relative efficiencies of the different ways of achieving the same objective. Thus in practice management tends to be experimental and more data are required from experiments in the reclaimed land habitat to improve the efficiency and precision of future programmes of management.

Some types of management with examples

Control of water-table

As implied in an earlier section the drainage and hydrological regime of reclaimed land is a key factor in its ecology. In fact, the ability to control the water-levels within the reclamation or, at least, to ameliorate the effects of unseasonal flooding is probably the single most important aspect of reclaimed land management. Certainly if the full potential of an area is to be exploited it must be independent of the (increasingly lower) water-tables in the surrounding farmland. This is not to say that the land is doomed for wildlife if no control of groundwater is possible—for example, the dry grassland habitat is often highly interesting and may attract large numbers of herbivorous animals, such as geese. However, for maintenance of the marshland habitat and production of the full range of potential wetland habitats a degree of control is essential.

In general, water-level fluctuations should be gradual and, in temperate areas, seasonal in character. The effects of unseasonal water-levels can be seen in the Veersemeer (see above) where in summer the water-level is 70–80 cm higher than in winter (a situation found naturally only in glacier-fed lakes) and a zone of bare mud, uncolonized by vegetation after 10 years, fringes the lake margins and islands (Beeftink et al., 1971). Prolonged periods of flooding during the summer become possible in those areas where there is increasingly efficient drainage of the surrounding land. Duffey (1971) describes the disastrous effects which a period of exceptional summer rainfall and flooding in 1968 had on the populations of many species in Woodwalton Fen National Nature Reserve, Huntingdonshire.

It is sometimes possible to have virtually complete control of water levels using a relatively simple and inexpensive system of flood control dams and sluices and perhaps a small float-switch operated pump. Where this has been possible programmes of management have been very successful. An example is provided by the Harderdijk meadow bird reserve in eastern Flevoland (the IJsselmeerpolders), an area of 98 ha (245 acres) established in 1965 (van Duin, 1969). Here water levels are kept high to produce flooded grassland during the winter and early spring. This attracts large numbers, and many species, of waterfowl. The water is gradually lowered during the spring, providing a suitable breeding area for meadow birds, including waders such as black-tailed godwit (*Limosa limosa*), lapwing (*Vanellus vanellus*), redshank (*Tringa totanus*) and ruff (*Pinlomachus pugnax*), and duck such as mallard (*Anas platyrhynchos*), gargeney (*A. querquedula*), gadwall (*A. strepera*) and shoveller (*A. clypeata*). Some of these have nested in exceptional densities, notably 105 pairs of black-tailed godwits in 1970 (de Jong, 1972). Part of the reserve is used as hay meadow (55 ha), the grass being cut after 15 June to increase the chances of survival of the ground-nesting birds, and the other part (43 ha) is grazed at a density of 1–1·5 cows per hectare. Fewer birds breed in the pasture than in the hay meadow. The reserve has a 'buffer zone' of 108 ha managed as grassland.

Also in eastern Flevoland is the waterfowl reserve at Kamperhoek where again control of water-levels with sluices and pumps has enabled the creation of a wetland area comprising deep pools, reedbeds, marshy ground, and scrub.

Provision of habitat

Where the principal management objective is to preserve, increase or attract specific desirable species [objective (3) above] or to diversify the area generally [objective (1)] it may be necessary to extend or create *de novo* the appropriate habitat. The existence of the habitat of a species in an area may itself be sufficient to attract that species. Two outstanding examples are Leighton Moss near Silverdale in Lancashire and Minsmere in Suffolk. At Leighton Moss, land originally reclaimed in 1840 and cultivated as arable farmland up to the end of the First World War has regressed to marshland and reedbed since that time. At the present day, the reedbeds, now part of a nature reserve managed by the Royal Society for the Protection of Birds, support a stable nesting population of the bittern (*Botaurus stellaris*) and, more recently, a small number of nesting bearded tits (*Panurus biarmicus*), both species being some 200 miles from their British breeding headquarters in Norfolk and Suffolk. This area is also the north-western limit of breeding reed warblers (*Acrocephalus scirpaceus*). At Minsmere, which dates from the Second World War and is also an R.S.P.B. reserve, considerable success in attracting birds to an area by providing their habitat has been achieved, in particular with nesting avocet (*Recurvirostra avosetta*) and common, sandwich and little terns (*Sterna hirundo, S. sandvicensis*, and *S. albifrons*).

A spectacular example of habitat provision attracting bird species is provided by the work of Dr. J. Harrison and his colleagues at the Bradbourne Pit Wildfowl Reserve at Sevenoaks, Kent. Whilst it is not an example of reclaimed land management, the report of this work (Harrison, 1971) is an important blueprint for wetland management for waterfowl in general. In addition to providing islands, loafing areas, nesting rafts and mudbanks, the managers of the reserve—who, potential managers of reclaimed land will be encouraged to learn, describe themselves as 'a small party of amateur enthusiasts' (Harrison, 1971)—adopted the principle of providing favoured food plants. For example, stomach contents of locally shot duck were studied (Harrison, 1960) and the planting programme planned accordingly. Particularly interesting in this context is the successful transplanting, survival, and seeding of *Scirpus maritimus*, a common plant of brackish wetland on reclaimed land, and referred to earlier as an important food plant. Olney (1970) lists other preferred food species of wildfowl in Britain. A list of techniques in the management of wetland areas for waterfowl based on case-studies, many of which are applicable in coastal marshland, is being produced by the International Waterfowl Research Bureau (1972).

Among the food plants planted at Sevenoaks were the tree species alder (*Alnus glutinosa*) and hawthorn (*Crataegus monogyna*) both of which occur, and may be planted, on reclaimed land. These, and other species such as willows (*Salix* spp), provide cover both for nesting birds and for migratory birds making their first landfall on what may be an otherwise featureless coast. The problem of planting trees in areas subject to periodic flooding is reviewed by Gill (1970) who lists a number of suitable species.

In most cases the provision of an appropriate habitat will require a programme of constant maintenance and repair, as the newly created habitat changes in time and space. For example, the provision of habitats suitable for the ephemeral element of the reclaimed marsh flora described earlier will require the periodic creation of bare mud areas by shallow dredging of parts of creek and fleet edges or zoned seasonal heavy use of vehicles. Other examples of maintenance include the control of reedbeds along shallow open water margins and the eradication of willows in rush and reedbed areas.

The extent to which appropriate habitats can be created successfully depends on our knowledge of the exact habitat requirements of the species we hope to encourage or attract. This in turn depends on careful observation and research. For example, the studies of Owen (1973) and Newton *et al.* (1973) in Britain and of Lebret (1965) in the Netherlands have provided detailed information on the food and habitat requirements of herbivorous wildfowl, particularly geese. Based on such research, goose-grazing habitats have been successfully created by sowing grasses, for example in the Veersemeer (Lebret, 1972). In addition, these studies have highlighted some of the major problems in managing areas for grazing wildfowl. One of these is illustrated by the study of geese on farmland in Scotland by Newton and Campbell (1973) who found that the size and situation of a field was the main factor affecting its use by geese, and that roads, buildings, tall trees, and hedges reduced the attractiveness of the area. Owen (1973) and Lebret (1972) also regard disturbance as the major factor influencing the choice of grazing area, suggesting that unpreferred or inferior feeding areas will be used by birds where their first choice of feeding area is subject to a degree of disturbance. Thus the provision of the habitat alone may not be sufficient to induce the species to occupy it. The landscape of the surrounding areas, the proximity of buildings and other structures are important factors affecting the chances of success which vary from species to species.

Grazing as a management tool

The traditional use of much of the reclaimed land in Europe has been as pasture. As indicated earlier, the major threat to the flora and fauna stems from the conversion of permanent pasture to arable land or short term leys and the retention of diversity will in many cases depend on the continuation of traditional grazing practices.

Where grazing is not practical, or it is not possible to adjust stocking rates to suit the condition of the pasture, an effective alternative may be mowing. The experiments of Spedding and Large (1957) showed that swards which had been mowed were similar in height and density to those on which the sheep stocking rate had been varied. In a study of permanent grasslands on Romney Marsh, Kent, Chippindale and Merricks (1956) demonstrated how gang-mowing could be used as an alternative to varying the stocking rates, concluding that it was particularly useful for restoring the 'bite' to pasture which had been understocked and on which weed invasion and rank growth had occurred. Cadwalladr and Morley (1973) have recently suggested on the basis of their mowing experiments on salting pasture at Bridgwater Bay, Somerset, that the area suitable for grazing by wigeon (*Anas penelope*) can be extended by mowing, without affecting the objectives of the grazier's sheep husbandry.

The effects of grazing or mowing on grassland wildlife are complex and there is a need for research on the specific effects in the reclaimed grassland habitat. Morris's studies of the invertebrate populations on grazed and ungrazed chalk grassland (see e.g. Morris, 1969) led him to suggest that a rotational system of grazing is the optimum one for the conservation of these populations in general (Morris, 1971). The salt-marsh and paramaritime plant species can persist for long periods in grazed reclaimed land (salt-marsh plants for up to 30 years or more (Gray, 1970) and paramaritime plants probably for centuries) and may disappear very quickly after cessation of grazing.

Certainly where it is possible, and until such time as the effects of changes in the grazing regime can be assessed, a sound policy would be to continue the traditional grazing in an area of reclaimed land which is known to be interesting for its wildlife.

Diversifying the habitat

A number of examples of ways in which reclaimed land habitats may be diversified have been mentioned above—the planting of trees, the creation of bare mud areas, and so on. The habitat lends itself well to activities such as excavating pools, creating small mounds and islands or enclosing areas with earth-banks, particularly where control of water table is possible.

There is also considerable scope for diversifying the habitat in time as well as in space by varying important factors such as water-table, grazing, and general access to maximize the seasonal usage of the area by wildlife. Examples are provided by the Harderijk meadow bird reserve referred to earlier and the Ouse Washes, Cambridgeshire and Norfolk (Gray, 1976).

The management plan

It has not been possible in the foregoing section to do more than outline a few examples of the type of management regimes which have been used or are suitable for use on reclaimed land. These examples indicate that the habitat in general has a high potential for management aimed at conserving wildlife. Indeed a fuller range of general management options is given in Table 19.

Table 19 The PREPLAN scheme. A list of general management techniques applicable to reclaimed land.

Code	Key words	Management objective — Action	Examples
P	Protection	Protection of the habitat from threat	Control of shooting, public access. Prevention of drainage and ploughing
R	Repair	Repair and rehabilitation of damaged habitats	Restoration of grazing, repair of flood-control system
E	Elimination	The total elimination of unwanted species	Shooting carrion crows, cutting down invasive scrub
P	Populations	Altering the balance of species populations	Mowing to maintain grazed sward species, extending wetland subhabitats. Control of competitors
L	*Laissez-faire*	Deliberate 'leaving alone' of the habitat	Monitoring colonization of newly reclaimed area
A	Augmentation	Augmenting species populations	Planting trees
N	Niches	Niche diversification by creating new habitats	Provision of habitat such as open water. Flattening old sea-walls, digging new pools

The notation PREPLAN (Protection, Repair, Elimination, Populations, *Laissez-faire*, Augmentation, and Niches) is a deliberate (if rather contrived) reminder that management requires careful planning, and involves deliberate positive choice from a range of options. These options may be inextricably inter-related as in the case of diversifying and altering population balance, or mutually exclusive as with augmenting species populations and the *laissez-faire* approach. They are examples of the types of action available and have in common the need to be preceded by a survey of the type described in the *Methods* section above, and in many cases by a programme of research into their effects and interactions.

Whilst the range of possible management options on reclaimed land in general may be high, they may in fact be strictly limited on any particular piece of land. Having completed a preliminary survey and assessed the potential of the land for wildlife conservation, the manager is faced with deciding on an appropriate and, above all, realistic objective. This may be governed by the history of the land, its natural diversity, its size, its drainage patterns, the degree of control possible over its development, its freedom from disturbance, the state of adjoining land and coast, and many other factors.

Whatever the final set of objectives, it is important that accurate records are kept and the effects of all actions regularly monitored. In this way we can hope to improve on the precision of future management and understand the system more fully.

In this context, an extremely important document is *The Management Plan*. Management plans are written for many of the nature reserves in Britain, whether the body responsible for their management is national (Nature Conservancy Council) or local (e.g. county trusts, county councils). Two examples of management plans which are widely available are that of Eggeling (1964) for the Island of Rhum, and that of Usher (1973) for Aberlady Bay Local Nature Reserve, East Lothian. The latter author in addition to discussing at much greater length than is possible here many of the conflicts and problems of management for biological conservation gives a *pro forma* for a management plan as well as the full text of the Aberlady Bay plan.

In summary, then, the reclaimed land habitat despite its artificial origins is often ecologically diverse and offers the possibility of management which can combine the interests of the agriculturalist and the conservationist in particular and probably other interests as well. Whilst there is an identifiable need for more scientific research, particularly on the 'permanent' reclaimed grassland habitat in Europe, it seems clear that, managed with the correct blend of imagination and realism, reclaimed land has an outstanding contribution to make to the wealth and diversity of our coastal wildlife.

Acknowledgements

The author would like to acknowledge the help of Drs. D. S. Ranwell and R. L. Jefferies, who commented on the text.

Chapter 14

The submaritime fringe

G. DICKINSON

BACKGROUND

Of all the parts of the environment influenced by the sea, least attention is generally paid to the submaritime fringe. This is the area immediately inland of those zones which are occupied by distinctive assemblages of plants and animals directly and dominantly under marine influence. The specialist's attention is most often focussed upon those limited parts in which there is an extreme and often highly variable habitat stimulating a particular organic response. In contrast to these areas with their sharp interfaces and limited extent, the submaritime fringe may be, but is not exclusively, a relatively diffuse zone extending 2 km or more inland. Though less dramatic than the shoreline habitats, the complexities and subtleties of the ecological relationships, the importance of its organic resources, and the extent of the zone are major reasons for serious study of the submaritime fringe.

Examining first the nature of the habitat, description and analysis of coastal landforms presents a number of problems. Coastal geomorphologists have tended to concentrate their attention on the study of processes involved in generating coastlines, and to this end have been produced general classifications by Johnson (1919), Shepard (1937), Cotton (1952) and Valentin (1952). Detailed studies have tended to be concerned with examination of processes in the intertidal and shallow coastal zones and most have terminated shortly above the high-water mark. King (1972) provides a detailed review of recent research trends. Whilst appropriate to process-orientated geomorphological analysis, this approach yields little information on the nature of the submaritime fringes, and therefore provides only a limited basis for planning and management. The recent series of coastal studies sponsored by the Countryside Commission for Scotland (e.g. Richie, 1971) has shown that a more broadly-based, descriptive morphological approach is more suitable for these latter aims. It is commonly recognized that the submaritime fringe is a product of both maritime and terrestrial processes, and that the exact morphology of any coastal zone depends upon the balance between these process types as well as on the nature of other variables, such as the material involved, the time over which the processes have been operating, and the role of man. Within the coastal zone, it is generally assumed that landforms further from the sea are more stable and tend to be older, and these premises are axiomatic in several classical models of coastal development. However, the value of such models is limited to only the most general level, and detailed local studies often reveal patterns which are the result of complex interactions of processes, often on a cyclic and interacting basis. Such studies tend to promote the view that the submaritime fringe is a

271

recognizable and discrete unit of the coastal environment, with its own process regime and its own characteristic assemblage of landforms.

Whatever approach is adopted, the geomorphological character of the coastal zone is normally shaped by both marine and non-marine processes, and the resultant complexity of its nature strongly influences both the plant and animal ecology of the area. Ecological approaches to the submaritime fringe may either be based on the concepts of habitat or community. In the former case, widely-understood terms such as grey dunes or dune scrub (Salisbury, 1952) are employed to characterize elements of the habitat which share relatively common environmental circumstances. In the second case, the existence of a distinctive plant community helps to demarcate the zone. If the community approach is chosen, there is a wide range of techniques available to the ecologist, but, except in studies of very restricted sites, analyses are rarely accompanied by detailed and accurate spatial information. Frequently, surveys employ both habitat and community approaches in some compound system (Tansley, 1939; Burnett, 1964). Most current studies are concerned with the dynamic aspects of coastal ecology and focus attention on interaction between individual species and between species and their environment, are less concerned with fitting the data (often collected with great diligence in large quantities) into classification schemes. Neither approach can be advocated without qualification, and any approach chosen must depend largely on the nature of the problem, research resources, and purpose of study.

In general, therefore, the consensus regarding the limits and general nature of the submaritime fringe is more notional than real. The geomorphologist approaching the study of an area is likely to be most interested in processes affecting the coastline and their attendant morphological fetatures. The ecologist is most concerned with species–habitat relationships and the constant changes occurring in these. However the nature of the submaritime fringe is generally regarded as having two diagnostic characteristics. Firstly it is located above the high-water mark of all normal tides, and secondly it extends inland to that point where marine influences, and the biotic communities associated with these, become negligible components in the ecosystem. The actual limits are variable and depend on prevailing side conditions, though in some cases they may be fairly clearly delimited, particularly on the landwards' edges, by the extent of certain plant species: e.g. *Smyrnium olustratum* (alexanders) in southern Britain, *Plantago maritima* (sea-plantain) in northern Britain are possible candidates. But rarely, unless study is taken to ecological race (ecotype) level, do these plants locate the zone with any precision. Many of the common indicator plants, such as those mentioned or *Armeria maritima* (sea-pink), have alternative habitats such as cultivated ground or mountains, and distribution patterns at the species level are unclear. It is convenient and appropriate under some circumstances to consider the submaritime fringe as a transitional zone between tidal and wholly terrestrial environments, but the zone may also be properly regarded as distinctive in its own right, distinguished from its neighbours on the balance and nature of environmental factors and the biological response, rather than forming part of a rigorous scheme or classification. It is from this standpoint of identity that the submaritime fringe is considered here.

SPECIAL FEATURES OF THE SUBMARITIME FRINGE

Of all coastal habitats, that provided by the submaritime fringe is usually most stable and least subject to periodic variation in ecological conditions. This has most important ecological consequences in that it may act as a buffer zone for some species, particularly those of the higher trophic levels, the habitats of which are essentially littoral but which are unable to withstand unfavourable conditions for sustained periods in the intertidal zones. Thus the submaritime fringe is important to several sea-birds as a nesting and breeding ground, even though their principal habitat is closer to the shore or actually marine. Terneries, for instance, may extend inland for some distance, and the limits of this terrestrial territory may be vigorously defended against all intruders during the breeding season. Further, the submaritime fringe provides an important source of food for species such as gulls, which have a varied diet. Animal-commuting to and from this zone is an important aspect of its ecology. For example in the Hebrides the huge colonies of rabbits, which feed on the relatively rich grassland vegetation of the submaritime dune pastures, are the major source of food for predators such as buzzards and scavengers such as hooded crows, the feeding ranges of which extend beyond the coastal zone. The extent to which the submaritime fringe supports the higher trophic levels depends on its primary productivity and on the range of habitats available, and in many cases the submaritime zone compares favourably in these respects with adjacent marine and terrestrial ecosystems.

The submaritime fringe provides a large pool of plant species available for colonization of suitable habitats. This is most important in areas in which coastal accretion is occurring and the narrow tidewater zones are moving seawards, leaving new areas in which the balance of processes has been substantially altered. In areas where erosion is occurring, the existence of a well developed submaritime vegetation may slow the rate to a level where sudden and detrimental changes are avoided and an ecological equilibrium is maintained. The vegetation of this zone is characterized by a wide diversity of species, which is a reflection of the less extreme ecological conditions and the variety of niches available. The absence of inundation by seawater, for example, allows a much wider range of species to grow in the area, and mosses, normally rarely present in the littoral zone, are often important components of the vegetation. Characteristic too are a number of species with wide ecological tolerances and consequently with catholic habitat preferences, but which in the submaritime fringe are at the seawards limits of their range. Often such species, e.g. *Calluna vulgaris* (ling, heath), are dominant over considerable areas.

Although not subjected to the kind of gross environmental variation which occurs along the shoreline, the habitat is nevertheless a demanding one. Wind transport of salt-spray, as every gardener within 1 km of the sea will testify, is a problem and though accumulations of ions in the soil are relatively low in most humid areas, due to leaching, the problem is more acute where evapotranspiration greatly exceeds precipitation (Goreham, 1958). The favourable moisture regime of north-western Europe is to some extent offset by the frequent occurrence of storms during which very large amounts of spray are carried considerable distances inland. Although short in duration, the long-term effects of storms are great and their importance as an ecological factor in the coastal zone is considerable. However most pressures on the habitat are due directly or indirectly to man. Removal of the higher trophic levels, the natural predators, by extermination or elimination of the habitat, may result in a population explosion at lower levels. Introductions can have a similar effect, causing in both cases formidable pressure on the vegetation of the zone. The use of the submaritime fringe for either pastoral or arable agriculture imposes a severe strain on a fragile soil ecosystem which is often deficient in nutrients and inherently unstable. Throughout the zone, soil organic matter is scarce and this is a most important factor influencing not only nutrient supply but also soil structure, water-holding capacity, and resistance to erosion. It clearly follows that husbandry of these areas must be of a high standard.

Man's most intense pressures, however, are exerted directly through use of the submaritime fringe for industrial or recreational purposes. The coastal zone often provides a suitable location for those types of industrial land-use which require extensive and level sites, and which require to be sited near a transhipment point. International movements of bulky raw materials, such as iron ore or crude oil, are normally by sea, and to minimize the costs incurred in redistribution a coastal location has become the optimal site for many refining industries. Many tracts of the coastal zone are able to provide the expansive level terrain necessary for large petroleum refineries and integrated steel works. By no means all submaritime fringes provide physically suitable sites, and the locational decision is further complicated by other considerations such as deep-water handling facilities, markets, and labour supply. However some of the most ecologically significant submaritime fringes are under severe pressure from these types of land-use, which though made less noxious and ecologically destructive are normally incompatible with other land-uses, and over all are detrimental to the ecosystems involved. Recreational land-use is the other major source of pressure and this is growing rapidly with increasing affluence and mobility. Physical destruction of vegetation, disturbance of animal communities, and in general a reduction in the complexity and an impairment in the functioning of ecosystems are the principal detrimental effects. These are concentrated at sites for which the main attractive quality is a landscape unmodified by other human activities. The agencies responsible for this type of pressure are trampling, which when localized to paths or other contact zones can cause acute damage, effects of motor vehicles (again normally a highly concentrated factor), spread of litter and other noxious and non-degradable pollutants, and physical disruption of animal communities.

METHODS

Significant ecological information may be obtained either directly from field surveys or indirectly by evaluation of existing sources of data. Before describing and analysing each of these, some general consideration of the aims of study are necessary. Three factors above all else govern the way in which essential information should be gathered. These are, in no significant order, the aims of the study, logistical resources, and time constraints. Quantitative as well as qualitative evaluation of these factors must be made to obtain a thorough understanding of the particular problem. All three are especially important in applied studies, where the survey is principally a means to an end, rather than an intellectual exercise. This does not mean that the standards or merit of studies relating to practical problems should be or are, in fact, any the lesser, but that the information collected is oriented towards definite requirements. Existing surveys can sometimes be utilized but they rarely provide the whole answer to the questions raised by management proposals, again indicating that the most important consideration is the applied study's requirements, rather than comparability with other projects or utilization of a praticular method or system. Often simple empirical and descriptive techniques will yield the data most useful for applied work, and they may be more suitable for this purpose than more sophisticated approaches. Generally the survey finally carried out is the result of a compromise, the exact nature of which is dictated by the balance of the three controlling elements discussed above. Thus it is not possible to give a prescription for obtaining significant information on the ecology or pressures of the submaritime fringe; but the manner of approach has certain general principles from which a study programme may be developed.

The most common means of obtaining ecological information is by direct field investigation. There is a very wide range of methods available, but for applied purposes in the submaritime fringe with its wider range of plants, animals and habitats, field studies will most often take the form of an inventory, seeking to provide information on distributions rather than to provide explanations. For environmental aspects of the coastal zone, the most appropriate approach is likely to be one which concentrates on mapping or otherwise describing the spatial distribution of minor morphological features, paying particular attention to factors, such as slopes, elevations, exposures, and stability of surface and substrate, which have maximum influence on the communities found in the resultant habitats. The same inventory approach can be usefully applied to vegetation studies, in this case concentrating on description and general explanation of plant communities. In this instance, a simple approach is often the most appropriate, and care should be exercised not to choose an inappropriately complex quantitative technique which has been developed to solve detailed and specific ecological questions and which would not be suited to the requirements of management and planning. In coastal areas where most of the important ecological gradients run from the shore inland, the line transect may be particularly useful as it gives the maximum information about habitat and community change when following such ecological gradients. However, maritime influences decrease with distance from the sea in a manner which is exponential rather than linear, so that in the submaritime fringe, other methods may be deemed preferable, particularly if the zone is being evaluated separately from those closer to the shore. Sampling by quadrats of various size may be used. These can be laid out in a variety of ways, but objective rather than subjective approaches are generally preferable. Examples of various approaches are discussed by Dickinson, Mitchell and Tivy (1971), and full evaluation of the wide range of techniques currently available is contained in the reviews of Greig-Smith (1964) and Shimwell (1971). In many cases of applied work, hybrid methods may be required.

Information on animal populations in the submaritime zone is very often scant and there is an urgent need for basic inventories in all areas where man's pressure is significant. The gathering of precise information is generally best left to animal ecologists (who unfortunately are usually in short supply, since the techniques involved require familiarity both with the type of specialist methods reviewed by Southwood (1966) and Macfadyen (1963), and with the species involved. However much useful data may be obtained by discussion with local amateurs and natural history societies, whose fund of knowledge is frequently very considerable, if not in an easily accessible or usable form for applied work. Careful consideration should be given to the need for this type of information, to the evaluation of potential sources, and to consultation with interested and informed parties with regional or national conservation briefs, such as county wildlife trusts or the Nature Conservancy Council, before commissioning field surveys.

The constant change which occurs in man-made pressure on the coastal zone means that field study of pressure-effects is not always the best method of obtaining information. At the most detailed level, a considerable amount of work has been carried out in recent years and findings not relating specifically to coastal areas may be relevant to the submaritime fringe, e.g. Streeter, 1971; Bayfield, 1971b. Direct observations normally concentrate on measuring ecological pressure through monitoring change in biological communities, and in particular on micro- or meso-scale variations in one or more species which have a perceptible response to pressure. For example, the lichen flora of an area may provide a measure of atmospheric pollution much more readily obtainable, than by use of atmospheric monitoring equipment alone.

Direct effects by man, vehicles or domestic animals are a major source of ecological pressure. With these, information is probably best obtained by detailed field study of sample or critical sites. Available techniques include the use of profile sections across the ecological gradients to measure effects on vegetation—a particularly useful method for work on trampling by man or animals. A quantitative measure of numbers of trampling agents may be obtained by the use of electronic beam or reed switch counters, and an estimate of impact on surfaces gained by use of thin wire 'tramplometers' (Bayfield, 1971a). Most work on pressure-effects on animal populations has been concerned with investigation of changes in fauna (particularly invertebrate) and in population size, though some studies concerned with higher animals, especially birds, have considered the impact on a wider range of ecological activities (Murton, 1971).

Few studies of ecological pressure have examined in any detail effects on the environment. Interest in pressure has tended to concentrate on its impact on plant and animal communities, and furthermore information on the nature of pressure-effects on the environment is extremely difficult to obtain, due to the number of factors involved and the way in which they interact. This is in part a result of the current balance of interests and priorities in conservation management. The technical difficulties of collecting data on such topics as soil structure, compaction, or changes in nutrient availability may be solved by using simple empirical parameters. The need for information on habitat pressure in the coastal zone is urgent, since most human impact on biological populations has been caused by man's modification of their environment, and to be effective management policies must focus on causal factors.

Two of the most important recent developments in methodology have concerned remote sensing devices and aerial photography. Access to sources of these data relating to the coastal zone is becoming widely available, and it can provide information on both the ecology of, and pressure upon, areas under study. Remote sensing, a rapidly developing field, incorporates such devices as infra-red photography (both so-called black and white and false-colour types) and multi-spectral photography (that is imagery produced by simultaneous exposure of photographic plates sensitive to different ranges of the electromagnetic spectrum). The most recent developments are sideways-looking airborne radar (SLAR) and infra-red line-scan imagery; the former utilizes the relatively long wavelength radar waves, whilst the latter picks up emitted infra-red radiation (as opposed to the reflected infra-red waves used in conventional photography) which may be used to produce a type of 'heat' map. A general résumé of these techniques is given by Harris and Cooke (1969), some ecological applications are discussed in greater detail by Johnson (1969), and these techniques have been evaluated in coastal areas by several workers under the sponsorship of the N.E.R.C. (Greenwood, 1974). The value of infra-red radiation to vegetation studies is being increasily realized and infra-red data are often regarded not merely as an adjunct to other more conventional sources, but as a priority basic source in itself. The value of such information lies in the fact that the leaves of plants reflect relatively low levels of visible light, but have relatively high infra-red reflectance. This allows recognition of plant communities at a detailed level, besides observations on plant pathology, and studies of shallow-water vegetation, all of which may be of importance in coastal studies. In the submaritime fringe, where patterns are often complex and processes difficult to distinguish, infra-red imagery can be of considerable value.

Conventional black and white photography, however, should not be underestimated as a source of information. Aerial photographs are fairly cheap, quite widely available, and coverage exists for much of the coasts of Britain and western Europe. Simple interpretations can be made by most users, and quantitative distance measurements are possible using appropriate photogrammetric techniques. The aerial photograph, often used directly in the field, has become a familiar tool of the ecologist, geomorphologist, and soil scientist, and is widely used in planning and management work (Howard, 1970). Specialized maps of considerable value in coastal studies can be produced; an example of this is the vegetation map of the island of Rhum, Inner Herbirdes, produced by the Nature Conservancy (1970). Another particular value of aerial photographs is that they can be produced at a wide variety of scales. Photographs covering thousands of square kilometres in a single frame may be taken from orbiting satellites at a distance of several hundred kilometres from the earth's surface. At the other end of the scale, photographs at a very large scale may be taken at a range of a few metres, if the camera is elevated on a pole or underneath a captive balloon. Change may be monitored by repeated photography at intervals which may vary from several years, for long-term changes of vegetation, to a few minutes in time-lapse photography for such purposes as investigation of the movements of large herbivores, or distribution of holidaymakers. As with field study, the exact choice of method must depend on the nature of the problem under investigation, but such is the value and flexibility of remote sensing and aerial photography that it is likely that these techniques will be used at some point in most surveys, especially in the reconnaissance and primary data collection stages.

LAND-USE IN THE SUBMARITIME FRINGE

The range of uses to which the submaritime fringe may be put is far wider than in any other part of the coastal environment, and this area has a wider range of land-use potential (often termed land-use capability) than almost any other analogous zone. Land-use lies at the heart of management problems in the coastal zone and reconciliation of conflicting interests is likely to be a most taxing process.

Educational and scientific uses

The very problem outlined above gives the area one of its main educational resources. Throughout Europe, but particularly along the crowded coastlines bordering the North Sea and its approaches, conflict between various forms of land-use is a problem of the first order. This, together with possible solutions, should be presented to as wide a range of people as possible, from a very general and simple picture for young schoolchildren, to detailed case-studies and research work for postgraduate students. Included should be all those likely to be involved in decisions which influence the land-use of the zone, this latter comprising a broad spectrum of people and professions. At first sight, the complexity of the processes controlling the habitats might seem to preclude the use of the submaritime fringe for basic ecological teaching, but besides the important plant and animal communities located there, the zone provides a convenient and compact site for study of such ecological factors as plant community patterns, soil development, role of drainage as a component of various habitats, influence of soil pH and consequences of concentration of certain ions.

As a centre for fundamental scientific research, the submaritime fringe demands answers to many vital questions relating to its nature, and provides an excellent base for a study of general ecological systems. Of the wide range of problems relating to the submaritime fringe, a representative selection in terms of type and scope includes basic inventories of invertebrate fauna along the remoter coasts of Britain and Europe, studies of the detailed post-glacial evolution of the zone in different areas, and close investigation of the ecology of certain critical plant species. In the realms of fundamental work this zone, generally lacking great structural complexities in its vegetation, is an appropriate location for the further of knowledge relating to the dynamics of plant communities.

Recreational use of the coastal zone

The submaritime fringe was one of the first landscapes to be used for recreation. The spas so often associated with seaside developments of the eighteenth and nineteenth centuries in Britain were often located on the submaritime fringe, and as sea-bathing replaced taking the waters, 'promenades' extended along the seafronts. Although traditional holiday-resorts in Britain have experienced some decline in the past two decades, the submaritime sites are still the best endowed and most fully developed, and the advent of cheap travel has placed a heavy strain on the zone in sunnier parts of Europe. Though having the most profound effects and generally leaving little trace of former patterns, such urban-type developments are normally highly concentrated in specific sites and are restricted in extent. Other types of recreational land-use have modified the environment less and are more amenable to harmonious integration with effective ecological management, but they may cover much larger areas. For more than one hundred years, golf courses have been located on low-lying sandy plains and on other types of submaritime fringes around the coasts of Scotland. Though at first these were little managed and caused minimal impact, growth in the game's popularity saw both continuous and rapid developments which have spread throughout the world. Many of the famous names of British golf—Carnoustie, St. Andrews, Muirfield, Troon, Lytham St. Annes—are located around the coast, where the diverse landscapes of the coastal zone provide a physiography and vegetation ideal for the sport, whilst the exposed aspect of these courses contribute to their unique challenge. Management of these areas is now a sophisticated science and Ranwell has rightly reminded ecologists that: 'No one has yet brought together the very considerable practical experience obtained by golf links management on sand-dunes.' (1972a, p. 220). In spite of considerable pressures, efficient course management can minimize damage, and will aid resource conservation in other spheres of recreation. It is to be hoped that the numerous golf courses being developed on many of the coasts of Europe can fulfill the same valuable role as those in Britain.

Most other types of recreational use are related to unorganized and spontaneous recreation. In terms of absolute numbers, the largest share of this belongs to the casual day-visitor travelling from adjacent urban centres, though in the remoter areas the major contributors are holidaymakers. The motor car and post-war affluence, together resulting in the vast personal mobility which typifies western Europe, have made enormous increases in this type of use throughout the coastal zone. The concomitant has been the rise of problems and acute pressures on coastal ecosystems, requiring effective management policies to be developed without delay. A particularly graphic example is the use of the submaritime fringe for caravan parks. Suitable sites readily accessible to population concentrations have seen mushrooming developments, resulting in landscape blighted by high densities of caravans. Many examples in Britain are well known, and the resulting deterioration in ecological conditions and intrinsic landscape qualities have been fully described (Patmore, 1970). The often quoted cases of the Welsh coasts are cited as being extreme situations, but these seem to be in the process of being emulated in parts of Yorkshire, Devon, and Cornwall. Proper management policies must be instituted to ensure adequate screening, sanitation, refuse-disposal arrangements, and rigorous control of numbers. Progressive authorities seem to be learning from earlier mistakes and, as in the county of Argyll where recent studies have shown that ecological damage will inevitably occur unless caravan use of its beautiful but fragile bays and coasts is restrained, the equally inevitable conclusion is the need for imposition of controls. In Europe the growth of caravan holidays has been less rapid, but camping enjoys great popularity and hugh sites, with tents often at formidably high densities (each one accompanied by its attendant motor vehicle), are a commonplace sight on many of the most attractive coasts of Spain, France, the Low Countries, and Germany. In either camping or caravanning, pressures are caused by site pollution, by trampling, by erection of service facilities, and most seriously by motor vehicles driven over open ground. Each one of these may be tackled quite easily if proper provisions are made, and though it may be expensive, these must be capable of handling maximum pressures without deterioration. With screens of shrubs and trees, such sites can tolerate high densities with limited detrimental effects, privacy for users, and very little visual impact on the landscape.

Economic uses

In the strict sense all uses are economic, since the submaritime fringe is a relatively scarce resource which can only be allocated to certain uses at any one point, but in this case the term economic implies those uses which are primarily aimed at the production of monetary profit. The number and diversity of such uses is prolific. In sylviculture, amenity and ecological stability though considered important are secondary to commerical timer production in the policy of the Forestry Commission in Britain. During the last 50 years, extensive coniferous plantations particularly of Scots pine (*Pinus sylvestris*) have become a feature of the submaritime landscape in many parts of Britain. Agriculture is the most widespread single land-use in the submaritime fringe, and is one of great economic importance. Grazing is carried out very widely, and in the absence of another incompatible land-use it is to be expected to occur. Although problems in maintaining the stability of the substrate may be found in certain areas, the zone has many advantages for pastoral agriculture. The relatively exposed coastal location and its effects on plant growth are offset by moderating influences on climate due to proximity to the relatively stable-temperatured sea, which result in an extended growing season. Even without the use of domesticated grasses, the submaritime fringe often provides rich grazings due to the preponderance of useful species and the relatively high carrying-capacity of coastal grasslands. Cultivation is also quite widely practised though in this case the problems of erosion, exposure, and salt-spray burning are more critical. However, where conditions beyond the submaritime fringe are inimical to crop husbandry, as exist in many parts of the highlands and Islands of Scotland, the zone may form the primary resource for cultivation.

But it is in terms of industrial use that the highest demand for land and the greatest profits per unit area are made. The submaritime fringe offers two kinds of general advantage when considered as a potential industrial site. Firstly the general site charac-teristics are favourable to those types of industry which demand fairly extensive tracts of level ground in an accessible location. Other zones may fulfill these requirements, but generally competition for land-use from other industrial users, agriculture or housing is so acute as to force price and rentals beyond the market capabilities of the extensive land-users. Added to this, coastal sites may be at a convenient distance from centres of population—neither too close to increase costs by the expensive environmental engineer-ing necessary to fit into a modern urban community, nor too far to hinder easy movements by the labour force. Secondly the location of the zone means that particular advantages are conferred on those industries which require the import of bulky raw materials for processing or refining. The increasing trends to site major oil refineries or integrated steel plants in coastal areas close to deep-water anchorages indicate the benefits which accrue to industry by minimizing transhipments and maximizing economies of scale with bulk carriers and super-tankers.

The problems which can result from this pressure are immense. Pollution has been and continues to be a threat to life in the proximity of industrial installations. Prior to the recent growth in interest in conservation, public pressure on industrial companies was minimal, but in recent years a much more widespread realization of the need for concern has substantially change this situation. New developments are now subject to rigorous scrutiny and promises of action to minimize detrimental effects on the vulnerable coastal zone are sought and given. Whilst giant corporations with major projects may produce large absolute problems and attract the most attention, they have been more willing to tackle seriously their problems than many smaller secondary enterprises. The latter, often indigenous, company is frequently less able to develop in a manner compatible with conservation interests, either due to lack of financial resources or to the nature of the business. The problems of the Seine estuary, Fos or Europort are in part due to this already widely-experienced British situation, and similar circumstances continue to arise throughout the industrial world.

POLICIES FOR MANAGEMENT AND CONSERVATION

This final section comprises four main components. The first three are case-studies dealing with different submaritime areas, which are put to different uses, present a variety of problems both in kind and scale, and illustrate a corresponding range of planning and management techniques. All three examples are located in Scotland; a fact which in part reflects the author's experience, but they are also exemplary cases with a wide range of applicability, reflecting that country's large share of Europe's coastline and its diverse character (see Steers, 1973). The fourth part discusses general approaches to the problems of management and conservation in the submaritime fringe and proposes the theory that whilst each problem requires a unique solution there can be a systematic approach to selection of methods.

Machair in the Outer Hebrides

Machair is the Gaelic name for a landscape found in many places along the west coasts of the Highlands and Islands of Scotland. Though imprecise in meaning, the term commonly encompasses a calcareous-sand plain, up to 2 km in breadth, immediately inland of the shoreline, a suite of aeolian geomorphological features such as dunes, slacks, cuestas, and blowouts composed of the same lime-rich shelly sand, together with its semi-natural calcicole grassland vegetation. The vegetation is treeless, due to the combined effects of exposure and grazing, and consists of a herb-rich sward dominated by *Ammophila arenaria* in the less stable areas, and by *Festuca rubra, Poa pratensis*, and certain sedges (*Cyperaceae*) in the more stable areas, these latter tending to be at a greater distance from the sea. (Dickinson, 1974). Conventional models of dune development (yellow dunes, grey dunes etc.) are inapplicable as there are no further contributions of fresh sand to continue building, but local movements and reworkings are considerable. The variety of habitats, the lime-rich substrate and the effects of man have combined to make the vegetation diverse and floristically rich. In comparison with the harsh peaty moorlands inland, this area forms one of the principal areas of organic resource in the west coast of Highland Scotland—a resource widely utilized by indigenous crofting communities from their earliest times. Man may also be regarded as an integral component in this ecosystem since human involvement in the area has been continuous throughout its evolution. Machair, more than most coastal areas, merits the use of the term landscape (applied in a similar, though less cogent manner to the German *Landschaft*), acknowledging that its existence depends on the physical environment, the habitats provided by this, the vegetation and animal life of the area, man's activities and all the manifestations of human culture in an interacting 'socio-ecosystem'.

The area provides many valuable and detailed examples of ecological problems, and yet may readily be set in a wider context. The most serious local ecological problem is the disturbance and disruption of substrate materials resulting from pressure on vegetation cover. This pressure may be temporary or permanent depending on location, and on the nature and extent of the agencies involved. Consequent damage to vegetation may be the result of natural erosion and deposition processes, and though this may occur at any point, it tends to occur most frequently in areas close to the shore, or in the highest parts of the zone. The configuration of the coastline, the form and depth of underlying rocks, weather conditions and changes in vegetation caused by other factors all contribute to the amount and type of natural pressure exerted on any machair area.

However, most problems are man-made, and pressure results from a wide range of human activities. Rabbits, introduced by man to the Outer Isles only a few hundred years ago, have increased in numbers to Malthusian levels due to the scarcity of predators other than raptorial birds such as buzzards; and the consequence has been formidable pressure on vegetation in several areas. Sporadic shooting has virtually no effect due to the breeding surplus available to make up numbers; the effects of myxomatosis (also introduced by man as a control), which were drastic a decade ago, have now been overcome and the disease currently has little effect as a result of the onset of natural immunity; and use of cyanide gas would have to be systematically applied to be effective, a procedure which is precluded by expense. The rabbit problem is likely to remain serious until concerted action is taken. Farming practices are the cause of much damage and modern methods may be as much to blame as traditional approaches. Pressure on the machair resulting from agricultural use was most acute about 150 years ago when the local population reached its peak (in many areas more than double the present figure), and the decline in crofters has meant that there has been a certain lessening of impact of the zone. However, being more suitable for all forms of agriculture than the interior moors, the area has continued to support much agricultural activity, so that reduction in pressure has been minimal. It is ironic that efforts made to counteract depopulation may result in ecologically detrimental effects. Nowadays machair is cultivated by machinery which is often responsible for damage in vulnerable sites, where erosion once initiated may rapidly spread. The use of traditional organic fertilizers, such as seaweed, has been largely replaced by more convenient artificial types, which though providing correct mineral balance make little addition to the restricted humus resources which are critical in maintaining soil stability, tilth, and water-holding capacity. Several recent changes illustrate how agricultural problems may be tackled. The great profitability of beef-cattle rearing has caused expansion in pastoral farming and in increase in farm income, both of which promote the use of machair as an agricultural resource. Brucellosis has been eliminated from many areas thereby enhancing the breeding qualities of cattle reared there. Progressive and conservative methods have been adopted, including fencing of land to control grazing efficiently, and the use of fences has been accelerated by land-tenure reforms which have allowed erstwhile common land to be apportioned amongst the crofters of a township, making for sounder ecological management as well as more profitable farming. Sterling work by the North of Scotland college of Agriculture has shown how conservative methods of grazing and cultivation, and in particular the introduction of new grass strains, can minimize pressures and improve cash returns. The same type of improvements have occurred in arable crop production. The main problems in agriculture in this submaritime area are lack of scientific knowledge, resistance of local people to new methods, and difficulties in communication of new ideas.

Other man-made problems have a more recent origin. The growth of tourism and recreation in the Highlands and Islands have caused considerable pressure on coastal areas, which form one of the main attractive areas. In Uist, the remoteness of the region and the strong local counter-attractions of splendid beaches and wild mountains mean that there are relatively few visitors to the machair—an area which is unspectacular even if possessed of a distinctive charm. However machair is increasingly used to park caravans, both touring and more permanent types, the latter often owned by crofters who use them to supplement farm income. In themselves these cause little damage, but pollution is often associated with the caravans, as even the most basic facilities are lacking, and the constant movement of motor vehicles to and from sites soon causes quite severe erosion. Part of the area is used as a defence installation and quite a large part of the machair of South Uist is under Ministry of Defence control, but beyond a very limited number of buildings and other installations there has been surprisingly little impact, and over-all effects have been beneficial since the kinds of pressure discussed above which may be ecologically harmful have been severely limited.

Certain problems relate to a wider scale, and solutions in correspondingly broader context of approach must be sought. For conservation interests, the pressures on the area due to the value of the resources and their economic use must be weighed carefully against the case for preservation. The strength of a conservation case depends largely on the scale of overview. At a local level, an ecosystem frequently appears to be more important than at a regional scale, but only in the latter case can a really strong argument for the conservationist viewpoint be established. The machair areas on the islands of North and South Uist and Benbecula, together forming a spatial entity, illustrate this situation. Machair is an unusal and ecologically interesting landscape in the European context, and it therefore has a high claim to priority in conservation. The most important sites in Uist are those which, besides having the major components of the machair ecosystem, are the breeding habitats of absolutely rare species. In the coastal zone of the Uists there are major breeding grounds of birds such as the grey lag goose (*Anser anser*), corncrake (*Crex crex*) and red-necked phalarope (*Phalaropus lobatus*), which are scarce species in Britain, and to a considerable extent also in Europe. The policy of two conservation organizations involved in the Uists has been to concentrate action on these critical sites. Machair comprises a major part of nature reserves set up by the Nature Conservancy Council and the Royal Society for the Protection of Birds, and the areas chosen fulfill both of the above requirements. Each organization has management plans and policies specifically designed for the reserves and having three general aims, management and conservation of the environment in particular as a habitat for the important species, protection of such species, and furtherance of scientific research. Other aims, such as public access for forms of recreation compatible with conservation, or education of a broad section of the public, must have a lower priority and are subordinated to the principal objectives of the reserves.

Currently this coastal zone is at an important crossroads in its utilization. In spite of many changes over the last 25 years or so, the landscape has not been significantly degraded as a biological resource, and in fact its importance and value as an area of scientific and conservation interest has been heightened by its relative stability in the face of increasing pressures on other submaritime areas. But this position can no longer be maintained without a complete analysis and evaluation of the current situation, and the development of strategic planning for the future within a broad spatial and temporal context. Co-ordination has not been much in evidence in past planning work, though local authorities have made efforts which have been thwarted by the complexities and lack of communication between existing agencies involved in the land-use of machair. Local government reorganization may aid rational planning in the future, but the pressures likely to be exerted on coastal lands, even, as in this case, in one of the most remote and inaccessible parts of western Europe, means the need for strategic planning is urgent.

County planning: tourism and conservation in East Lothian

Some 15 km east of Leith, the coastline of the county of East Lothian swings north-eastwards into the Firth of Forth to form a broad promentary. On the northernmost point of this foreland is the resort town of North Berwick, at its south-western limit are the suburbs of Edinburgh and at the south-eastern apex is the historic town of Dunbar. Along this 50 km of coastline are a succession of wide sandy bays sporadically broken by low rocky headlands and backed by dunes and sand plains. Inland of the coastal zone lies a low fertile plain underlain by diverse igneous and sedimentary rocks of lower Carboniferous age, which in the relatively dry and sunny climate prevailing make this the premier agricultural district in Scotland and one of the best in Britain. The coasts of East Lothian are also amongst the leading in landscape quality and amenity in lowland Britain, and their justifiable fame is widely appreciated by the tens of thousands of day visitors who use the area each summer, as well as by the smaller but significant number of tourists who come from all parts of the world to enjoy family holidays at its resorts or to play golf on its celebrated courses. Ecological pressures throughout this coastal zone are high, but as is to be expected they are concentrated in certain locations, whilst other areas remain relatively untouched. Thus the 64 ha coastal National Nature Reserve at Aberlady Bay and the 8 km stretch of coastline around Tyninghame and Tyne Mouth Bay, which are less accessible and protected by national and local statute from activities detrimental to their ecology and quality of landscape, are areas relatively free from acute pressures. At the opposite extreme, it is estimated that 100,000 or more visitors come to the village of Gullane annually, being attracted by the fine beach, the spectacular dunes and its golf courses, and the impact of people on this part of the submaritime fringe is profound. The resort of North Berwick, with a population of 4,400, has a peak daily demand of more than 2,000 car parking spaces. In these and similar areas, ecological pressures are severe, and though many visitors do not leave the built-up areas, the ecosystems of the coastal zone are under a load which cannot be borne without the most careful management.

In this submaritime area several different kinds of ecological pressures can be recognized. Car parking, which though confined to a narrow band by roadsides, is a serious problem in this and other popular coastal areas where roads run close to the shore. In any areas where severe regulation is not imposed, complete destruction of roadside plant and animal communities will occur in a very short period. Fortunately vehicles cannot be taken to beaches, but at many points footpaths run across the area, often radiating from convenient access points to popular beaches. Here again a band of less profound but still serious ecological disturbance several metres in width is found. All types of recreational pollution form patterns which replicate those of road and path and show the effects of damage to plant and animal communities, disfiguration of the landscape, not to mention hazards to farm livestock and machinery and potential dangers to visitors' enjoyment and health. Probably the most difficult ecological problem is that of resolving the inevitable conflict of interests which must arise in such an area. Land-use management techniques employed by the Nature Conservancy Council, County Council, greenkeepers or farmers are likely to be different in both objective and practice, and difficult to reconcile. Each agency expects and obtains a considerable degree of autonomy and only by negotiation and compromise can a viable strategy for the whole area be obtained: in East Lothian it is a compliment to the participants that there is such a degree of success.

In several aspects of countryside management, the work of the East Lothian County Council has been recognized as being amongst the best in Britain. The dune stabilization programme, which has received considerable attention, has been discussed fully elsewhere (Tindall, 1967), but more appropriate to the present exposition are other aspects of management relating to the submaritime fringe. The most important single factor was the County Council's preparation of a strategic plan for the development of recreation in the county. It was a forward-looking document with a primary aim of creating harmony between competing land-uses and development in an ecologically sound manner. This plan recieved Governmental approval in 1965 and since that time has been implemented at a rapid rate. This action follows and extends the initial basic guidelines laid down earlier by Parliamentary Act of 1936 and in the county development plan of 1953. A feature of the work since inception has been liaison and co-operation between the various organizations involved in the land-use of the coastal zone, and as a result of this, the area has gained greatly from the wide range of expertise brought to bear upon its problems.

In detail, the ideas embodied in the country recreational plan can best be examined by analysis of the coastal park at Yellowcraigs, near the village of Dirleton at the northernmost tip of the East Lothian coast, some 2 km west of North Berwick and 37 km by road from Edinburgh. The park, rather less than 1 km^2 in area though almost 1 km in depth at its maximum point, extends from the shore across several lines of dunes through a broad area of links (which is penetrated by the low crag of the fossil volcanic vent giving the park its name), to merge at the landwards edge with an extensive area of post-glacial raised-beach material now forming fine arable farmland. The park was established on land donated to the county in 1944, and wartime debris and damage was cleared by voluntary labour in 1960. Further work carried out in the next decade included construction of an access road with an all-weather surface and provision of a well laid-out car park. This, though grass covered, as is shown in Figure 90, has sufficient entry points to allow heavily

Figure 90. The well laid-out grass car park with hard-surface access road
at Yellowcraigs, East Lothian

worn areas of turf to be rested when necessary and is well supplied with litter bins. The result is a facility which is unobtrusive, quite inexpensive, practical, and most importantly which has caused little damage to the environment both during establishment and during operation; showing that man-made features can, if carefully planned and managed, fit into an area where conservation and amenity are the prime interests.

A little further inland, a beautifully laid-out caravan park was established in 1968 by the Caravan Club of Great Britain. The site forms a narrow triangle which is well screened by trees and shubbery both externally from the road and coast, and internally where divisions by arcs of thicket promote an aspect of privacy. In spite of the maintenance of amenity value of this site, it is essentially functional and its unobtrusive external character and pleasant internal environment should be a model (one, alas sorely needed in nearly every part of Britain) for such coastal developments. Figure 91, showing the view westwards

Figure 91. Yellowcraigs links and dunes from Yellowcraigs Hill. Note the unobtrusively located public convenience screened by trees

from the top of the low hill at the centre of the park, illustrates the characteristic unmodified environment. Though considerable parts of the submaritime fringe in this area are under farmland or are used for golf courses, the kind of landscape shown in Figure 91 is quite extensive and thus all tracts do not require to be rigorously protected. Nevertheless management techniques at Yellowcraigs ensure that modification to vegetation is minimized and a wide range of habitats is provided for a variety of species. This has led, in co-operation with the Scottish Wildlife Trust, to the provision of a nature trail. But besides such formal facilities, day-visitors and picnickers are welcomed and the provision of a barbecue site is an imaginative feature. In all, this is an area meant to be used and enjoyed by a wide cross-section of the public and this aim is manifestly successful as measured by the 50,000 or so visitors the park receives annually.

In conclusion, this case-study illustrates the advantages of positive planning in a coastal area. Inheriting a sizeable legacy of problems, and having a high tourist resource value which has been increasingly utilized in the past two decades, the area has provided an excellent testing ground for management ideas. In many other cases such opportunities have been lost and future problems have been allowed to grow to levels where drastic restorative action becomes essential and planning aimed at holding and repair operates rather than conservative management. The East Lothian example shows how early action and appropriate methods can reconcile the effects of development and the interests of ecology, for the action of the county has been environmentally successful and is good for its tourist industry.

Industrial development on the Clyde

The final case study presented here is not so much a record of success or failure, but more a review of circumstances in an area likely to be the focus of major developments in the future. The area is the Hunterston peninsula on the west bank of the Firth of Clyde, some 45 km west-south-west of Glasgow. This tract of land, about 8 km^2 in extent, forms an approximate rectangle of low ground extending from the hills which border most of the northern part of the estuary into the Firth, here some 35 km wide and opening out to the Northern Atlantic trade route. Two small islands, 2 km offshore of the promentary, provide a sheltered anchorage which is more than 20 m in depth and thus ships of the very largest size could be accommodated in these roads. Adequate communications already exist with the industrial heart of Scotland and these could readily be improved if developments were to take place. Around the site are several existing towns which could provide the infrastructure necessary for rapid large-scale industrial projects. The potential of this area of submaritime fringe was fully realized during the mid-1960s, and in the intervening period, though there has been only limited construction, there has been much debate and inquiry into various proposals, culminating in 1971 with Government sanction for the rezoning of 920 hectares of land on the peninsula for industrial use. The past two years has seen four major development options put forward, involving expenditure of up to £180m for the construction of steel mills, stockyards, oil-refineries and the like (Campbell, 1973). Over the same period, opposition to development both specific and general has grown, and at the time of writing has reached a high level. It is not the aim of this section to evaluate these developments or their ecological impact, which are complex and detailed issues, but to briefly examine the ethos of the situation—a situation which illustrates many of the problems arising from industrial pressures on the submaritime fringe in Europe and North America.

As a recreational resource the Hunterston peninsula is of some value, though the original potential has been rather diminished by the presence of a large nuclear electricity generating station to the north-west of the area. On the other hand, the Hunterston promentary has a certain amenity value which is enhanced by the fact that most of the upper reaches of the accessible south bank of the Clyde estuary are already quite intensively developed. Within this rather narrow context it is of some ecological importance, but this is purely local, as it has neither been designated a Site of Special Scientific Interest (S.S.S.I.), nor have any local conservation organizations sought to acquire it. A few years previously, land at Ardmore Point, some 30 km to the north, had seemed threatened by a similar industrial project and was vigorously defended by the Scottish Wildlife Trust. But at Hunterston ecological interest has been mainly directed towards ensuring that the pollution associated with the new industry will be minimized, and detrimental effects on the surrounding land and estuary limited as much as possible.

Ecologically two aspects of this case-study are important. Firstly having achieved acceptable standards of impact on the environment, significant work can be carried out in monitoring changes in environment, habitats and biological populations. With recording of these data and detailed evaluation of their significance, there can be valuable feedback into future projects. Secondly the strategy of planning is worth closer study. Impact on the area has been considered in both the context of what is acceptable within the area, and what priority the area has for conservation in a regional and national context. Whilst minimizing the effects of industry is an axiom in all conservation work, many ecologists would agree that new projects in hitherto undeveloped areas are not merely inevitable but necessary. The manifest advantages of the submaritime fringe for commercial (and other) use and man's pressure on the zone, already acute in many areas, are likely to increase. In future developments, the role that the ecologist should play is to provide essential information, to state priorities, and to defend the most important sites by every means at his disposal including acting in a positive manner at the planning stage by suggesting which sites may be developed. Hunterston is not a priority site, but the ecologist has much to learn from the pattern of its evolution.

Conclusions

Most of the general principles in the case-studies discussed above can be applied to coastal zones in many countries. In France and Spain, many problems have developed in submaritime areas of the Costa Blanca, Costa Brava, Marseilles, and Rhône delta regions. The coastal lands of southern Europe are particularly vulnerable, since natural marine recycling processes have a limited capacity for handling pollution, the landlocked Mediterranean Sea acting as a virtually closed system. Severe problems have been encountered in the extensive coastal zones bordering the North Sea in the Low Countries, France, and Germany, where the estuaries of several of the great rivers of Europe provide some of the most favoured sites for industry and commerce in the world. The consequences have been loss of superb farmland, loss of amenity, and in some areas formidable pollution problems. Although some of the most prosperous communities in the world are found in these areas, the environmental and social costs paid in achieving this have considerably offset the monetary wealth generated in these coastal lands. In future, enterprises located in the submaritime fringe must be considered both in an ecological context and in a broad regional context, as well as conforming to much higher standards of emission control.

To sustain our heritage of coastal lands, the ecological perspective must be given a prominent role in future work, and those concerned with maintaining and conserving the submaritime fringe as a living habitat for important and interesting biological populations must involve themselves in planning. Few ecologists (but it is only fair to note that there are a few) are set against all development, and therefore positive planning must seek to locate projects in sites which are least important in a regional, national or even continental ecological context, to minimize the impact of development by establishment and enforcement of minimum standards in pollution control and similar practices, and to protect at all costs the most important sites. The case-studies quoted above show the progress which has already been made in building up experience and an armoury of techniques for the detailed management of the submaritime fringe when used for agriculture and recreation, and how these form part of an effective conservation programme. This vital work should be improved, extended and constantly fed back into practice wherever and whenever possible.

All of this demands that there must be a framework of approach to ecological management of the submaritime fringe. Due to the biological and human complexities of this zone and its use, such a framework must be a compromise between flexibility to the specific problem, and certain general elements which will enable the ecological case to be presented consistently and clearly. The approach, different facets of which have been illustrated by the above examples, can be summarized as follows. Firstly an inventory of existing conditions and the current nature of the ecosystems under review must be made. This need not be complex and should be broadly-based rather than deep, and its objective—the provision of an over-all description of the area—must constantly be borne in mind. Secondly there must be an appraisal of the area, including its priority for conservation in a wide spatial context, the vulnerability of its various communities and habitats to change, and types and resources of management available. These must form the basis for input of ecological information to the third component, the policy towards the area. This is the critical element, yet is the one most often absent. Policy towards an area must include the main strategies to be put forward to land owners, developers, local or central government, or any other groups involved in the land-use which will shape the ecological future of the area. For obvious reasons, human interest in the submaritime fringe is much greater than in other littoral zones, so that strategy in policies and communication of these is of fundamental importance. Fourthly from this a plan must be developed. If the policy is concerned with over all strategy, then planning can be envisaged as being concerned with the tactics of how this may be carried out. Specific techniques should be detailed and the entire emphasis of the plan should be directed towards practicality in operation, including the financial aspect. Fifthly management must be implemented. Although this sounds obvious, it may be more difficult to achieve than first appears since essential infrastructure or administrative capacity may be absent, and before work can commence such facilities must be created. At this final stage the ecologist should not relinquish his interest, but make sure that this work is not wasted by weak implementation. From this too will come the essential feedback without which advances in the care and maintenance of the submaritime fringe will be hard won in the future.

PART IV

CONCLUSION

Chapter 15

Coastline management:
Some general comments on management
plans and visitor surveys

MICHAEL B. USHER

INTRODUCTION

The coastline, as has been noted in the preface, is not a single habitat but rather a heterogeneous mixture of habitats ranging from horizontal structures such as mud-flats through various degrees of slope to vertical structures such as rocky cliffs. Even within such a heterogeneous collection a further division may be recognized on the basis of the proportion of time that the habitat is flooded—a *Zostera* (eelgrass) bed is flooded for nearly, if not completely, 100% of the time whilst the *Festuca* (fescue) grassland with *Scilla verna* (spring squill) is never likely to be submerged though it might be subjected to salt-spray. Furthermore, the chapters dealing with salt-marshes have shown that the plant composition is likely to be related to the time that the particular area is flooded. This relation is partly of a physical nature, since some plants are more efficient at trapping water-borne debris whilst others would be killed by the continual deposition of mud on them. It might be noted that *Spartina* spp (cordgrasses), plants of the frequently inundated communities, have stiffly erect leaves whilst many plant species of less frequently inundated communities have leaves held in a more horizontal direction.

The habitats of the coastline are thus extremely varied, and ecologically they are often in a state of flux. With fully terrestrial habitats the concept of the climax or edaphic climax community is well established. It is perhaps no accident that many studies on the successional process have used habitats of the coastline, the end point of the succession often being that community furthest from the sea, subject, of course, to this community not being influenced by human activity such as agriculture or forestry. Management of the coastline is thus often going to be concerned with management of seral communities. Before, however, investigating any techniques, it is pertinent to ask if these communities have any special ecological characteristics. Two such characteristics are, perhaps, obvious.

First, there is an ecology of catastrophes. In many chapters the processes of erosion or accretion have been referred to—the territorial limits of one community may be decreasing, allowing more salt-spray to reach communities that were previously beyond the splash zone and hence climax or near-climax communities will find themselves very close to the sea. Alternatively, the territorial limits of another community might increase as sand is deposited in front of sand-dunes or as more fine particles are trapped at the seaward margin of a salt-marsh. Not only are these large-scale processes constantly occurring but there are less spectacular catastrophes. Any cliff, either of hard or soft parent material, is in a state of constant change, the crevices and sheltered ledges are frequently changing and these can provide habitats for such a species as *Asplenium marinum* (sea spleenwort) on the west coast of Scotland or nesting sites for *Delichon urbica* (house martin) in Kincardineshire and Berwickshire, Scotland.

Secondly, many of the coastal communities are species-poor. Certainly, when walking out on mud-flats the dominance of *Spartina* spp is striking—indeed the community is virtually a monoculture. There are a few animal species in the *Spartina* community, but it is evident that the fauna of salt-marshes has been less fully documented than the flora (Ranwell, 1972a, p. 111). Perhaps, unfortunately, it is almost a truism that where wildlife management is concerned the maxim is 'if you look after the plant community the animal community will look after itself', with the result that research has concentrated on the management of the plants. However, even a casual observer would conclude that there is some correlation between the number of species of plants and the number of species of animals. But, whichever groups are studied, it is apparent that there is a general rule that the further one goes (up to a maximum of 100–200 m) from the actual land–sea margin the greater the species diversity. The study of coastal ecosystems has, of course, been very much a contributory factor to the theory that ecological diversity increases as the successional process proceeds.

In thinking of the management of a parcel of coastal land, three ecological questions must be considered:

(1) At what stage in the ecological succession is the parcel of land that is to be managed?
(2) What are the dynamics of the present species diversity—are some species expected to become locally extinct (relics of previous seral communities), or are others overdue in their colonization?
(3) What are the chances of a catastrophe?

As well as these ecological questions there is going to be a multitude of sociological and socio-ecological questions dealing with a definition of objectives, attempting to define what sort of harvest the parcel of land is expected to yield, and the general relations between 'man' and the 'natural' species.

To anyone who is concerned with management of natural or semi-natural coastline communities, the basis is often the *management plan*, a confidential document setting forth the management policy and the methods recommended for achieving the manager's aims. Such documents are discussed in the next section. All too often someone or some authority is going to question the management policy asking 'Is it worth it?' In our society most comparisons on land-use are based on economic considerations—how much money will agricultural and forestry development yield, how many houses can be built or how many jobs will be created if such and such a development takes place ? If land is being reclaimed from the sea for agricultural use, then economic arguments can be used provided that some discounting percentage can be agreed to equate the costs of reclamation work today with the benefits of agriculture in the future. However, if a natural and semi-natural coastline is to be conserved, is there any economic value? Some approaches to this problem will be considered in the third section.

THE MANAGEMENT PLAN

A *pro forma* for a management plan

The *pro forma* that follows is very similar to those given by Stamp (1969) and Usher (1973), and it closely follows the pattern that has been used by the Nature Conservancy (now the Nature Conservancy Council). This *pro forma* has been adapted to the coastal situation, introducing some features which are particularly appropriate to ecosystems of this type.

Introduction

This normally sets the area to be managed into its geographical situation, and compares it with series of similar sites elsewhere along the coast and with terrestrial habitats elsewhere in the country or vicinity.

Chapter 1. *Name and general information*

(a) Name
(b) Location
(c) Brief description
(d) Area and boundaries
(e) Access
(f) History of establishment
(g) Bye-laws
(h) Permits
(i) Grid references
(j) Maps
(k) Collections of museum material
(l) Collections of photographs

Chapter 2. *Reasons for establishment*

(a) General. Reasons associated with the over-all management policy of the responsible authority.
(b) Specific. Reasons associated with the establishment and management of this particular area of land.

Chapter 3. *Surveys and scientific information*

(a) Topography
(b) Drainage and hydrological regime
(c) Characteristics of erosion and accretion processes
(d) Geology
(e) Climate
(f) Soils
(g) Vegetation
(h) Fauna
(i) Land-use history
(j) History of catastrophes
(k) Archaeology and ancient monuments
(l) Research projects
(m) Public and recreational interest
(n) Statutory bodies
(o) Sporting rights
(p) Pest control

Chapter 4. *Aims of management*

The major aim of a nature reserve will usually be wildlife conservation, with subsidiary aims such as education, research, recreation, and amenity. It may be necessary to specify short-term aims of public relations where there is hostility to the form of management, or of legal changes in boundaries, leases, etc. In coastal areas the subsidiary aims might become as important as the conservational aim: recreation might form an important aspect of management, and proper planning and control of recreational facilities can considerably enhance the conservation potential of a site. Alternatively, in areas of pasture and reclaimed land the establishment of suitable agricultural practice may be of prime importance, as for example in machair grasslands (Knox, 1974).

Chapter 5.   Management programme

   (a) Scientific management. An interpretation of the aims of management into practice.
   (b) Estate management. Details of the maintenance of the estate; e.g., buildings, sea-defences, fences, paths, bridges, walls, and signs. Conservationally desirable structures such as information points, hides, shelters, nest-boxes, etc., would also be included in this section.

Chapter 6.   Public access

   (a) Bye-laws
   (b) Public rights of way and access to the shore
   (c) Permits
   (d) Any other rights or privileges

Chapter 7.   Wardening

Details of the present policies regarding the use of full- or part-time wardens, and the likely demand for, or changes in, wardening facilities in the future.

Chapter 8.   Time schedule and finance

The items on the management programme will need to be costed, and in the light of any limitation on financial resources a time schedule with priorities will be drawn up.

Chapter 9.   Division of responsibilities

The responsibility for implementing the whole plan or specific parts of it will be detailed.

Chapter 10.   Advice and records

Structure of committees responsible for management decisions and advice, finance, etc. Records will include periodic or progress reports, research activity designed to monitor the effects of management practice, and surveys of rates of erosion, accretion, etc. It is also important to record detailed statements of management decisions that have been taken, paying particular attention to recording the reasons why all decisions were made.

Chapter 11.   Renewal, authorship of the plan and references

A date will usually be specified for the renewal of the plan. The references might include all literature quoted in the plan as well as a bibliography of the site itself. Authorship of each section (where the plan has been written by several people) should be given. Acknowledgements may also be included.

Appendices

Lists of plant species; lists of animal species; map(s) showing items detailed in the scientific management programme; map(s) showing items detailed in the estate management programme.

Many other items could be included in the appendices, such as copies of legal documents relating to the land; tables of income and expenditure; short reports on research, etc., too bulky for inclusion in the body of the plan; charts of animal habitats; details of soil profiles; weekly, monthly, or yearly records of physical processes like erosion; etc.

Some general considerations

The *pro forma* for a management plan can be seen to consist of four parts:

(1) Descriptive (introduction and chapters 1–3)
(2) A statement of aims (chapter 4)
(3) Prescriptive (chapters 5–11)
(4) Collections of data (appendices).

In order to illustrate some of the general points of preparing such plans for coastal areas, reference will be made to Spurn Nature Reserve. Spurn is a narrow peninsula, about 5 km in length, situated in the mouth of the River Humber on the east coast of England (approximately 53° 34′ N, 0°07′ E). This thin peninsula, covering a land area of only about 130 ha (see Figure 92), is owned by the Yorkshire Naturalists' Trust Ltd. and managed primarily as a nature reserve. However, besides being an exceptionally well documented coastal feature (its history is reviewed by de Boer, 1964) and being renowned for the diversity of its wildlife (Elton, 1966), its very geographical position makes it an important area for recreation (Usher *et al.*, 1974). In terms of catastrophes, de Boer (1964) has shown that there is a 250-year cycle with the peninsula being built-up, breached, almost completely washed away and then reformed several hundred metres west of its previous location.

The descriptive part of a management plan is perhaps the simplest of the four parts to compile, especially when the area is well-known scientifically. For Spurn there is a wealth of published material, mostly in *The Naturalist*, with papers and notes appearing virtually every year from the 1880s till the present time and no doubt continuing into the future. However, for most coastal sites there will not be this quantity of published information, and simple ecological survey work will be required to determine the most important components of the soil and vegetation. Regional solid and drift geological and soil survey publications (reports, handbooks and maps) will probably indicate the salient features of the geology and pedology respectively. The Meteorological Office maintains records of climatic data, and, in Britain at least, it is likely that climatic records will be available for the actual or a neighbouring area. The fauna might prove more difficult if there are no published accounts, but a lot of qualitative information can often be obtained from amateur naturalists living locally. This is particularly true for birds since ornithologists often maintain notebooks recording birds sighted, nests, etc. Data on mammals, reptiles, amphibians and higher plants can often be gained from such sources; but information is often sparse for the lower plants and most groups of invertebrates. However, there is in Britain a number of nature reserves for which there is no faunal list.

The coastline, perhaps more than any other terrestrial ecosystem, is subjected to a wide variety of human influences (Figure 93), and the management plan will endeavour to find

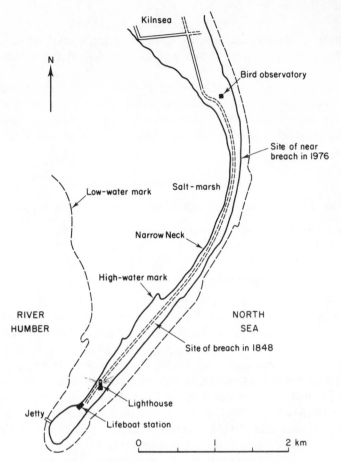

Figure 92. A plan of Spurn Nature Reserve, indicating the
location of features mentioned in the text

Figure 93. A view of the distal kilometre of Spurn Peninsula, showing the lighthouse,
jetty and some of the buildings erected during the former military occupation of the
peninsula

and recommend ways of avoiding conflicts between these influences. Parties interested in Spurn are shown in Table 20. It will be seen that some of these parties could be expected to have interests in any terrestrial habitat (for example the County, District and Local Councils and the Nature Conservancy Council). However, there are several authorities that are concerned solely with the coastal environment (such as H.M. Coastguards, Trinity

Table 20   A list of parties interested in Spurn Nature Reserve, a reserve owned by the Yorkshire Naturalists' Trust Ltd. (all parties are given in alphabetical order)

General, conservational and wildlife interests (of a continuing nature)

British Transport Docks Board	Leaseholder—jetty, navigation beacon and radio navigational station. Pilotage.
British Trust for Ornithology	Annual bird ringing course
Countryside Commission	Designation of 'Heritage Coast' in conjunction with County Council
Department of the Environment	Feasibility study of reclamation of Spurn Bight
H.M. Coastguards	Left Spurn in 1965, but now returned
Humberside County Council	Formerly East Riding County Council. Matters dealing with planning, oil pollution, etc.
Nature Conservancy Council	Spurn is an S.S.S.I. Consultation on planning matters
Oil Pollution Prevention Centre	At Immingham. A liaison body.
Ordnance Survey	Triangulation Point
Rothamsted Experimental Station	Moth trap
Royal National Lifeboat Institution	Lifeboat station established in 1810. A full-time crew lives at Spurn
Trinity House	Leaseholder—lighthouse and surrounding ground
Universities of Hull and Leeds and Hull College of Education	Leaseholders—buildings for use as field centres
Yorkshire Naturalists' Union	Leaseholder—of buildings and land for Spurn Bird Observatory, sea-watch huts and Heligoland traps

Other interests, interests of only a short-term nature, and essential services

Central Electricity Generating Board
Charitable Trusts
Conservation Corps (both national and local based on Hull)
Council for the Preservation of Rural England
Easington Parish Council
East End Sea Angling Club
East Yorkshire Conservation Council
East Yorkshire Local History Society
General Post Office
H.M. Customs and Excise
Holderness District Council
Hull Water Department
Ministry of Defence (including the Army—Northern Command—and Navy)
National Farmers Union
Royal Society for the Prevention of Cruelty to Animals (R.S.P.C.A.)
Social Science Research Council
Yorkshire Electricity Board
Yorkshire Insurance Company
Yorkshire Water Authority

House, R.N.L.I., and the British Transport Docks Board). An assessment and understanding of the role of all these authorities is an important component of the descriptive part of the management plan.

This descriptive part really establishes the basis on which management proposals can be formulated. Thus, although the primary aim of management is likely to follow closely the main reason for establishing management in the first instance, it will be tempered with constraints that themselves might be secondary aims. Thus, in the published plan of Aberlady Bay Nature Reserve (Usher, 1967), although the primary aim was wildlife conservation, the constraint was that a local authority was managing the area and had, in a manner, to account to its ratepayers. The reserve, although fulfilling a conservational function, was thus also fulfilling recreational and educational roles for the community living within the area. In theory, Spurn is owned by a Trust devoted to conservation, and thus the constraint of accountability to financial sponsors is removed. However, other management constraints are imposed, perhaps the most important being public opinion. Although on strictly conservational grounds it might be beneficial to close Spurn to all but the essential users, yet such action would almost certainly lose public support for a measure of conservation management for this habitat and might in the long run lead to the loss of an area that is acknowledged to be conservationally important (Countryside Commission, 1969).

The aims of management thus endeavour to set realistic guidelines for management prescriptions. Ranwell (1972a) has reviewed such uses as golf links, forestry, sand extraction, recreation, and conservation. In his concluding section he considers multiple use, the means whereby at least two forms of use are planned side by side. In a world with a more or less constant length of coast and an increasing population and an increasing amount of leisure time, it is important that the aims of management should always consider multiple usage, either for the present or the future. It is very often true that the 'aims of management' chapter is the most difficult to write, but the most important to prepare thoroughly.

The third part of the plan is prescriptive, dealing with the practical implementation of the aims. In general there are four influences to consider:

(1) Scientific management
(2) Estate management
(3) Financial implications
(4) People.

The emphasis placed upon each of these four influences will vary from coast to coast, but if the coast is to be managed as a golf course influences (2) and (3) might be most important, whilst for a coastal nature reserve influences (1) and (4) are of importance. Influence (4) works in two ways, both of planning the needs of people visiting the coast and in specifying responsibility for management or certain aspects of management. These influences can best be considered by looking at some examples of management in practice.

Some examples of planning

Five examples from coastal areas of Britain will be used to illustrate the relation between the aims of management and planning as shown in the third section of management plans. Rather frequently there are insufficient scientific data available for precise management prescriptions to be formulated, and hence it is frequent for a management plan to endeavour to establish more data. Such is the case at Tentsmuir National Nature Reserve, situated on the south side of the Firth of Tay in Eastern Scotland. One of the aims of management is: 'To measure coastal accretion or erosion and investigate the reasons for it' (Tentsmuir Management Plan, 1963). Although much is already known about accretion on the east-facing shore, the reasons for it are not clearly understood. A survey of coastal changes in Chapter 3 of the Tentsmuir management plan indicates that sediment brought down by the River Tay is one factor, but perhaps more important are the movements of banks of sand below high water. However, the continuation of the accretion process is important in maintaining the series of plant communities in the succession from bare sand to stable dune communities. Since relatively little is known about the accretion process, the management plan stipulates:

> An important long-term research project is the measurement of coastal accretion and study of shore line changes. The topographical survey provides a basis for future physiographical studies, which will include:
>
> (a) Continued monthly measurement of the sand level against two lines of wooden posts erected in 1965 across the foreshore at the north end of the Reserve at right angles to the high-water mark.
> (b) Biannual levelling across the foreshore from the concrete controls of the trigonometrical survey. These lines will also be levelled after periods of easterly gales.
> (c) Aerial photography of the Reserve at low-tide springs to plot long-term changes of foreshore and sandbanks.

A second example relates to the salt-marsh area of Newborough Warren National Nature Reserve in Anglesey, North Wales. This reserve has a wide range of habitats, in which grow several rare species of plants. Some of the aims of management are concerned with the maintenance of this diversity of habitats. Among the conservation objects stated in the Newborough Warren management plan (1964) are:

> (i) To promote the natural succession of the plant and animal communities characteristic of the various types of habitat of the reserve.
> (ii) To maintain representative examples of the present range of habitats and their associated flora and fauna.

The salt-marsh is situated in the estuary of the Afon Cefni. The zonation of salt-marsh plants is well-known, species known from Newborough Warren being listed in one of the appendices to the plan. The management prescription takes account of three factors: knowledge of the existing vegetation, the experimental artificial pool built to encourage feeding shore birds, and the unknown consequences of colonization by an 'alien' grass *Spartina* × *townsendii*.
The prescription is:

> This area is the largest area of salt-marsh on any existing or proposed reserve in North Wales. In 1958, it was invaded at the seaward end of the estuary by *Spartina townsendii*. Colonization by this species is limited as yet but, following further study, the decision will have to be taken whether to attempt its control or let the invasion take the natural course with the risk that the natural salt-marsh community will be entirely replaced by a *Spartina* sward.
>
> The artificial pool created on the salt-marsh in 1958–59 has proved successful in increasing the density of birds in the area, but periodic clearing of the developing hydroseral vegetation will have to be undertaken if the area of water is not to be greatly reduced.

Scientific management is thus a blend between gathering further data and either undertaking tasks or managing by 'controlled neglect' based on data already available. People, however, generally require a greater degree of planning since facilities have to be established. At Barns Ness on the East Lothian coast, on the east coast of Scotland, the County Council was given 4 km of coastline. Although there is no formal management plan in the sense previously described, the County Council states:

> 'The development of these lands is a good example of co-operation between governmental, commercial, and voluntary bodies of multi-purpose uses.' (Barns Ness, undated).

The booklet on the area shows that several definite management practices have taken place to cater for the human influences on these coastal ecosystems. The main planning decisions, following the establishment of the area, were:

(1) The County Council obtained a 'Countryside Grant' to provide water and drainage;
(2) Together with the Camping Club of Great Britain and Ireland, the County Council has established a camp site;
(3) Together with the Scottish Wildlife Trust, the County Council has set aside a wildlife area which is a resting place for migrant birds attracted to the lighthouse and which provides winter grazing for sheep;
(4) A trail has been established, mainly to demonstrate the geology of the area, although the booklet also includes details of the plant and animal life.

The aim of this area is essentially multi-usage of a coastal strip situated within one hour's travelling time of a large centre of population (Edinburgh). Although the aims of management have not been specified nor prescriptions drawn up in the form of a plan, the published material relates to the educational values of these habitats. This is not the place to give a guide to the preparation of nature trails: however, introductory material for trails can be found in *Projects for Environmental Studies* (1970) and Usher (1973).

Recreational planning can be as important as educational planning. Indeed, Barns Ness described above provides not just for a wildlife sanctuary and a trail, but also for a camping site. Burton (1967, 1971) has drawn attention to the different needs of 'overnight' as opposed to 'day-visitors'. Most managers of coastal ecosystems are likely to be faced with day-visitors, as at Spurn. On this reserve, one of the aims of management is:

(a) To provide such amenities for visitors as are compatible with the foregoing (conservation and research) aims.
(b) To conserve and if possible enhance the appearance of Spurn as landscape by preventing inappropriate development or activities, and by removing as far as possible and desirable the derelict buildings left from former military occupation (Spurn Management Plan, 1976).

The management plan for this coastal area has thus catered for the recreational interest. The use of the reserve and the interests in various recreational pursuits have been surveyed by Usher *et al.* (1974). This survey showed that a majority of visitors came for picnicking (29%), sunbathing (30%), walking (65%), or swimming (16%) whereas a minority was interested in birdwatching (23%) or other natural history studies (15%). Of those interviewed, 82% indicated that 'peace and quiet' was one of the attractions of this habitat and 51% mentioned the sandy beaches, while only 38% indicated that the wild birds were an attraction (Figure 94).

Figure 94. The entrance to a Heligoland trap. The study of birds was one of the main
reasons for the establishment of Spurn Nature Reserve

The level of public recreational interest in Spurn has thus been determined. The
management plan prescribes ways of integrating this recreational use with wildlife
conservation. Perhaps most importantly the recreational interest is associated with 'peace
and quiet' and hence conservation and recreation are likely to be compatible with each
other. Zonation of recreational use is likely to be used in this context: survey work has
established which sections of the 10 km of shoreline are most suited to recreation, and
records of natural historians have indicated the areas on the peninsula most suited to
wildlife or harbouring the rarer species. Management will thus be concerned with
providing car parking facilities in the areas most appropriate for recreational use and
making it difficult or impossible to park a car elsewhere. More data are, however, required
to assess the 'carrying capacity' of the recreational zones: when this is known an effective
limit can be placed on the number of cars entering the reserve area. It would clearly be
beneficial to adjust car parks to the appropriate carrying capacity of the beaches rather
than to limit the use of the beaches by the size of existing car parks. In any case, the location
of car parks will channel recreational use into various areas which can be watched for signs
of erosion damage. If this occurs the appropriate action, such as the provision of walkways
(Countryside Commission, 1969; Ranwell, 1972a) or the creation of alternative paths, can
be taken.

These examples have concentrated on the scientific, educational and recreational use of
coastal ecosystems and of integrated forms of management between two of these forms of
land-use. However, these uses occupy only a proportion of the British coastline (Countryside Commission, 1969): agricultural use is perhaps the most important. A fifth
example of management, where both conservation and agriculture are important, concerns a community that is confined to the north of Britain, the sand-dune machair. Again it
is unlikely that any of the machair areas have formal management plans, but some general
points relating to their management are discussed by several authors in a volume edited by
Ranwell (1974). Machair is a community resembling a dune pasture and, due to the high

shell content of dunes and blown sand in the west and north of Scotland, it is usually calcareous. It is floristically rich, containing many species characteristic of calcareous grasslands (Gimingham, 1974) and is usually subjected to management by grazing.

Management essentially follows traditional lines of balancing exploitation and preservation of the habitat. Over-grazing, as well as other factors such as rabbits, can lead to severe wind erosion (Knox, 1974). Ploughing is carried out late in the spring to minimize the risk of erosion. Arable crops, mostly oats, are grown for a period of three years and then the area is sown out to grass for a long 'fallow' period. Thus, although the machair is used for both grazing and crops, there are two factors to be balanced in considering its exploitation: grazing pressure should be sufficiently light to preserve the vegetation cover of the sandy soil and cropping on any area should be for sufficiently short a period to minimize erosion and loss of fertility. This latter is also important since the machair receives approximately 1,300 mm of rain per year. Traditionally the machair has been fertilized with seaweed.

These examples, which have been drawn fairly randomly and are not exceptional in relation to much of the British coast, demonstrate clearly an essential feature of management of the majority of coastal communities: how does one conserve the community whilst at the same time use it? A balance will always have to be struck between conservation and utilization, and indeed, if this book does no more, the means of finding this balance for a whole series of coastal communities have been indicated in the preceding chapters.

THE VISITOR SURVEY

The Questionnaire

People, as has been seen in the previous section, are an important aspect of the management of coastal systems. The length of coast available to people is constant or nearly so, and hence as human populations have increased, as people have become more mobile with the general availability of the car, and as the amount of leisure time has increased, the human pressures on a limited length of coast have also increased. If integrated forms of management are required then it is desirable to have data not only on the ecology of the environment but also on the sociology of the people using that environment. Of necessity, this book has been concerned with the ecology and applied ecology of coastal systems, but in concluding it we should also think of the users.

There are several methods of estimating the view and reactions of the user. It is, for example, possible to take the view that everything is all right if people use the area—in a free society they could have gone somewhere else instead. Alternatively, it is possible to react to every criticism that is voiced either directly to the manager or indirectly via the media (e.g. letters in newspapers). Furthermore, one could rely solely on one's own reactions and those of a management committee if such a committee exists. All these are subjective assessments of the user. Perhaps, it is better to take a sample of users and to ask a few questions (assuming that the user is prepared to spend a few minutes giving replies). This is the questionnaire approach, and provides data on which some statistical comparisons can be made.

The design of the questionnaire is not as simple as it might appear: a general discussion on the subject is given by Burton (1971). A few of the essential considerations in designing a questionnaire are:

(1) Determine beforehand what you want to know, and then ask a question that does just that.

(2) Consider the length of the questionnaire since one that is too long might lose the interest of the interviewee. A short questionnaire is more economical of the interviewer's time, since more short questionnaires can be completed per day than long ones.

(3) Consider carefully the use of 'open' and 'closed' forms of questions. The 'closed' question requires an answer of the 'yes' or 'no' form or for the interviewee to select one or two answers from a list of possible answers—this is easy for both interviewee and the interviewer to record. The 'open' question requires the interviewee to think out an answer, and takes time for the interviewer to record (it is also much more difficult to subject such data to statistical analysis after the survey has been completed).

(4) Consider the desirability of using personal questions. The majority of interviewees do not mind answering questions about what activities they have been doing or whether they would use a nature trail; far fewer would wish to divulge their income or the socio-economic groups to which they belong.

(5) Consider the use of 'repeat' questions. At times it might be inappropriate to ask essentially the same question in two different ways. However, if one wants to make a check on the reliability of answers it might be useful to have these 'repeat' questions, making sure that the original and repeat questions are at different points in the questionnaire.

(6) Consider all sources of bias. 'Closed' questions, particularly, can introduce some bias. The showing of a card with several possible replies might have a bias towards replies higher up the list since these are the ones first seen by the interviewee. Hence, several cards with a random arrangement of replies can be useful. Also, as the questionnaire is carried out *in situ* on the coastal system, the replies are biased since you are questioning only those people who have come—what about the views of the many people who have not come? It might then also be appropriate to question a sample of people in the catchment area of the coastal system.

(7) Is it desirable to have a pilot survey? Often, a draft questionnaire can be tried out during a short pilot survey, and almost always such a pilot survey will demonstrate faults in the draft questionnaire. If there is time to run a pilot survey, it is probably always desirable to do so.

The analysis of survey data

A survey of the users of Spurn was carried out in the summer of 1970. The aims were threefold: to determine the recreational pattern of the visitors; to determine the interest in natural history in general; and to collect data for the economic analysis of demand for that recreational resource. The first two of these are discussed here, the third is left till the next section. The questionnaire used in the Spurn survey is shown by Usher (1973) and some of the recreational results are discussed in Usher *et al.* (1974).

Visitors were handed a card and asked the 'closed' question: 'Could you tell me which of these activities you have done while you were here today?' Following this question the interviewee was asked: 'Which of these would you say was the main object of your visit?' The replies to this question are given in Table 21. This Table demonstrates two facets of the statistical treatment of the data. First, it is far simpler for comparisons if the data is changed to a percentage basis. One of the main uses of these surveys is to repeat them periodically and to determine trends in recreational patterns. Surveys at different times or in different places are unlikely to be the same size, and hence percentage data is more easily compared. Secondly, it is often useful to know the accuracy of your data, so that in

Table 21   The main reason for visiting Spurn Nature Reserve

Reason	Number of interviewees	Percentage of responding groups	Approximate 95% confidence intervals
Picnicking	139	11	9·2–13·1
Birdwatching	124	10	8·3–12·0
Other nature studies	29	2	1·4–3·1
Swimming	26	2	1·3–3·0
Sunbathing	69	6	4·7–7·6
Walking	161	13	11·1–15·2
Fishing	95	8	6·0–10·4
Peace and quiet	75	6	4·7–7·6
Curiosity	331	27	24·3–29·8
Other	176	14	12·0–16·2

making comparisons statistically significant differences can be determined. The percentages given in Table 21 have 95% confidence limits, although these limits are not easy to calculate (Sokal and Rohlf, 1969). They are based on the binomial distribution, which, if in a sample size of n there is a probability of a particular outcome of p (the probability of all other outcomes is $q = 1 - p$), predicts a mean of np outcomes of the particular type with a variance of npq. Fortunately, tables have been drawn up of the confidence intervals (Rohlf and Sokal, 1969). These two statistics, the percentage and its confidence interval, are thus the basic descriptive statistics for presenting the results of a single question. However, it is often instructive to compare the results of a pair of questions. A hypothetical case is shown in Table 22 where one has the replies to two closed questions, which is appropriate for a χ^2 contingency-table analysis (most elementary books on statistics include such an analysis). A χ^2 calculated for the data in the table is 27·843, a value larger than the tabulated values with two-degrees of freedom. Inspection of the Table shows that there is an association between membership and an interest in natural history. In a χ^2 analysis the data are actual counts of people and it is inappropriate to use percentage data.

Table 22   Results of two 'closed' questions: (1) Are you a member of the organization responsible for this coastal system (yes, no)? (2) Why have you visited the areas today (picknicking, natural history, other)? The numbers in brackets in the Table represent expected frequencies given the null hypothesis of no association

Question (2)	Question (1) Yes	No	Total
Picnicking	21 (31·5)	189 (178·5)	210
Natural History	14 (4·5)	16 (25·5)	30
Other	10 (9·0)	50 (51·0)	60
Total	45	255	300

Such analyses can thus be used to present the data of single questions or to relate questions in surveys or to relate similar questions between surveys. The Spurn survey found in general that a relatively small proportion of people visited the nature reserve for natural history reasons and that the majority were recreational users wanting to use the beach (Figure 95). The areas of greatest recreational potential were identified in the survey as were other areas on the reserve of virtually no recreational potential, as for example the salt-marsh (Figure 96).

Figure 95. A view of Spurn Peninsula. The photograph was taken looking north-wards from near the lighthouse, and the Narrow Neck region of the reserve can be seen near the skyline. The beaches, the main recreational attraction, can be seen particularly on the North Sea side of the peninsula

Figure 96. The salt-marsh situated on the Humber side of the Spurn Pensinsula: an area not frequented by the recreational user

'Demand' versus 'Management'

Having completed the survey work, one aspect of an analysis is the mathematical relationship between visitation rate and the cost of a visit. The visitation rate can be taken to equate to the 'demand', in an economic sense, for the coastal system. Generally, it has been found that a graph of the logarithm of the visitation rate ($\log_e V$) against the logarithm of the cost ($\log_e C$) gives a statistically satisfactory fit using a regression analysis technique

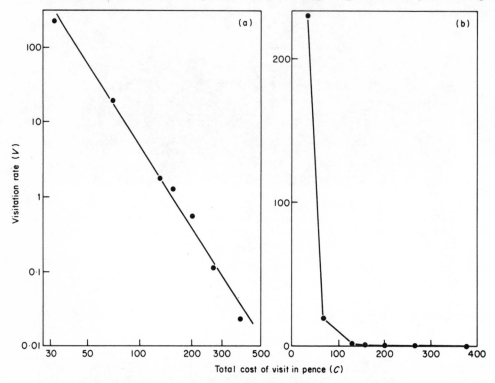

Figure 97. The relation between visitation rate (V, the number of visitors per thousand population of the zone) and the total cost of a visit (the travel cost plus a gate charge of 25p). (a) Both axes are logarithmic, and the line is the regression equation quoted in the text, equation (1); (b) The data plotted on arithmetic axes, showing the extremely steep nature of the 'demand curve'

Table 23   Basic data for analysis of the 'demand' for the recreational experience of Spurn. The zones have geographical meaning and are shown in Figure 98, and the costs include a gate charge (25p) and the estimated travel costs at 1970 levels. Data taken from Usher (1973)

Zone	Number of groups	Cost (p)
0	138	31·62
1	151	69·16
2	634	130·98
3	49	157·48
4	61	201·64
5	70	267·88
6 and further	114	403·07

[see Figure 97(a)]. Thus, using the data from the Spurn survey (Table 23), Usher (1973) derived the equation

$$\log_e V = 18 \cdot 261 - 3 \cdot 633 \log_e C \qquad (1)$$

or

$$V = 8 \cdot 522 \times 10^7 \times C^{-3 \cdot 633}$$

where V is the visitation rate from a zone expressed as numbers per 1,000 of that zone's population, C is the cost of a visit in pence and the zones are as shown in Figure 98. If this

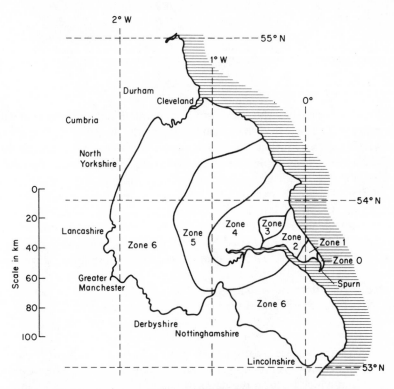

Figure 98. The location of Spurn Nature Reserve on the East Coast of England, showing the seven zones (numbered 0–6) that have been used in the analyses in the text

curve is drawn on arithmetic axes [Figure 97(b)], it will be seen that the demand curve never actually cuts either the ordinate or the abscissa. Thus, in any modelling exercise, however much the cost is, there will always be an arbitrarily small visitation rate, whereas in the real world there must be some point at which the visitation rate reaches zero. There can be three approaches to the resolution of this problem, one of which is here extended so that it has a less subjective basis.

First, it is possible to arrange an arbitrary cut-off point on the visitation rate. Thus, if the number of visitors from a zone is always rounded to the nearest integer, if this figure drops below 0·5 then the visitation rate is scored as zero. Using this approach, the predicted number of visitors and income with a 50p gate charge, using equation (1), is shown in Table 24, where it can be seen that the predicted number of visiting groups is 606 giving a predicted income of £303·00.

Table 24   The predicted number of visiting groups and income if the demand curve expressed by equation (1) is used where there is no cut-off on the visitation rate except that of rounding to the nearest whole number of visiting groups

Zone	Cost (p)	Population of zone	Predicted visitation rate	Predicted number of groups visiting	Predicted income (£)
0	56·62	600	36·472	22	11·00
1	94·16	7,800	5·747	45	22·50
2	155·98	366,000	0·9185	336	168·00
3	182·48	39,200	0·5194	20	10·00
4	226·64	108·800	0·2363	26	13·00
5	292·88	633,400	0·0931	59	29·50
6+	428·07	4,172,600	0·0234	98	49·00
Total	—	—	—	606	303·00

Secondly, a marginal cost can be fixed, and it is assumed that anyone faced with paying this sum or more will be deterred from visiting. Using the Spurn data, Usher (1973) and Usher *et al.* (1974) fixed a margin at £4·03 since this was the average cost of the 10% of most costly visits (see, for example, Trice and Wood, 1958). Assuming this margin and a gate charge of 50p, the predicted number of visitors and income using equation (1) is similar to that in Table 24, except that zone 6+ is now more expensive than the margin. Hence the visitation rate for this zone is zero, giving a predicted total of 508 visiting groups with an income of £254·00. This method suffers from the disadvantage that a margin cannot be exactly determined, and that any margin is likely to vary between socio-economic classes and hence it is inaccurate to treat the population as a homogeneous whole.

Thirdly, the actual demand equation can be adjusted so that it actually cuts the appropriate axis. Thus, Smith and Kavanagh (1969) fit the equation

$$\log (V+1) = a - k \log (C) \tag{2}$$

to their data on visitation rate and cost of a trout fishery in Grafham Water, England (where a and k are positive regression constants). They quote Knetsh's (1964) study of another reservoir in which he used the demand equation

$$\log (V+0·80) = a - k \log (C). \tag{3}$$

There appears to be no *a priori* reason for accepting 1 or 0·80 as suitable constants, and indeed these two examples are specific cases of the general demand equation

$$\log_e (V+\lambda) = a - k \log_e (C) \tag{4}$$

where λ is a positive non-zero constant which causes the equation to cut the C-axis when $V=0$ at C_{max}

$$C_{max} = (e^a/\lambda)^{1/k} \tag{5}$$

where C_{max} could be considered as the marginal cost at which the visitation rate becomes zero. The major problem is to determine an objective method of estimating the 'best' value of λ. One possible method is to choose λ such that it maximizes the variance ratio in testing the fit of the regression line (this is similar to methods discussed for transformations by Box and Cox, 1964). Using the Spurn data in Table 23, the variance ratios for several values of

Table 25 Estimating the value of λ in fitting a demand equation of the form $\log_e (V+\lambda) = a - k \log_e C$

λ	k	a	Variance ratio
0	3·6331	18·2611	455·7
0·04	3·3422	17·0333	1,550·5
0·043	3·3277	16·9725	1,560·1
0·05	3·2958	16·8396	1,529·0
0·25	2·8304	14·9294	250·0
0·5	2·5685	13·8874	122·1
1	2·2799	12·7720	69·9

Table 26 The predicted number of visiting groups and income when a constant (+0·043) has been included in the equation for the demand curve, equation (6). The number of visiting groups has also been rounded to the nearest whole number

Zone	Cost (p)	Population of zone	Predicted visitation rate	Predicted number of groups visiting	Predicted income (£)
0	56·62	600	34·451	21	10·50
1	94·16	7,800	6·305	49	24·50
2	155·98	366,000	1·141	417	208·50
3	182·48	39,200	0·6592	26	13·00
4	226·64	108,800	0·2984	32	16·00
5	292·88	633,400	0·1024	65	32·50
6+	428·07	4,172,600	0	0	0
Total	—	—	—	610	305·00

λ are shown in Table 25, where it will be seen that the 'best' value of λ is 0·043. Using this value and a gate charge of 50p, the predicted number of visitors and income are shown in Table 26. Unfortunately this rule of selecting a value of λ to maximize the variance ratio, which also maximizes the correlation coefficient between $\log_e (V+\lambda)$ and $\log_e C$, cannot always be applied. Thus, no positive value of λ could be found for the data given in Table 7 of Smith and Kavanagh's (1969) paper on visitation rates at Grafham Water, since all variance ratios with positive λ were less than that with λ equal to zero.

Having derived a mathematical model for relating the visitation rate or 'demand' to the cost of a visit, a demand schedule can be prepared (Smith and Kavanagh, 1969). Essentially this is a table setting out the predicted number of visitors for a large number of increasing, hypothetical 'gate' charges until there is some gate charge at which the predicted visitation rate reaches zero.

Two forms of mathematical relation have been used to analyse the Spurn data. First, the untransformed data have been used with equation (1) and a marginal cost of a visit has been assumed at £4·03. The second used the transformation of equation (4) before carrying out the regression analysis, and with the 'best' value of λ being 0·043 this gives

$$\log_e (V+0·043) = 16·972 - 3·328 \log_e C \qquad (6)$$

Demand schedules for these two equations are given in Table 27, and they are shown graphically in Figure 99. It might be possible to use the two regression equations to estimate the number of visitors if the gate charge were to be reduced to less than the 25p

Table 27   A Clawson demand schedule for Spurn Nature Reserve. At the time of the survey the actual gate charge was 25p per car entering the Reserve. The coefficients of the two equations are given in the text

Gate charge p	Estimated number of visitors from all zones, using	
	$\log_e V = a + b \log_e C$ equation (1)	$\log_e (V + \lambda) = a + b \log_e C$ equation (6)
(25)	(1,225)	(1,225)
30	1,056	1,108
40	794	829
50	508	610
75	296	357
100	187	222
125	128	143
150	91	91
175	50	56
200	36	36
250	19	14
300	0	2
350	0	0

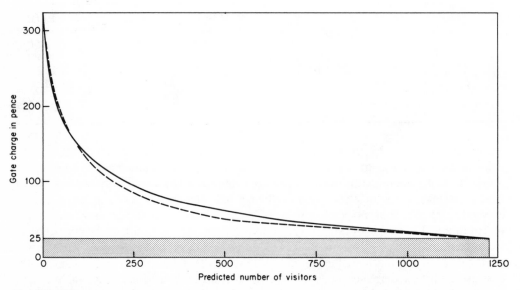

Figure 99. The Clawson demand schedule for a sample of 1,225 visiting groups. The stippled area represents the actual number of visitors paying a gate charge of 25p. The dashed line represents the estimated number of visitors if equation (1) is used together with a margin at £4·03; the continuous line represents estimates based on equation (6)

which was actually charged, but such an estimation would be an extrapolation in the most sensitive region of the demand curve [Figure 97(b)] and is probably unjustified.

The area under the demand curve gives an estimate of the benefits of the recreational experience. It will be seen from Figure 99 that these benefits have two components. First, there is the amount of money actually received at the gate to Spurn (£306, which is the same for either of the equations, and is based on 1,225 groups each paying 25p. This area is stippled in Figure 99). Second, there is the amount that visitors would have been prepared to pay over and above the 25p gate charge (estimated, from the area between the curve and the stippled area in Figure 99, as £499 when there is a marginal cost and £552 when *V* is transformed). These gives the total benefits to the sample group as £805 and £858 for the two methods respectively. Pitt (1971) indicated that whilst the sample of 1,225 visitors was being taken approximately 3,420 groups visited Spurn, and thus that the above data represent approximately 36% of the users during the months July to September inclusive. Taking the benefits as £858 for the sample, this indicates that all the recreational users had benefits of approximately £2,380 over the three-month summer period. During this period Spurn is more heavily used by recreational visitors than at other periods of the year, and hence this figure would suggest that the total recreational benefits are of the order of £5,000 per annum (at 1970 prices).

Such a study can set some sort of value on the benefits of the recreational or conservational experience of the coastal ecosystem. It does show the value to the community of maintaining such an environment, and a comparison of such values between several areas could form a basis for larger scale land-use or land-capability studies. However, such a study is particularly suited to a coastal system being used for recreation, since the more visitors the larger the estimated benefits will be. It is perhaps poorly adjusted to areas being managed for conservation where the benefits take a non-tangible form, such as peace and quiet or a lack of disturbance. For conservation areas it might be more suitable to estimate the benefits by counting the people who did not come to the area concerned.

CONCLUSIONS

In concluding both this final chapter and the book, one can reflect on the diversity of both organisms and systems that have been referred to in the preceding pages. One important point emerges throughout, and that is that the system must be considered as a whole. The coastal system includes the physical processes of the sea in modifying and moulding the system, the availability or excess of nutrients and energy, the autotrophic and hetero-trophic forms of life, and the impact of human activities. In the chapters of this book various coastal ecosystems have been described, but it seems that all too frequently not enough is known about the working of the whole system. Perhaps it is indicative that this 'handbook' has had to rely heavily upon the descriptive approach to so many of the systems: at this stage still not enough is known to quantity them. It is perhaps going to be one of the most spectacular developments of the next decade that more will be understood of the totality of the coastal systems and that this knowledge will find its way into the formulation of mathematical or simulation models that have a predictive value for the coastal manager.

This book summarizes the present position, showing the art of coastal management as it is now. Let us hope that this demonstrates where the weaknesses are, and that the next such handbook will be able to include a greater measure of precision in forecasting or predicting what management is required. The sea will, however, always ensure that complete precision is impossible, the ecology of catastrophes is likely always to remain with the coastal manager.

References

Adams, L. (1959) An analysis of a population of snowshoe hares in Northwestern Montana, *Ecol. Monogr.*, **29**, 141–170.

Adriani, M. J. (1958) Halophyten, in: *Handbuch der Pflanzenphysiologie*, ed. Ruhland, W., Springer, Berlin, 709–736.

Adriani, M. J., and Maarel, E. van der (1968) *Voorne in de Branding*, Stichting Wetenschappelijk Duinonderzoek, Osstvoorne.

Adriani, M. J., and Terwindt, J. H. J. (1974) *Sand Stabilization and Dune Building*, Rijkswaterstaat Communications No. 19, The Hague.

Alabaster, J. S. (1970) Testing the toxicity of effluents to fish, *Chem. Ind.*, **24**, 759–764.

Alexander, W. B., Southgate, B. A., and Bassindale, R. (1936) Summary of the Tees estuary investigations, *J. mar. biol. Ass. U.K.*, **20**, 717–724.

Allen, E. J., and Todd, R. A. (1902) The fauna of the Exe estuary, *J. mar. biol. Ass. U.K.*, **6**, 295–335.

Allen, S. E., Carlisle, A., White, E. J., and Evans, C. C. (1968) The plant nutrient content of rainwater, *J. Ecol.*, **56**, 497–504.

Anderson, R. R. (1969) Temperature and rooted aquatic plants, *Chesapeake Sci.*, **10**, 157–164.

Anderson, S. S. (1972) The ecology of Morecambe Bay, II. Intertidal invertebrates and factors affecting their distribution, *J. Appl. Ecol.*, **9**, 161–178.

Arber, M. A. (1940) The coastal landslips of south-east Devon, *Proc. Geol. Assoc.*, **51**, 257–271.

Arber, M. A. (1941), The coastal landslips of west Dorset, *Proc. Geol. Assoc.*, **52**, 273–283.

Ardré, F., Cabañas Ruesgas, F., Fischer-Piett, E., and Seoane, J. (1958) Petite contribution á une monographie bionomique de la Ria de Vigo, *Bull. Inst. Océanogr. Monaco*, **55**, 1–56.

Aubert, M., and Aubert, J. (1973) Pollutions marines et aménagement des rivages, *Rev. intern. Océanogr. med.*, **1973** (suppl.), 309 pp.

Backlund, H. O. (1945) Wrack fauna of Sweden and Finland, *Opusc. ent.*, **1945** (suppl.).

Bagnold, R. A. (1941) *The Physics of Blown Sand and Desert Dunes*, Methuen, London.

Baker, J. M. (1970) The effects of oil pollution and cleaning on salt marsh ecology, *Annu. Rep. Oil. Poll. Res. Unit, Orielton*, **1969**, 3–26.

Baker, J. M. (1971) Studies on salt-marsh communities. Contributions to the effects of oil spillage, in: *The Ecological Effects of Oil Pollution on Littoral Communities*, ed. Cowell, E. B., Institute of Petroleum, London, pp. 16–98.

Baker, J. M. (1973) Recovery of salt marsh vegetation from successive oil spillages, *Environ. Pollution*, **4**, 223–230.

Bakker, C., Nienhuis, P. H., and Wolff, W. J. (1973) *Biologische en milieuhygienische evaluatie van zoute bekkens in de afgesloten Oosterschelde*, Rapport Delta Inst. Hydrobiol. Res., Yerseke.

Bakker, D. (1958) The significance of nitrogen for the natural vegetation of the Yssel-lake polders (in Dutch), *Von Zee Tot Land*, **26**, 53–66.

Bakker, D. (1960a) A comparative life-history study of *Cirsium arvense* L. Scop. and *Tussilago farfara* L., the most troublesome weeds in the newly reclaimed polders of the former Zuiderzee, in: *The Biology of Weeds*, ed. Harper, J. L., Oxford University Press, Oxford.

Bakker, D. *Senecio congestus* (R. Br.) DC. in the Lake Yssel polders, *Acta Bot. Neerl.*, **9**, 235–259.

Bakker, D., and Zweep, W. van der (1957) Plant-migration studies near the former island of Urle in the Netherlands, *Acta Bot. Neerl.*, **6**, 60–73.

Ball, D. F., and Williams, W. M. (1974) Soil development on coastal dunes at Holkham, Norfolk, England, *Proc. 10th Int. Cong. Soil Science*, Moscow, VI, 380–386.
Bannister, P. (1966) The use of subjective estimates of cover abundance as the basis for ordination, *J. Ecol.*, **54**, 665–674.
Barkman, J. J., Doing, H., and Segal, S. (1964) Kritische Bemerkungen und Vorschlage zur quantitativen Vegetationsanalyse, *Acta Bot. Neerl.*, **13**, 394–419.
Barnes, H. (1959) *Oceanography and Marine Biology: A Book of Techniques*, Macmillan, London.
Barnes, R. S. K. (1973) The intertidal lamellibranchs of Southamption Water, with particular reference to *Cerastoderma edule* and *C. glaucum. Proceedings malacol. Soc. London*, **40**, 413–433.
Barnes, R. S. K. (1974) *Estuarine Biology*, Arnold, London.
Barnes, R. S. K., Coughlan, J., and Holmes, N. J. (1973) A preliminary survey of the macroscopic bottom fauna of The Solent, with particular reference to *Crepidula fornicata* and *Ostrea edulis*, *Proc. malacol. Soc. Lond.*, **40**, 253–275.
Barnes, R.·S. K., Dorey, A. E., and Little, C. (1971) An ecological study of a pool subject to varying salinity (Swanpool, Falmouth). An introductory account of the topography, fauna and flora, *J. Anim. Ecol.*, **40**, 709–734.
Barnes, R. S. K., and Green, J. (eds.) (1972) *The Estuarine Environment*, Applied Science, London.
Barnes, R. S. K., and Jones, J. M. (1974) Observations on the fauna and flora of reclaimed land near Calshot, Hampshire, *Proc. Hants. Field Club Archaeol. Soc.*, **29**, 81–91.
Barns Ness (undated) East Lothian County Council, Haddington.
Barrett, J. M., and Yonge, C. M. (1958) *Pocket Guide to the Sea Shore*, Collins, London.
Barrett, M. J. (1972) Predicting the effect of pollution in estuaries, *Proc. Roy. Soc. (B)*, **180**, 511–520.
Bassindale, R. (1938) The intertidal fauna of the Mersey estuary, *J. mar. biol. Ass. U.K.*, **23**, 83–98.
Bassindale, R. (1941) Studies on the biology of the Bristol Channel, IV. The invertebrate fauna of the Bristol Channel and Severn Estuary, *Proc. Bristol nat. Soc.*, **9**, 143–201.
Bassindale, R. (1943a) Studies on the biology of the Bristol Channel, XI. The physical environment and intertidal fauna of the southern shores of the Bristol Channel and Severn Estuary, *J. Ecol.*, **31**, 1–29.
Bassindale, R. (1943b) A comparison of the varying salinity conditions of the Tees and Severn estuaries, *J. Anim. Ecol.*, **12**, 1–10.
Battaglia, B. (1965) Advances and problems of ecological genetics in marine animals, in: *Genetics Today* Vol. 3, ed. Geerts, S. J., Pergamon Press, Oxford, pp. 451–462.
Bayfield, N. G. (1971a) A simple method for detecting variations in walker pressure laterally across paths, *J. Appl. Ecol.*, **8**, 533–535.
Bayfield, N. G. (1971b) in *The Scientific Management of Animal and Plant Communities for Conservation*, ed. Duffey, E., and Watts, A. S., Blackwell, Oxford, pp. 469–485.
Beadle, L. C. (1943) Osmotic regulation and the faunas of inland waters, *Biol. Rev.*, **18**, 172–183.
Beaufort, L. F. de (1954) *Veranderingen in de Flora en Fauna von de Zuiderzee na de Afluiting in 1932*, Uitgegeven door, de Ned. Dierk. Ver., Amsterdam, pp. 1–359.
Beeftink, W. G. (1965) De zoutvegetatie van ZW-Nederland beschouwd in Europees verband, *Meded. Landbouwhogeschool Wageningen*, **65**, 1–167.
Beeftink, W. G. (1966), Vegetation and habitat of the salt marshes and beach plains in the south-western part of the Netherlands, *Wentia*, **15**, 83–108.
Beeftink, W. G. (1968) Die Systematik der europaischen Salzpflanzengesellschaften, in: *Pflanzensoziologische Systematik*, ed. Tüxen, R., Junk, The Hague, pp. 239–263.
Beeftink, W. G. (1972) Übersicht über die Anzahl der Aufnahmen europäischer und nordafrikanischer Salzpflanzengesellschaften für das Projekt der Arbeitsgruppe für Datenverarbeitung, in: *Grundfragen und Methoden in der Pflanzensoziologie*, ed. Tüxen, R., Junk, The Hague, pp. 371–396.
Beeftink, W. G. (1973) Ecologie en vegetatie met betrekking tot het Deltaplan, in: *De Gouden Delta, Symposium Gent, 24th. November, 1971, Pudoc, Wageningen, pp. 81–109.
Beeftink, W. G. (1975a) The ecological significance of embankment and drainage with respect to the vegetation of the south-west Netherlands, *J. Ecol.*, **63**, 423–458.
Beeftink, W. G. (1975b) Vegetationskundliche Dauerquadratforschung auf periodisch überschwemmten und eingedeichten Salzböden im Südwesten der Niederlande, in: *Sukzessionsforschung*, ed. Schmidt, W., Cramer, Vaduz, pp. 567–578.
Beeftink, W. G., Daane, M. C., and Munck, W. de (1971), Tien jaar botanischoecologische verkenningen langs het Veerse Meer, *Natuur en Landschap*, **25**, 50–63.

Beeftink, W. G., Daane, M. C., Liere, J. M. van, and Nieuwenhuize, J. (1977) Analysis of estuarine soil gradients in salt marshes of the southwestern Netherlands with special reference to the Scheldt estuary, *Hydrobiologia*, **52**, 93–106.

Bellamy, D. J. (1968) The kelp project, *Triton*, **12**, 16–17.

Bellamy, D. J., John, D. M., Jones, D. J., Starkie, A., and Whittick, A. (1972) The place of ecological monitoring in the study of pollution of the marine environment, in: *Marine Pollution and Sea Life*, ed. Ruivo, M., FAO/Fishing News Books, London, pp. 421–425.

Bellan, G. (1972) Effects of an artificial stream on marine communities, *Mar. Poll. Bull.*, **3**, 74–77.

Bird, E. C. F. (1969) *Coasts*, M.I.T. Press, Massachusetts and London.

Bird, E. C. F. (1976) *Shoreline changes during the past century: a preliminary review*, Department of Geography, University of Melbourne.

Bird, E. C. F., and May, V. J. (1976) *Shoreline changes in the British Isles during the past century*, Division of Geography, Bournemouth College of Technology.

Birks, H. J. B. (1973) *Past and Present Vegetation of the Isle of Skye:Palaeoecological Study*, Cambridge University Press, London.

Black, C. A. (1965) *Methods of Soil Analysis*, University Press, Madison, Wisconsin.

Blom, C. W. P. M. (1976) Effects of trampling and soil compaction on the occurrence of some *Plantago* species in coastal sand-dunes, *Oecol. Plant.*, **11**, 225–241.

Blom, C. W. P. M., and Willems, R. J. B. M. (1971) De Westduinen op Goeree, *De Levende Natuur*, **74**, 219–225.

Blom, C. W. P. M., and Blom-Steinbusch, A. M. (1974) Een vegetatiekartering van het Natuur-monument 'Quackjeswater' in de duinen van Voorne, *De Levende Natuur*, **77**, 3–15.

Böcher, T. W. (1969), Stenstrandens plantverden, *Danmarks Natur*, **4**.

Boer, G. de (1964) Spurn Head: its history and evolution, *Trans. Inst. Br. Geogr.*, **34**, 71–89.

Boerboom, J. H. A. (1960) De plantengemeenschappen van de Wassenaarse duinen, *Meded. Landbouwhogeschool, Wageningen*, **60**, 1–135.

Boerboom, J. H. A. (1963), Het verband tussen bodem en vegetatie in de Wassenaarse duinen, *Boor en Spade*, **13**, 120–155.

Boerboom, J. H. A. (1964) Microclimatological observations in the Wassenaar dunes, *Meded. Landbouwhogeschool, Wageningen*, **64**, 1–28.

Boney, A. D. (1965) Aspects of the biology of the seaweeds of economic importance, *Adv. Mar. Biol.* , **3**, 105–253.

Bourne, W. R. P., and Dixon, T. J. (1974) The Sea-birds of the Shetlands, in: *The Natural Environment of Shetland*, ed. Goodier, R., Nature Conservancy Council, Edinburgh, pp. 130–144.

Bowden, K. F., and Sharaf el Din, S. H. (1966) Circulation, salinity and river discharge in the Mersey Estuary, *Geophys. J. Roy. Astr. Soc.*, **10**, 373–384.

Box, G. E. P., and Cox, D. R. (1964) An analysis of transformations, *J. R. Statist. Soc. (B)*, **26**, 211–252.

Boyce, S. G. (1954) The salt spray community, *Ecol. Monogr.*, **24**, 29–67.

Boyden, C. R., and Little, C. (1973) Faunal distribution in soft sediments of the Severn Estuary, *Estuarine coastal Mar. Sci.*, **1**, 203–223.

Brahtz, J. F. P. (ed.) (1972) *Coastal zone management: multiple use with conservation*, Wiley, New York and London.

Brattegard, T. (1966) The natural history of the Hardangerfjord, 7. Horizontal distribution of the fauna of rocky shores, *Sarsia*, **22**, 1–54.

Braun-Blanquet, J. (1964) *Pflanzensoziologie*, 3rd Ed., Springer, Wien.

Brehm, K. von (1971) *Seevogel–schutzgebiet Jauke-Haien-Koog*, Tier Und Umwelt, Heft, 617.

Brett, J. R. (1956) Some principles in the thermal requirements of fishes, *Quart. Rev. Biol.*, **31**, 75–87.

Brightmore, D., and White, P. H. F. (1963) *Lathyrus japonicus* Willd., *J. Ecol.*, **51**, 795–801.

Briquet, A. (1931) *Le littoral du nord de la France et son évolution morphologique*, Paris.

Brooke, H. (1967) *Hengistbury Head, Hampshire: some problems in the use of a public open space*, unpublished M.Sc. (Conservation) thesis, London University.

Brown, I. C. (1943) A rapid method of determining exchangeable hydrogen and total exchangeable bases in soils, *J. Soil Sci.*, **56**, 353–357.

Bruun, P. (1960) quoted by Christiansen, S., Wave power and the Djursland coast, *Guidebook Denmark*, International Geographical Conference, Copenhagen.

Buller, A. T., McManus, J., and Williams, D. J. A. (1971) *Investigations in the estuarine environments of the Tay. Physical aspects*, Tay Estuary Research Centre, Dundee University.

Burden, R. F., and Randerson, P. F. (1972) Quantitative studies on the effects of human trampling on vegetation as an aid to the management of semi-natural areas, *J. Appl. Ecol.*, **9**, 439–458.

Burnett, J. H. (ed.) (1964) *The Vegetation of Scotland*, Oliver and Boyd, Edinburgh.

Burton, T. L. (1967) The basic elements of planning for recreation, in: *The Biotic Effects of Public Pressures on the Environment*, ed. Duffey, E., Monks Wood Experimental Station Symposium No. 3, Nature Conservancy, pp. 156–162.

Burton, T. L. (1971) *Experiments in Recreation Research*, George Allen and Unwin, London.

Buxton, N. E., Gillham, R. M., and Pugh Thomas, M. (1976) Some aspects of the biology of the Dee estuary, in: *Problems of a Small Estuary*, ed. Nelson-Smith, A., and Bridges, E. M., University College of Swansea Institute of Marine Studies, pp. 9:3/1–9:3/12.

Cadwalladr, D. A., and Morley, J. V. (1973) Sheep grazing preferences on a saltings pasture and their significance for wigeon (*Anas penelope* L.) conservation, *J. Br. Grassland Soc.*, **28**, 235–242.

Cambers, G. (1976) Temporal scales in coastal erosion systems, *Trans. Inst. Br. Geogr.* (*N.S.*), **1**, 246–256.

Campbell, G. (1973), Options at Hunterston, *The Scotsman* (1/6/73), Edinburgh.

Carey, A. E., and Oliver, F. W. (1918) *Tidal Lands*, Blackie, London.

Carlson, R. E. (1972) Lakeshore physiography and use, *Shore Beach*, **40**, 3–14.

Carr, A. P. (1969) The growth of Orford spit: cartographic and historical evidence from the sixteenth century, *Geogr. J.*, **135**, 28–39.

Carr, A. P. (1970) The evolution of Orford Ness, Suffolk before 1600 A.D.: geomorphological evidence, *Zeit für geomorph. N.S.*, **14**, 289–300.

Carr, A. P., and Baker, R. E. (1968) Orford, Suffolk: evidence for the evolution of the area during the Quaternary, *Trans. Inst. Br. Georgr.*, **45**, 107–123.

Carriker, M. R. (1951) Ecological observations on the distribution of oyster larvae in New Jersey estuaries, *Ecol. Monogr.*, **21**, 19–38.

Carson, M. A., and Kirkby, M. J. (1972) *Hillslope Form and Process*, Cambridge University Press, Cambridge.

Carter, L. (1973) An industrial approach to waste disposal in estuaries, *J. environ. Planning*, **1**, 5–11.

Caspers, M. (1959) Die Einteilung der Brackwasser-Regionen in einem Aestuar, *Arch. Oceanogr. Limnol.*, **11** (suppl.), 155–169.

Cavé, A. J. (1960) Muizen en hun bestrijding in Oostelijk Flevoland, *Flevoberichten*, **A19**, 1–16.

Cavers, P. B., and Harper, J. L. (1967) The comparative biology of closely related species living in the same area, *IV Rumex*. The nature of adaptation to a seashore habitat, *J. Ecol.*, **55**, 73–82.

Chambers, M. R., and Milne, H. (1975a) Life cycle and production of *Nereis diversicolor* (O. F. Müller) in the Ythan Estuary, Scotland, *Estuarine coastal mar. Sci.*, **3**, 133–144.

Chambers, M. R., and Milne, H. (1975b) The production of *Macoma balthica* (L.) in the Ythan Estuary, *Estuarine coastal mar. Sci.*, **3**, 443–455.

Chapman, C. R. (1966) The Texas Basins Project, in: *A Symposium on Estuarine Fisheries*, Am. Fish. Soc. Spec. Publ. No. 3, Washington, D.C., pp. 83–92.

Chapman, V. J. (1934) The plant ecology of Scolt Head Island, in: *Scolt Head Island. The Story of its Origin: the Plants and Animal Life of the Dunes and Marshes*, ed. Steers, J. A., Heffer, Cambridge, pp. 85–163.

Chapman, V. J. (1938–59) Studies in salt marsh ecology. Sections I–IX, *J. Ecol.*, **26**, 144–179; **27**, 160–201; **28**, 118–152; **29**, 69–82; **47**, 619–639.

Chapman, V. J. (1947) *Suaeda fruticosa* Forsk, *J. Ecol.*, **35**, 303–310.

Chapman, V. J. (1960a) The plant ecology of Scolt Head Island, in: *Scolt Head Island*, 2nd Edn., ed. Steers, J. A., Heffer, Cambridge.

Chapman, V. J. (1960b) *Salt Marshes and Salt Deserts of the World*, Hill, London.

Chapman, V. J. (1964) *Coastal Vegetation*, Pergamon, Oxford and London.

Chappell, H. G., Ainsworth, J. F., Cameron, R. A. D., and Redfern, M. (1971) The effects of trampling on a chalk grassland ecosystem, *J. Appl. Ecol.*, **8**, 869–882.

Chester, C. G. C., *et al.* (1956) Studies of the decomposition of seaweeds and seaweed products by micro-organisms, *Proc. Linn. Soc. Lond.*, **166**, 87–97.

Chippendale, H. G., and Merricks, N. W. (1956) Gang mowing and pasture management, *J. Br. Grassland Soc.*, **2**, 1–9.

Christiansen, W. (1937) Beobachtungen an Dauerquadraten auf der Lotseninsel Schleimunde, *Schr. Naturwiss. Ver. Schleswig-Holstein*, **22**, 69–88.

Clark, M. J. (1971) The coast of south-west Hampshire, in: *Field Studies in South Hampshire*, Southampton Branch Geographical Association, pp. 98–106.

Clark, M. J., Ricketts, P. J., and Small, R. J. (1976) Barton does not rule the waves, *Geogr. Mag.*, **48**, 580–588.

Clark, R. B., and Panchen, A. L. (1971) *Synopsis of Animal Classification*, Chapman & Hall, London.

Cobb, R. T. (1956) *Shingle Street: a study of the changes which have taken place on this part of the Suffolk coast, with some suggestions as to their cause*, M.Sc. thesis, University of Sheffield.

Cole, H. A. (ed.) (1971) A discussion on biological effects of pollution in the sea, *Proc. Roy. Soc. B*, **177**, 275–468.

Colenbrander, H. J. (1975) Agro-hydrologisch onderzoek in relatie tot het structuurschema voor de toekomstige drinkwatervoorziening, *Versl. Meded. Comm. Hydrol. Onderz. T.N.O.*, **16**, 33–52.

Colloque sur l'unification des méthodes d'analyse des eaux saumâtres méditerranéennes (1966), *Comm. Int. Sci. Mer Méditerranée*, Messine, Monaco.

Colombo, G. (1972) Primi risultati delle ricerche sulle residue Valli di Comacchio e piani delle ricerche future, *Boll. Zool.* (Italy), **39**, 471–478.

Coode, J. (1853) Description of the Chesil Bank, with remarks upon its origin, the causes which have contributed to its formation, and upon the movement of shingle generally, *Minut. Proc. Inst. civ. Engrs.*, **12**, 520–546.

Corlett, J. (1948) Rates of settlement and growth of the 'pile' fauna of the Mersey estuary, *Proc. Trans. Liverpool biol. Soc.*, **55**, 2–28.

Cottam, C. (1935) The eelgrass shortage in relation to waterfowl, *Game Conf. Trans.*, **20**, 272–279.

Cotton, C. A. (1952) Criteria for the classification of coasts, *17th Int. Geog. Cong. Abs. of Papers*, p. 15.

Coughlan, J. (1966) *The Blackwater Oyster Investigations 1959–65*, Central Electricity Research Laboratory, Bradwell-on-Sea, Essex, reports RD/L/R 1351 and 1387.

Countryside Commission (1968) *The Coasts of England and Wales* (several volumes), H.M.S.O., London.

Countryside Commission (1969) *Coastal Preservation and Development. A study of the coastline of England and Wales. Nature Conservation: Special study reports*, Vol. 2, H.M.S.O., London.

Countryside Commission (1970a) *The Coastal Heritage—a conservation policy of coasts of high quality scenery*, H.M.S.O., London.

Countryside Commission (1970b) *The Planning of the Coastline*, H.M.S.O., London.

Cowell, E. B. (ed.) (1971) *The Ecological Effects of Oil Pollution on Littoral Communities*, Institute of Petroleum, London.

Cramp, S., Bourne, W. R. P., and Saunders, D. (1974) *The Sea-birds of Britain and Ireland*, Collins, London.

Crapp, G. B. (1971a) Monitoring the rocky shore, in: *The Ecological Effects of Oil Pollution on Littoral Communities*, ed. Cowell, E. B., Institute of Petroleum, London, pp. 102–113.

Crapp, G. B. (1971b) Chronic oil pollution, in: *The Ecological Effects of Oil Pollution on Littoral Communities*, ed. Cowell, E. B., Institute of Petroleum, London, pp. 187–203.

Crapp, G. B. (1973) The distribution and abundance of animals and plants on the rocky shores of Bantry Bay, *Irish Fish. Invest. B*, **9**, 1–35.

Crapp, G. B., Withers, R. G., and Sullivan, C. E. (1971) Investigations on sandy and muddy shores, in: *The Ecological Effects of Oil Pollution on Littoral Communities*, ed. Cowell, E. B., Institute of Petroleum, London, pp. 208-216.

Crawford, R. M. M., and Wishart, D. (1971) A multivariate analysis of the development of dune-slack vegetation in relation to coastal accretion at Tentsmuir, Fife, *J. Ecol.*, **53**, 729–744.

Crofts, R. (1973) The utilization of sandy coasts in the Scottish Highlands, paper presented at *Inst. Br. Geogr. Conf.*, Birmingham.

Cumming, N. M. (1925), Notes on strand plants, 1. *Atriplex babingtonii*, *Trans. Proc. bot. Soc. Edinburgh*, **29**, 171–185.

Dalby, D. H. (1970) The salt-marshes of Milford Haven, Pembrokeshire, *Field Studies*, **3**, 297–330.

Dalby, R. (1957) Problems of land reclamation. Salt-marsh in The Wash, *Agricultural Review*, **2**, 31–37.

D'Ancona, U. (1959) The classification of brackish water with reference to the north Adriatic lagoon, in: Symposium on the classification of brackish waters, *Arch. Oceanogr. Limnol.*, **11** (suppl.), 93–109.

D'Ancona, U., and Battaglia, B. (1962) La lagune salmastre dell'Alto Adriatico, ambiente di popolamento e di selezione, *Publ. Staz. Zool. Napoli*, **32** (suppl.), 315–335.

Darby, H. C. (1940) *The Medieval Fenland*, Cambridge University Press, London.

Darby, H. C. (1956) *The Draining of the Fens*, 2nd Edn., Cambridge University Press, London.

Dartington Amenity Research Trust (1973) *Southampton Water*, Sports Council, London.

Davies, J. L. (1972) *Geographical Variation in Coastal Development*, Oliver & Boyd, Edinburgh.

Davies, M. S., and Snaydon, R. W. (1973) Physiological differences among populations of *Anthoxanthum odoratum* L. collected from the Park Grass Experiment, Rothamsted, I. Response to calcium, *J. Appl. Ecol.*, **10**, 23–32.

De Boer, G., and Carr, A. P. (1969) Early maps as evidence for coastal changes, *Geogr. J.*, **135**, 17–39.

Department of the Environment (1972) *Out of Sight, Out of Mind* (Report of a working party on sludge disposal in Liverpool Bay), H.M.S.O., London.

Department of the Environment (1973) *Report of a survey of the discharges of foul sewage to the coastal waters of England and Wales*, H.M.S.O., London.

De Turville, C. M. (1975) *Thermograph Temperatures in Milford Haven 1966–1973*, Central Electricity Generating Board, Portishead, Bristol, report SSD/SW/75/N37.

Dickinson, G. (1974) *A Geographical Study of Machair Vegetation in the Uists*, unpublished Ph.D. thesis, University of Glasgow.

Dickinson, G., Mitchell, J., and Tivy, J. (1971) The application of phytosociological techniques to the geographical study of vegetation, *Scott. Geogr. Mag.*, **87**, 83–102.

Dicks, B. (1973) Changes in salt-marsh vegetation around a refinery effluent outfall at Fawley, *Annu. Rep. Oil Poll. Res. Unit, Orielton*, **1972**, 14–20.

Dijk, C. van (1974) The distribution of the root-nodule endophyte of *Alnus glutinosa* (L.) Gaertn. in the field, *Verhandelingen der Koninklijke Nederlandse Akademie van Wetenschappen Afd. Natuurkunde, Tweede Reeks*, **63**, 110–114.

Doing, H. (1974) Lanschapoecologie van de duinstreek tussen Wassenaar en IJmuiden, *Meded. Landbouwhogeschool, Wageningen*, **74**, 1–111.

Doing Kraft, H. (1954) L'Analyse des carrés permanents, *Acta Bot. Neerl.*, **3**, 421–424.

Dorey, A. E., Little, C., and Barnes, R. S. K. (1973) An ecological study of the Swanpool, Falmouth, II. Hydrography and its relation to animal distributions, *Estuarine Coastal Mar. Sci.*, **1**, 153–176.

Drinnan, R. E. (1957) The feeding of the oystercatcher (*Haematopus ostralegus*) on the edible cockle (*Cardium edule*), *J. Anim. Ecol.*, **26**, 441–469.

D.S.I.R. (Department of Scientific and Industrial Research) (1960) *Notes on Water Pollution, No. 10. Oxygen demand tests for effluents*, Water Pollution Research Lab., Stevenage.

Duffey, E. (1971) The management of Woodwalton Fen: a multidisciplinary approach, in: *The Scientific Management of Animal and Plant Communities for Conservation*, ed. Duffey, E., and Watt, A. S., B.E.S. Symposium **11**, 581–597.

Duffey, E., and Watt, A. S. (eds.) (1971) *The Scientific Management of Animal and Plant Communities for Conservation*, Blackwell, Oxford.

Duin, I. R. H. A. van (1969) *Natuurbouw in Flevoland*, Rijksdienst voor de IJsselmeerpolders, Zwolle.

Dunn, J. N. (1972) *A general survey of Langstone Harbour with particular reference to the effects of sewage*, Report commissioned by Hampshire River Authority, unpublished.

Eales, N. B. (1961) *The Littoral Fauna of Great Britain*, Cambridge University Press, London.

Edwards, R. S., and Claxton, S. M. (1964) The distribution of airborne salt particles of marine origin in the Aberystwyth area, *J. Appl. Ecol.*, **1**, 253–263.

Eggeling, W. J. (1964) A nature reserve management plan for the Island of Rhum, Inner Hebrides, *J. Appl. Ecol.*, **1**, 405–419.

Eisma, D. (1968) Composition, origin and distribution of Dutch coastal sands between Hoek van Holland and the island of Vlieland, *Neth. J. Sea Res.*, **4**, 123–267.

Eklund, O. (1924) Strandtyper i Skargardshavet, *Terra*, **36**.

Eklund, O. (1931) Über die Ursachen der regionalen Verteilung der Schärenflora Südwest-Finlands, *Acta bot. fenn.*, **8**, 1–133.

Elliot, J. (1847), Account of Dymchurch wall, *Minut. Proc. Inst. civ. Engrs.*, **6**, 466–480.

Ellis, E. A. (1960) The purple (hybrid) Marram, *Ammocalamogrostis baltica* (Fluegge) P. Fourn. in East Anglia, *Trans. Norfolk Norwich Nat. Soc.*, **19**, 49–51.

Elton, C. (1966) *The Pattern of Animal Communities*, Methuen, London.

Eltringham, S. K. (1971) *Life in Mud and Sand*, English Universities Press, London.

Emery, K. O. (1969), *A Coastal Pond Studied by Oceanographic Methods*, Elsevier, New York.

Emery, K. O., and Stevenson, R. E. (1957) Estuaries and Lagoons, I. Physical and chemical characteristics, *Geol. Soc. America Mem.*, **67** (fasc. 1), 673–693.

Feekes, W. (1936) De ontwikkeling van de natuurlijke vegetatie in de Wieringermeerpolder, de eerste groote droogmakerij van de Zuiderzee, *Ned. Kruidk. Arch.*, **46**, 1–295.

Feekes, W. (1943) De Piamer Kooiwaard en Makkumerwaard, *Ned. Kruidk. Arch.*, **53**, 288–330.

Feekes, W., and Bakker, D. (1954) De ontwikkeling van de natuurlijke vegetatie in de Noordoost-polder, *Van Zee Tot Land*, No. 6.

Ferrari, C., Pirola, A., and Piccoli, F. (1972) Ricerche idrobiologiche nelle Valli di Comacchio, II. Saggio cartografico della vegetazione delle Valli di Comacchio, *Ann. Univ. Ferrara. Sez. i Ecologia*, **1** (2), 35–54.

Fischer-Piette, E. (1931) Sur la pénétration des diverses espèces marines sessiles dans les estuaires et sa limitation par l'eau douce, *Ann. Inst. Océanogr. Monaco*, **10**, 214–242.

Fischer-Piette, E., and Seoane-Camba, J. (1962) Écologie de la ria-type; la Ria del Barquero, *Bull. Inst. Océanogr. Monaco*, **58**, 1–36.

Fischer-Piette, E., and Seoane-Camba, J. (1963) Examen écologique de la Ria de Camariñas, *Bull. Inst. Océanogr. Monaco*, **61**, 1–38.

Fisher, O. (1866) On the disintegration of a chalk cliff, *Geol. Mag.*, **3**, 354–356.

Fisher Cassie, W., Simpson, J. R., Allen, J. H., and Hall, D. G. (1962) Final report of hydraulic and sediment survey of the estuary of the River Tyne, *Bull. Univ. Durham Dept. civ. Engng.*, **24**, 47 pp.

Flevoland Facts and Figures (1972) Rijksdienst voor de IJsselmeerpolders, Zwolle.

Flinn, D. (1974) The coastline of Shetland, in: *The Natural Environment of Shetland*, ed. Goodier, R., Nature Conservancy Council, Edinburgh, pp. 13–23.

Food and Agriculture Organization (1965) *Soil erosion by water: some measures for its control on cultivated lands*, FAO Agricultural Development Paper No. 81, FAO United Nations, Rome.

Frank, E. C., and Lee, R. (1966) *Potential solar beam irradiation on slopes*, U.S. Forest Service Research Paper RM-18, U.S. Forest Service, Rocky Mountains Forest Experimental Station.

Fraser, J. H. (1932) Observations on the fauna and constituents of an estuarine mud in a polluted area, *J. mar. biol. Ass. U.K.*, **18**, 69–85.

Friedrich, H. (1965) *Meeresbiologie*, Borntraeger, Berlin.

Freijsen, A. H. J. (1975) Some experiments on the calcicolous plant *Cynoglossum officinale* L., *Verhandelingen der Koninklijke Nederlandse Akademie van Wettenschappen Afd. Natuur-kunde, Tweede Reeks*, **66**, 97–99.

Fuller, R. (1972) *The assessment of coastal habitat resources from aerial photography, 1. A quick and easy method for assessment of the major habitat resources*, Coastal Ecology Research Paper No. 1, Norwich.

Fuller, R. (1973) *The assessment of coastal habitat resources from aerial photography, 2. Mapping and assessment of salt marsh vegetation resources*, Coastal Ecology Research Paper No. 3, Norwich.

Fuller, R. M. (1975) The Culbin shingle bar and its vegetation, *Trans. Proc. bot. Soc. Edinburgh*, **42**, 293–305.

Gabriel, P. L. (1973) *The plankton of polluted waters*, Ph.D. thesis, University of Wales.

Gabriel, P. L., Dias, N. S., and Nelson-Smith, A. (1975) Temporal changes in the plankton of an industrialized estuary, *Estuarine coastal mar. Sci.*, **3**, 145–151.

Gameson, A. L. H., and Barrett, M. J. (1958) Oxidation, reaeration and mixing in the Thames Estuary, *U.S. Publ. Health Serv. Tech. Rep.*, **W58-2**, 63–93.

Géhu, J. M. (1960a) La végétation des levées de galets du littoral français de la Manche, *Bull. Soc. bot. N. France*, **13**, 141–152.

Géhu, J. M. (1960b) Quelques observations sur la végétation et l'écologie d'une station réputée de l'Archipel de Chausey: l'île aux Oiseaux, *Bull. Lab. marit. Dinard*, **46**, 78–92.

Géhu, J. M. (1960c) Une site célèbre de la côte Nord bretonne: Le Sillon de Talbert (C-du-N). Observations phytosociologiques et écologiques, *Bull. Lab. marit. Dinard*, **46**, 93–115.

Géhu, J. M. (1963) *Sarothamnus scoparius* spp *maritimus* dans le Nord-Ouest Français, *Bull. Soc. bot. N. France*, **16**, 211–222.

Géhu, J. M., and Géhu, J. (1959) Note phyto-écologique concernant la station de *Crambe maritima* L. de l'anse du Guesclin (Ille-et-Vilane), *Bull. Lab. marit. Dinard*, **45**, 56–62.

Geiger, R. (1965) *The Climate Near the Ground*, Harvard University Press, Cambridge, Mass.

Gierloff-Emden, H. G. (1961) Nehrungen und Lagunen, *Petermanns Geogr. Mitt.*, **105**, 81–92; 161–176.

Gill, C. J. (1970) The flooding tolerance of woody species—a review, *Forestry Abstracts*, **31**, 671–688.

Gillham, M. E. (1953) An ecological account of the vegetation of Grassholm Island, Pembrokeshire, *J. Ecol.*, **41**, 84–99.

Gillham, M. E. (1957) Vegetation of Exe Estuary in relation to water salinity, *J. Ecol.*, **45**, 735–756.

Gillner, V. (1960) Vegetations- und Standortsuntersuchungen in den Strandwiesen der Schwedischen Westküste, *Acta Phytogeogr. Suecica*, **43**, 1–198.

Gillner, V. (1965) Salt marsh vegetation in Southern Sweden, *Acta Phytogeogr. Suecica*, **50**, 97–104.

Gilson, H. C. (1966) The biological implications of the proposed barrages across Morecambe Bay and the Solway Firth, in: *Man-made Lakes*, ed. Lowe-McConnell, R. H., Academic Press, London, pp. 129–137.

Gimingham, C. H. (1964) Maritime and submaritime communities, in: *The Vegetation of Scotland*, ed. Burnett, J. H., Oliver & Boyd, Edinburgh, pp. 67–143.

Gimingham, C. H. (1974) Plant communities of the machair and floristic relationships with non-dune vegetation, in: *Sand Dune Machair*, ed. Ranwell, D. S., Institute of Terrestrial Ecology, Natural Environment Research Council, London, pp. 13–14.

Glopper, R. J. de (1969) *Shrinkage of subaqueous sediments of Lake IJssel (The Netherlands) after reclamation*, Rijksdienst voor de IJsselmeerpolders, Kampen, Overdruk No. 39.

Glude, J. B. (1972) Information requirements for rational decision-making in control of coastal and estuarine oil pollution, in: *Marine Pollution and Sea Life*, ed. Ruivo, M., FAO/Fishing New Books, London, pp. 622–624.

Glue, D. E. (1971) Salt-marsh reclamation stages and their associated bird life, *Bird Study*, **18**, 187–198.

Goldberg, E. D. (ed.) (1972) *A Guide to Marine Pollution*, Gordon & Breach, New York.

Goldsmith, F. B. (1973a) The vegetation of exposed sea-cliffs at South Stack, Anglesey, I. The multivariate approach, *J. Ecol.*, **61**, 787–818.

Goldsmith, F. B. (1973b) The vegetation of exposed sea-cliffs at South Stack, Anglesey, II. Experimental studies, *J. Ecol.*, **61**, 819–829.

Goldsmith, F. B. (1974) Ecological effects of visitors in the countryside, in: *Conservation in Practice*, ed. Warren, A., and Goldsmith, F. B., Wiley, London, pp. 117–231.

Goldsmith, F. B. (1975a) The sea-cliff vegetation of Shetland, *J. Biogeog.*, **2** (4), 297–308.

Goldsmith, F: B. (1975b) The evaluation of ecological resources in the countryside for conservation purposes, *Biol. Conserv.*, **7**, 89–96.

Goldsmith, F. B., and Harrison, C. M. (1975) Methods for the description and analysis of vegetation, in: *Methods in Plant Ecology*, ed. Chapman, S. B., Blackwell, Oxford.

Goldsmith, F. B., Munton, R. J. C., and Warren, A. (1970) The impact of recreation on the ecology and amenity of semi-natural areas: methods of investigation used in the Isles of Scilly, *Biol. J. Linn. Soc.*, **2**, 187–306.

Golterman, H. L. (1969) *Methods for Chemical Analysis for Fresh Waters*, I.P.B. Handbook No. 8, Blackwell, Oxford and Edinburgh.

Goodman, P. J. (1960) Investigations into 'die-back' in *Spartina townsendii* agg, II. The morphological stucture and composition of the Lymington sward, *J. Ecol.*, **48**, 711–724.

Goodman, P. J., Braybrooks, E. M., and Lambert, J. M. (1959) Investigations into 'die-back' in *Spartina townsendii*, I. The present status of *S. townsendii* in Britain, *J. Ecol.*, **47**, 651–677.

Goodman, P. J., and Williams, W. T. (1961) Investigations into 'die-back' in *Spartina townsendii* agg, III. Physiological correlates of 'die-back', *J. Ecol.*, **49**, 391–398.

Goreham, E. (1958) Soluble salts in dune sands from Blakeney Point in Norfolk, *J. Ecol.*, **46**, 373–9.

Gramolt, D. W. (1960) *The coastal marshlands of East Essex between the seventeenth and mid-nineteenth centuries*, M.A. thesis, University of London.

Gray, A. J. (1970) The colonization of estuaries following barrage building, in: *The Flora of a Changing Britain*, ed. Perring, F., Classey, Hampton, pp. 63–72.

Gray, A. J. (1972) The ecology of Morecambe Bay, V. The salt marshes of Morecambe Bay, *J. Appl. Ecol.*, **9**, 207–220.

Gray, A. J. (1974) The genecology of salt marsh plants, *Hydr. Bul.* (Amsterdam), **8**, 152–165.

Gray, A. J. (1976) The Ouse Washes and the Wash, in: *Nature in Norfolk, a Heritage in Trust*, ed. Norfolk Naturalists' Trust, Jarrold & Son, Norwich, pp. 123–129.

Gray, A. J., and Adam, P. (1974) The reclamation history of Morecambe Bay, *Nature Lancs.*, **4**, 13–20.

Green, B. H. (ed.) (1971) *Wildlife Conservation in the North Kent Marshes. Report of a Working Party*, Nature Conservancy S.E. Region, Wye.

Green, J. (1968) *The Biology of Estuarine Animals*, Sidgwick & Jackson, London.

Greenwood, J. G. W. (ed.) (1974) *Remote Sensing Evaluation Flights, 1971. Part 2. South Harris and Gairloch Area*, The Natural Environment Research Council, Publications Series C, No. 12.

Greig-Smith, P. (1957) *Quantitative Plant Ecology*, 1st Edn., Butterworth, London.

Greig-Smith, P. (1964) *Quantitative Plant Ecology*, 2nd Edn., Butterworth, London.

Grime, J. P., and Hunt, R. (1975) Relative growth rate: its range and adaptive significance in a local flora, *J. Ecol.*, **63**, 393–422.

Grimes, B. H., and Hubbard, J. C. E. (1969) The use of aerial photography in the Nature Conservancy, *Photogr. J.*, **109**, 5.

Grimes, B. H., and Hubbard, J. C. E. (1971) A comparison of film type and the importance of season for interpretation of coastal marshland vegetation, *Photogramm. Rec.*, **7**, 213–222.

Grimes, B. H., and Hubbard, J. C. E. (1972) Modern techniques of aerial photography, *Endeavour*, **31**, 130–134.

Grøntved, J. (1960) On the productivity of microbenthos and phytoplankton in some Danish Fiords, *Medd. Danm. Fisk. Havunders. N.S.*, **3**, 55–92.

Grøntved, J. (1962) Preliminary reports on the productivity of microbenthos and phytoplankton in the Danish Wadden Sea, *Medd. Danm. Fisk. Havunders.*, **3**, 347–378.

Guilcher, A. (1954) *Morphologie littorale et sous-marine*, Presses Universitaires de France, Paris.

Guilcher, A. (1967) Origin of sediments in estuaries, in: *Estuaries*, ed. Lauff, G. H., American Association for the Advancement of Science, Washington, pp. 149–157.

Haderlie, E. C., and Clark, R. B. (1959). Studies on the biology of the Bristol Channel, XIX. Notes on the intertidal fauna of some sandy and muddy beaches in the Bristol Channel and Severn Estuary, *Proc. Bristol nat. Soc.*, **29**, 459–468.

Haeck, J. (1969a) The immigration and settlement of carabids in the new IJsselmeerpolders, *Symp. Biol. Stat. Wijster. Wageningen*, **8**, 33–52.

Haeck, J. (1969b) Colonization of the mole (*Talpa europaea* L.) in the IJsselmeerpolders, *Neth. J. Zool.*, **19**, 145–248.

Haggett, P. (1964) Regional and local components in the distribution of forested areas in Southeast Brazil: a multivariate approach, *Geogr. J.*, **130**, 365–380.

Haltiner, G. J., and Martin, F. L. (1957), *Dynamical and Physical Meteorology*, McGraw-Hill, New York.

Hammond, R. C. (1967) Visitor pressure at Wye and Crundale Downs NNR, in: *The Biotic Effects of Public Pressure on the Environment*, ed. Duffey, E., Monkswood Experimental Station Symposium No. 3, Nature Conservancy, pp. 127–133.

Harden, V. P. (1963), *The Open Pits*, M.Sc. thesis, University of London.

Harper, J. L. (1971) Grazing, fertilizers and pesticides in the management of grasslands, in: *The Scientific Management of Animal and Plant Communities for Conservation*, ed. Duffey, E., and Watt, A. S., B.E.S. Symp. 11, pp. 15–31.

Harris, D. R., and Cooke, R. U. (1969) The landscape revealed by aerial sensors, *Geog. Mag.*, **42**, 1, 29–38.

Harrison, J. G. (1960) *A technique for removing wildfowl viscera for research*, Wildfowl Trust 11th Annual Report, pp. 135–136.

Harrison, J. G. (1971) *A Gravel Pit Wildfowl Reserve*, W.A.G.B.I. Publication, Chester.

Harrison, J. G. (1973) *Wildfowl of the North Kent Marshes*, W.A.G.B.I. Publication, Chester.

Hartley, P. H. T., and Spooner, G. M. (1938) The ecology of the Tamar estuary, I. Introduction, *J. mar. biol. Ass. U.K.*, **22**, 501–508.

Hartog, C. den (1959) The epilithic algal communities occurring along the coast of the Netherlands, *Wentia*, **1**, 1–241.

Hassouna, M. G., and Wareing, P. F. (1964) Possible role of rhizosphere bacteria in the nitrogen nutrition of *Ammophila arenaria*, *Nature*, **202**, 457–469.

Hawes, F. B. (ed.) (undated) *Hydrobiological studies in the River Blackwater in relation to the Bradwell nuclear power station*, Central Electricity Generating Board, London.

Haynes, J., and Dobson, M. (1969) Physiography, Foraminifera and sedimentation in the Dovey estuary (Wales), *Geol. J.*, **6**, 217–256.

Hedgpeth, J. W. (1957) Estuaries and lagoons, II. Biological aspects, *Geol. Soc. Amer. Mem. 67*, **1**, 693–734.

Hedgpeth, J. W. (1967) Ecological aspects of the laguna Madre. A hyperhaline estuary, in: *Estuaries*, ed. Lauff, G. H., American Association for the Advancement of Science, Washington, pp. 408–419.

Heerebout, G. R. (1970) A classification system for isolated brackish inland waters, based on median chlorinity and chlorinity fluctuation, *Neth. J. Sea Res.*, **4**, 494–503.

Hepburn, I. (1952) *Flowers of the Coast*, Collins, London.

Hewett, D. G. (1970) The colonization of sand-dunes after stabilization with Marram Grass (*Ammophilia arenaria*), *J. Ecol.*, **58**, 653–668.

Hey, R. W. (1967) Sections in the beach-plain deposits of Dungeness, Kent, *Geol. Mag.*, **104**, 361–370.

Heydermann, B. (1960) Verlauf und Abhängigkeit von Spinnensukzessionen im Neuland der Nordseeküste, *Verh. Deutsch. Zool. Ges. Bonn/Rhein*, **1960**, 431–457.

Hill, T. G., and Hanley, J. A. (1914) The structure and water content of shingle beaches, *J. Ecol.*, **2**, 21–38.

Holdgate, M. W. (1971) *The Seabird Wreck of 1969 in the Irish Sea*, Natural Environment Research Council, London.

Holme, N. A. (1949) The fauna of sand and mud banks near the mouth of the Exe Estuary, *J. mar. biol. Ass. U.K.*, **28**, 189–237.

Holme, N. A., and McIntyre, A. D. (1971) *Methods for the Study of Marine Benthos*, I.B.P. Handbook No. 16, Blackwell, Oxford.

Hope-Simpson, J. F., and Jefferies, R. L. (1966) Observations relating to vigour and debility in Marran grass (*Ammophila arenaria* (L.) Link., *J. Ecol.*, **54**, 271–274.

Horikawa, K., and Sunamura, T. (1967) A study of cliff erosion using air photos, *Coastal Engineering in Japan*, **10**.

Horikawa, K., and Sunamura, T. (1970) A study of erosion on coastal cliffs and submarine bedrocks, *Coastal Engineering in Japan*, **13**, 127–140.

Howard, J. A. (1970) *Aerial photo-ecology*, Faber, London.

Hoyle, J. W., and King, G. T. (1955) The lateral stability of shingle beaches, *J. Instn. munic. Engrs.*, **81**, 356–366.

Hubbard, C. E. (1968) *Grasses*, 2nd Edn., Penguin Books, Harmondsworth.

Hubbard, J. C. E. (1965) *Spartina* marshes in southern England, VI. Pattern of invasion in Poole Harbour, *J. Ecol.*, **53**, 799–813.

Hubbard, J. C. E. (1970) Effects of cutting and seed production in *Spartina anglica*, *J. Ecol.*, **58**, 329–334.

Hubbard, J. C. E. (1971) The use of aerial photography in the survey of coastal features, in: *The application of aerial photography to the work of the Nature Conservancy*, ed. Goodier, R., Nature Conservancy, Edinburgh, pp. 36–42.

Hubbard, J. C. E., and Grimes, B. H. (1972) The analysis of coastal vegetation through the medium of aerial photography, *Med. biol. Illust.*, **22**, 182-190.

Hubbard, J. C. E., and Ranwell, D. S. (1966) Cropping *Spàrtina* salt-marsh for silage, *J. Br. Grassland Soc.*, **21**, 214–217.

Hubbard, J. C. E., and Stebbings, R. E. (1967) Distribution, date of origin and acreage of *Spartina townsendii* (s.l.) marshes in Great Britain, *Proc. Bot. Soc. Br. Isl.*, **7**, 1–7.

Hulings, N. C., and Gray, J. S. (1971) A manual for the study of meiofauna, *Smithsonian Contr. Zool.*, No. 78, Smithsonian Institution Press, Washington, D.C.

Hutchinson, J. N. (1967) The free degradation of London Clay cliffs, *Proc. Geotechnical Conf.*, Oslo, **1**, 113–118.

Hutchinson, J. N. (1968) Field meeting on the coastal landslides of Kent, *Proc. Geol. Assoc.*, **79**, 227–237.

Hutchinson, J. N. (1969) A reconsideration of the coastal landslides at Folkestone Warren, Kent, *Geotechnique*, **19**, 6–38.

Hutchinson, J. N. (1970) A coastal mudflow on the London Clay cliffs at Beltinge, North Kent, *Geotechnique*, **20**, 412–438.

Hutchinson, J. N. (1971) Field and laboratory studies of a fall in Upper Chalk cliffs at Joss Bay, Isle of Thanet, *Roscoe Memorial Symposium*, Cambridge.

Hutchinson, J. N. (1976) Coastal landslides in the cliffs of Pleistocene deposits between Cromer and Overstrand, Norfolk, England, *Laurits Bjerrum Memorial Volume*, Oslo, 155–182.

Hydraulics Research Station (1975) The Wash water storage scheme—A historical review, *Rep. No. DE.26*, H.R.S., Wallingford.

Hydro Delft (1971) The Delta project, halfway to completion, *Hydro Delft*, **23/24**, 1–15.

Inglis, C. C., and Allen, F. H. (1957) The regimen of the Thames estuary as affected by currents, salinities and river flow, *Proc. Inst. Civ. Eng.*, **7**, 827–868.

Inglis, C. C., and Kestner, F. J. T. (1958) Changes in The Wash as effected by training walls and reclamation works, *Proc. Inst. Civ. Eng.*, **11**, 435–466.

Institute for Marine Environmental Research, *Reports* for 1971–73, 1973–74, 1974–75, Plymouth.

International Bird Census Committee (1969) Recommendations for an international standard for a mapping method in bird census work, *Bird Study*, **16**, 249–255.

International Waterfowl Research Bureau (1972) *Manual of Wetland Management*, Slimbridge.

Ishizuka, K. (1974) Maritime vegetation, in: *The Flora and Vegetation of Japan*, ed. Numata, M., Kodansha, Tokyo.

Jakobsen, B. (1954) The tidal area in south-western Jutland and the process of the salt-marsh formation, *Geogr. Tidsskr.*, **53**, 49–61.

Jakobsen, B. (1964) Vadehavets morfologi en geografisk analyse af vadelandskabets formudvikling med saerlig hensyntagen til Juvre Dybs tidecandsområde, *Folia Geogr. Danica*, **11**,1–176.

Jakobsen, B., and Jensen, K. M. (1956), Undersøgelser vedrørende landvindingsmetoder i Det danske Vadehav, *Geogr. Tidsskr.*, **55**, 21–61.

Jelgersma, S. (1966) Sea level changes during the last 10,000 years, *Roy. Met. Soc. Symposium on World Climate from 8,000 to 0 B.C.*, London.

Jelgersma, S., and Regteren Altena, J. F. van (1969) An outline of the geological history of the coastal dunes in the western Netherlands, *Geol. en Mijnb.*, **48**, 335–342.

Jewell, P. A. (1974) Managing animal populations, in: *Conservation in Practice*, ed. Warren, A., and Goldsmith, F. B., Wiley, London, pp. 185–198.

Johnson, C. S. (1972) Macro-algae and their environment, *Proc. Roy. Soc. Edinburgh (B)*, **71**, 195–207.

Johnson, D. W. (1919) *Shore Processes and Shoreline Development*, Wiley, New York.

Johnson, P. L. (ed.) (1969) *Remote Sensing in Ecology*, University of Georgia Press, Athens, Georgia, U.S.A.

Jones, E. B. G., and Eltringham, S. K. (ed.) (1971) *Marine Borers, Fungi and Fouling Organisms of Wood*, O.E.C.D., Paris.

Jones, E. B. G., and Farnham, W. (1973) Japweed: a new threat to British coasts, *New Scientist*, **60**, 394–395.

Jones, R., and Etherington, J. R. (1971) Comparative studies of plant growth and distribution in relation to waterlogging, IV. The growth of dune and dune-slack plants, *J. Ecol.*, **59**, 793–802.

Jong, H. de (1972) Het weidevogelreservaat in Oosterlijk Flevoland, *Limosa*, **45**, 49–57.

Jorde, I., and Klavestaf, N. (1963) The natural history of the Hardangerfjord, 4. The benthonic algal vegetation, *Sarsia*, **9**, 1–99.

Kamps, L. F. (1962) *Mud distribution and land reclamation in the eastern Wadden Shallows*, Rijkswaterstaat Communications, 4, The Hague, pp. 1–73.

Keefe, C. W. (1972) Marsh production: A summary of the literature, *Contr. Marine Sci.*, **16**, 163–181.

Kershaw, K. A. (1964) *Quantitative and Dynamic Ecology*, Arnold, London.

Ketchum, B. H. (ed.) (1972) *The Water's Edge: Critical Problems of the Coastal Zone*, M.I.T. Press, Massachusetts and London.

Ketner, P. (1972) *Primary production of salt-marsh communities on the island of Terschelling in the Netherlands*, Thoben Offset, Nijmegen, pp. 1–181.

Kidson, C. (1959) Uses and limits of vegetation in shore stabilization, *Geography*, **44**, 241–250.

King, C. A. M. (1959) *Beaches and Coasts*, 1st Edn., Arnold, London.

King, C. A. M. (1972) *Beaches and Coasts*, 2nd Edn., Arnold, London.

Knetsch, J. L. (1964) The economics of including recreation as a purpose of Eastern water projects, *J. Farm Econ.*, **46**.

Knox, A. J. (1974) Agricultural use of machair, in: *Sand Dune Machair*, ed. Ranwell, D. S., Institute of Terrestrial Ecology, Natural Environment Research Council, p. 19.

König, D. (1968) Biologische Auswirkungen des Abwassers einer Öl-Raffinerie in einem Vorlandgebeit an der Nordsee, *Helgo. wiss. Meeresuriters*, **17**, 321–334.

Koninklijk Nederlands Meteorologisch Instituut (1972) *Klimaatatlas van Nederland*, Koninklijk Nederlands Meteorologisch Instituut, Staatsuitgeveri, The Hague.

Krumbein, W. C., and Slack, H. A. (1956) The relative efficiency of beach sampling methods, *Tech. Memo. Beach Eros. Bd. U.S.*, **90**, 1–34.

Kühl, H. (1972) Hydrography and biology of the Elbe estuary, *Oceanogr. mar. Biol. annu. Rev.*, **10**, 225–309.

Kühl, H., and Mann, H. (1962) Über da Zooplankton der Unterelbe, *Ver. Inst. Meeresforsch. Bremerhaven*, **8**, 53–70.

Landsberg, S. Y. (1956) The orientation of dunes in Britain in relation to wind, *Geogr. J.*, **122**, 176–189.

Lance, G. N., and Williams, W. T. (1968) Note on a new information statistic classificatory program, *Computer J.*, **11**, 195.

Lauff, G. H. (ed.) (1967) *Estuaries*, American Association for the Advancement of Science, Washington.

Leach, J. H. (1971) Hydrology of the Ythan estuary with reference to the distribution of major nutrients and detritus, *J. mar. biol. Ass. U.K.*, **51**, 137–157.

Lebret, T. (1965) *The prospects for wild geese in the Netherlands*, Wildfowl Trust 16th Annual Report, pp. 85–91.

Lebret, T. (1972) Vogels van de Natuurreservaten in het Veerse Meer in de afsluiting 1961–1970, *Limosa*, **45**, 1–24.

Leeuwen, C. G. van (1966a) A relation theoretical approach to pattern and process in vegetation, *Wentia*, **15**, 255–46.

Leeuwen, C. G. van (1966b) Het botanische beheer van natuurreservaten op structuur-oecologische grondslag, *Gorteria*, **3**, 16–28.

Leeuwen, C. G. van (1969) Milieu-geografisch onderzoek aan plantensoorten, *Jaarverslag R.I.V.O.N.*, **1968**, 14–16.

Leney, F. M. (1974) The ecological effects of public pressure on picnic sites, *J. Sports Turf Res. Inst.*, **50**, 47–51.

Lewis, J. R. (1964) *The Ecology of Rocky Shores*, English Universities Press, London.

Lewis, J. R. (1972) Problems and approaches to baseline studies in coastal communities, in: *Marine Pollution and Sea Life*, ed. Ruivo, M., FAO/Fishing News Books, London, pp. 401–404.

Lewis, W. V. (1931) The effect of wave incidence on the configuration of a shingle beach, *Geogr. J.*, **78**, 129–148.

Lewis, W. V. (1932) The formation of Dungeness foreland, *Geogr. J.*, **80**, 309–324.

Lewis, W. V., and Balchin, W. G. V. (1940) Past sea levels at Dungeness, *Geogr. J.*, **96**, 258–285.

Liddle, M. J. (1975a) A selective review of the ecological effects of human trampling on natural ecosystems, *Biol. Conserv.*, **7**, 17–36.

Liddle, M. J. (1975b) A theoretical relationship between the primary productivity of vegetation and its ability to tolerate trampling, *Biol. Conserv.*, **8**, 251–255.

Liddle, M. J., and Greig-Smith, P. (1975a) A survey of tracks and paths in a sand-dune ecosystem, 1. Soils, *J. Appl. Ecol.*, **12**, 893–908.

Liddle, M. J., and Grieg-Smith, P. (1975b) A survey of tracks and paths in a dune ecosystem, 2. Vegetation, *J. Appl. Ecol.*, **12**, 909–930.

Little, C., Barnes, R. S. K., and Dorey, A. E. (1973) An ecological study of the Swanpool, Falmouth, 3. Origin and history, *Cornish Stud.*, **1**, 33–48.

Londo, G. (1966) *Rapport over een botanische studiereis naar Engelse duingebieden van 9 tot 23 juli 1966*, Rapport R.I.V.O.N, Zeist.

Londo, G. (1971) *Patroon en proces in duinvalleibegetaties langs een gegraven meer in de Kennemerduinen*, Diss. Nijmegen, Cuyk.

Londo, G. (1974), *Karteringseenheden op vegetatiekundige basis*, Rijksinstituut voor Natuurbeheer, Leersum.

Londo, G. (1975a) *Nederlandse lijst van hydro-, freato- en afreatofyten*, Rijksinstituut voor Natuurbeheer, Leersum.

Londo, G. (1975b) Infiltreren is nivelleren, *De Levende Natuur*, **78**, 74–79.

Londo, G. (1976) Over de Nederlandse lijst van hydro-, freato- en afreatofyten, *Gorteria*, **8**(2), 25–29.

Longbottom, M. R. (1970) The distribution of *Arenicola marina* (L.) with particular reference to the effects of particle size and organic matter of the sediments, *J. Exp. Mar. Biol. Ecol.*, **5**, 138–157.

Maarel, E. van der (1959) *Verslag van een onderzoek naar de begroeiing van de oevers van het meer in het Paardekoppenvlak*, Kennermeerduinen, Verslag Univ. van Amsterdam.

Maarel, E. van der (1966a) Dutch studies on coastal sand dune vegetation, especially in the Delta region, *Wentia*, **15**, 47–82.

Maarel, E. van der (1966b) *Over vegetatie structuren, -relaties, en -systemen*, Diss. Utrecht, Zeist.

Maarel, E. van der (1969) On the use of ordination models in phytosociology, *Vegetatio*, **19**, 21–46.

Maarel, E. van der (1971) Plant species diversity in relation to management, in: *The Scientific Management of Animal and Plant Communities for Conservation*, ed. Duffey, E., and Watt, A. S., Blackwell, Oxford, pp. 45–63.

Macdonald, J. (1954) Tree planting on coastal sand dunes in Britain, *Adv. Sci.*, **11**, 33–37.

Macfadyen, A. (1963) *Animal Ecology Aims and Methods*, 2nd Edn., Pitman, London.

Macnae, W. (1963) Mangrove swamps in South Africa, *J. Ecol.*, **51**, 1–25.

Macnae, W. (1966) Mangroves in Eastern and Southern Australia, *Austr. J. Bot.*, **14**, 67–104.

Macnae, W. (1968) A general account of the fauna and flora of mangrove swamps and forests in the Indo-West-Pacific region, *Adv. mar. Biol.*, **6**, 73–270.

Malloch, A. J. C. (1971) Vegetation of the maritime cliff-tops of the Lizard and Lands End peninsulas, West Cornwall, *New Phytol.*, **70**, 1155–1197.

Manohar, M., and Bruun, P. (1970) Mechanics of dune growth by sand fences, *The Dock and Harbour Authority*, **1970**, 243–252.

Marchesoni, V. (1954) Il trofismo della Laguna Veneta e la vivicazione marina, III. Ricerche sulle variazioni quantitative del fitoplancton, *Arch. Oceanogr. Limnol.*, **9**, 147–281.

Margalef, R. (1963) On certain unifying principles in ecology, *Amer. Nat.*, **97**, 357–374.

Marsh, A. S. (1915) The maritime ecology of Holme-next-the-Sea, Norfolk, *J. Ecol.*, **3**, 65–93.

Matthews, J. R. (1955), *Origin and Distribution of the British Flora*, Hutchinson, London.

May, V. J. (1964) *A study of recent coastal changes in south-east England*, unpublished M.Sc. thesis, University of Southampton.

May, V. J. (1966) A preliminary study of recent coastal changes and sea-defences in south-east England, *Southampton Res. Ser. in Geog.*, **3**, 3–24.

May, V. J. (1969) Reclamation and shoreline change in Poole Harbour, Dorset, *Proc. Dorset Nat. Hist. and Arch. Soc.*, **90**, 141–154.

May, V. J. (1971a) The retreat of chalk cliffs, *Geogr. J.*, **137**, 203–206.

May, V. J. (1971b) Hengistbury Head: a study of physical processes, in: *Field Studies in South Hampshire*, Southampton Branch Geographical Association, pp. 111–116.

May, V. J. (1976) Cliff erosion and beach development: the case of Shipstal Point, Dorset, *Proc. Dorset Nat. Hist. and Arch. Soc.*, **97**, 8–12.

McCrone, A. (1966) The Hudson river estuary. Hydrology, sediments and pollution, *Geogr. Rev.*, **56**, 175–189.

McIntyre, A. D. (1969) Ecology of marine meiobenthos, *Biol. Rev.*, **44**, 245–290.

McIntyre, A. D. (1971) The range of biomass in intertidal sand, with special reference to the bivalve *Tellina tenuis*, *J. mar. biol. Ass. U.K.*, **50**, 561–575.

McNulty, J. K. (1970) Effects of abatement of domestic sewage pollution on the benthos, volumes of plankton and the fouling organisms of Biscayne Bay, Florida, *Stud. trop. Oceanogr. Miami*, **9**, 107 pp.

Means, R. E., and Parcher, J. V. (1964) *Physical Properties of Soils*, Constable, London.

Meijer, J. (1972) An isolated earthworm population in the recently reclaimed Lauwerszeepolder, *Pedobiologia*, **12**, 409–411.

Melchiorri-Santolini, U., and Hopton, J. W. (ed.) (1972) Detritus and its role in aquatic ecosystems. Proceedings of an I.B.P. Unesco Symposium, *Mem. Ist. Ital. Idrobiol.*, **29** (suppl.).

Milne, A. (1938) Ecology of the Tamar estuary, III. Salinity and temperature conditions in the lower estuary, *J. mar. biol. Ass. U.K.*, **22**, 529–542.

Milne, A. (1940a) Some ecological aspects of the intertidal area of the estuary of the Aberdeenshire Dee, *Trans. Roy. Soc. Edinburgh*, **60**, 107–139.

Milne, A. (1940b) The ecology of the Tamar estuary, IV. Distribution of the fauna and flora on buoys, *J. mar. biol. Ass. U.K.*, **24**, 69–87.

Milne, H., and Dunnet, G. M. (1972) Standing crop, productivity and trophic relations of the fauna of the Ythan estuary, in: *The Estuarine Environment*, ed. Barnes, R. S. K., and Green, J., Applied Science, London, pp. 86–106.

Ministry of Housing and Local Government (1964) *Pollution of the Tidal Thames*, H.M.S.O., London.

Mitchell, J. K. (1968) A selected bibliography of coastal erosion, protection and related human activity in North America and the British Isles, *Natural Hazards Research Working Paper No. 4*, University of Toronto.

Mook, J. H. (1967) Habitat selection by *Lipara lucens* Mg. (Diptera, Chloropidae) and its survival value, *Archs. Néerl. Zool.*, **17**, 469–549.

Mook, J. H. (1969) Observations on the colonization of the new IJsselmeerpolders by animals, *Symp. Biol. Stat. Wijster. Wageningen*, **8**, 13–31.

Mook, J. H., and Haeck, J. (1965) Dispersal of *Pieris bassicae* L. (Lepidoptera, Pieridae) and of its primary and secondary hymenopterous parasites in a newly reclaimed polder of the former Zuiderzee, *Archs. Néerl. Zool.*, **16**, 293–312.

Moore, E. J. (1931) The ecology of the Ayreland of Bride, Isle of Man, *J. Ecol.*, **19**, 115–136.

Moore, P. D. (1971) Computer analysis of sand-dune vegetation in Norfolk, England, and its implications for conservation, *Vegetatio*, **23**, 323–338.

Morgan, J. P. (1967) Ephemeral estuaries of the Deltaic environment, in: *Estuaries*, ed. Lauff, G., American Association for the Advancement of Science, Washington, D.C., pp. 115–120.

Morgenstern, N. R., and Price, V. E. (1965) The analysis of the stability of general slip surfaces, *Geotechnique*, **1**, 79–93.

Morgenstern, N. R., and Price V. E. (1967) A numerical method for solving the equations of stability of general slip surfaces, *Computer J.*, **9**, 388–393.

Morris, M. G. (1969) Populations in invertebrate animals and the management of chalk grasslands in Britain, *Biol Conserv.*, **1**, 225–231.

Morris, M. G. (1971) The management of grassland for the conservation of invertebrate animals, in: *The Scientific Management of Animal and Plant Communities for Conservation*, ed. Duffey, E., and Watt, A. S., B.E.S. Symposium, **11**, 527–552.

Morris, R. F. (1960) Sampling insect populations, *A. Rev. Ent.*, **5**, 243–264.

Moyse, J., and Nelson-Smith, A. (1963) Zonation of animals and plants on rocky shores Dale, Pembrokeshire, *Field Studies*, **1**, 1–31.

Murton, R. K. (1971) *Man and Birds*, Collins, London.

Muus, B. I. (1967) The fauna of Danish estuaries and lagoons, *Medd. Danm. Fisk Havunders*, **5** (fasc. 1).

Myers, J. E. (1954) A survey and comparison of the natural and inned salt marshes at Leigh-on-Sea, Essex, *Essex Nat.*, **29**, 155–175.

Nair, N. B. (1962) Ecology of marine fouling and boring organisms of western Norway, *Sarsia*, **8**, 1–88.

National Parks Commission (1967) *The Coasts of South Wales and the Severn Estuary*, H.M.S.O., London.

Natural Environment Research Council (1972) *The Severn Estuary and the Bristol Channel: an assessment of present knowledge*, Natural Environment Research Council, London.

Nature Conservancy (1958) *Evidence of the Nature Conservancy for the public enquiry into the proposed nuclear power station at Dungeness, Kent, to be held at Lydd, New Romney, Kent on 16 December 1958*, Eyre & Spottiswoode, London.

Nature Conservancy (1970) Map: *Isle of Rhum Vegetation* 1/20,000, Crown Copyright, H.M.S.O., London.

Nature Conservation at the Coast (1969) Countryside Commission Coastal Preservation and Development, H.M.S.O., London.

Naylor, E. (1965a) Effects of heated effluents upon marine and estuarine organisms, *Adv. mar. Biol.*, **3**, 63–103.

Naylor, E. (1965b) Biological effects of a heated effluent in docks at Swansea, S. Wales, *Proc. Zool. Soc. London*, **144**, 253–268.

Nelson-Smith, A. (1965) Marine biology of Milford Haven: the physical environment, *Field Studies*, **2**, 155–188.

Nelson-Smith, A. (1967) Marine biology of Milford Haven: the distribution of littoral animals and plants, *Field Studies*, **2**, 435–477.

Nelson-Smith, A. (1968a) The effects of oil pollution and emulsifier cleansing on marine life in south-west Britain, *J. Appl. Ecol.*, **5**, 97–107.

Nelson-Smith, A. (1968b) Biological consequences of oil pollution and shore cleansing, *Field Studies*, **2** (suppl.), 73–80.

Nelson-Smith, A. (1971) Discussion, in: *The Ecological Effects of Oil Pollution on Littoral Communities*, ed. Cowell, E. B., Institute of Petroleum, London, pp. 235–236.

Nelson-Smith, A. (1972a) *Oil Pollution and Marine Ecology*, Elek Scientific, London.

Nelson-Smith, A. (1972b) Effects of the oil industry on shore life in estuaries, *Proc. Roy. Soc. (B)*, **180**, 487–496.

Nelson-Smith, A. (1975) Effects of long-term, low-level exposure to oil, in: *Petroleum and the Continental Shelf of North-West Europe*, Vol. 2, Applied Science, London.

Nelson-Smith, A., and Bridges, E. M. (1976) *Problems of a Small Estuary* (proceedings of a seminar on the Burry Inlet), University College of Swansea Institute of Marine Studies.

Newborough Warren Management Plan (1964) *Newborough Warren—Ynys Llanddwyn National Nature Reserve, Anglesey: Revised Management Plan*, Nature Conservancy.

Newell, R. C. (1964) Some factors controlling the upstream distribution of *Hydrobia ulvae* (Pennant) (Gastropoda, Prosobranchia), *Proc. Zool. Soc. London*, **142**, 85–106.

Newell, R. (1965) The role of detritus in the nutrition of two marine deposit feeders, the prosobranch *Hydrobia ulvae* and the bivalve *Macoma balthica*, *Proc. Zool. Soc. London*, **144**, 25–45.

Newton, I., and Campbell, C. R. G. (1973) Feeding of geese on farmland in east-central Scotland, *J. Appl. Ecol.*, **10**, 781–801.

Newton, I., Thorn, V. M., and Brotherston, W. (1973) Behaviour and distribution of wild geese in south-east Scotland, *Wildfowl*, **24**, 111–121.

Newton, L. E. (1965) *Taxonomic studies in the British species of Puccinellia*, M.Sc. Thesis, University of London.

Nicholson, E. M. (1960) Scolt Head as a National Nature Reserve, in: *Scolt Head Island*, ed. Steers, J. A., Heffer, Cambridge, pp. 6–9.

Nicholson, I. A. (1952) *A study of Agropyron junceum in relation to the stabilization of coastal sand and the development of sand dunes*, M.Sc. Thesis, University of Durham.

Nordhagen, R. (1940) Studien über die maritime Vegetation Norwegens, 1. Die Pflanzengesellschaften der Tangwalle, *Bergens Mus. Aarb.*, **1939–40**, 1–123.

O'Connor, F. B., Goldsmith, F. B., and Macrae, M. M. (1975) *Kynance Cove, Experimental Restoration Project*, University College, London.

Odum, H. T. (1956) Primary production in flowing waters, *Limnol. Oceanogr.*, **1**, 102–117.

Odum, E. P. (1962) Relationship between structure and function in the ecosystem, *Jap. J. Ecol.*, **12**, 108–118.

Odum, E. P. (1971) *Fundamentals of Ecology*, 3rd Edn., Saunders, Philadelphia.

Odum, E. P., and De La Cruz, A. (1967) Particulate organic detritus in a Georgia salt-marsh-estuarine ecosystem, in: *Estuaries*, ed. Lauff, G., American Association for the Advancement of Science, Washington, D.C., pp. 383–388.

Odum, E. P., and Smalley, A. E. (1959) Comparison of population energy flow of a herbivorous and deposit-feeding invertebrate in a salt marsh ecosystem, *Proc. Nat. Acad. Sci. U.S.A.*, **45**, 617–622.

Ogilvie, M. A., and Matthews, G. V. T. (1969) Brent geese, mudflats and man, *Wildfowl*, **20**, 119–125.

Ogilvie, M. A., and St. Joseph, A. K. M. (1976) Dark-bellied Brent geese in Britain and Europe, 1955–76, *Brit. Birds*, **69**, 422–349.

Oliver, F. W. (1912) The shingle beach as a plant habitat, *New Phytol.*, **11**, 73–99.

Oliver, F. W. (1913) Some remarks on Blakeney Point, Norfolk, *J. Ecol.*, **1**, 4–15.

Oliver, F. W. (1929) Blakeney Point reports, *Trans. Norfolk Norwich Nat. Soc.*, **12**, 630–653.

Oliver, F. W., and Salisbury, E. J. (1913a) Topography and vegetation of Blakeney Point, *Trans. Norfolk Norwich Nat. Soc.*, **9**, 502–542.

Oliver, F. W., and Salisbury, E. J. (1913b) Vegetation and mobile ground as illustrated by *Suaeda fruticosa* on shingle, *J. Ecol.*, **1**, 249–272.

Olney, P. J. S. (1970) Food habits of wildfowl in Britain, *The New Wildfowler in the 1970s*, 86–97.

Olson, J. S. (1958a) Rates of succession and soil changes on southern Lake Michigan dunes, *Botan. Gaz.*, **119**, 125–170.

Olson, J. S. (1958b) Lake Michigan dune development, 1. Wind velocity profiles, *J. Geol.*, **66**, 345–351.

Olsson-Seffer, P. (1909) Hydrodynamic factors influencing plant life on sandy sea shores, *New Phytol.*, **8**, 37–49.

Oosterveld, P. (1976) Beheer en ontwikkeling van natuurreservaten door begrazing, *Natuur en Landschap*, **29**, 161–171.

O'Sullivan, A. J. (1971) Ecological effects of sewage discharge, *Proc. Roy. Soc.* (*B*), **177**, 331–351.

O'Sullivan, A. J., and Richardson, A. J. (1967) The *Torrey Canyon* disaster and intertidal marine life, *Nature*, **214**, 448; *ibid.*, 541–542.

Ovington, J. D. (1950) The afforestation of the Culbin sands, *J. Ecol.*, **38**, 303–319.

Ovington, J. D. (1951) The afforestation of the Tentsmuir sands, *J. Ecol.*, **39**, 363–375.

Owen, M. (1973) The management of grassland areas for wintering geese, *Wildfowl*, **24**, 123–130.

Pahlsson, L. (1966) Vegetation and microclimate along a belt transect from the Esker Knivsas, *Botan. Notiser*, **119**, 401–418.

Pahlsson, L. (1974a) Relationship of soil, microclimate, and vegetation on a sandy hill, *Oikos*, **25**, 21–34.

Pahlsson, L. (1974b) Influence of vegetation on microclimate and soil moisture on a Scanian hill, *Oikos*, **25**, 176–186.

Patmore, J. A. (1970) *Land and Leisure*, David & Charles, Newton Abbot.

Paviour-Smith, K. (1956) The biotic community of a salt meadow in New Zealand, *Trans. Roy. Soc. N.Z.*, **83**, 525–554.

Pearse, A. S., *et al.* (1942) Ecology of sand beaches at Beaufort, N.C., *Ecol. Monogr.*, **12**, 135–190.

Pearson, T. H. (1970) The benthic ecology of Loch Linnhe and Loch Eil, a sea-loch system on the west coast of Scotland, 1. The physical environment and distribution of the macrobenthic fauna, *J. exp. mar. Biol. Ecol.*, **5**, 1–34.

Pearson, T. H. (1972) The effect of industrial effluent from pulp and paper mills on the marine benthic environment, *Proc. Roy. Soc. (B)*, **180**, 469–485.

Percival, E. (1929) A report on the fauna of the estuaries of the R. Tamar and the R. Lynher, *J. mar. biol. Ass. U.K.*, **16**, 81–108.

Pérès, J. M. (1961) *Oceanographic Biologique et Biologie Marine*, Vol. 1, Press Universitaires de France, Paris.

Pérès, J. M., and Picard, J. (1958) Manuel de Bionomie benthique de la mer Méditerranée, *Rec. Trav. St. mar. Endoume*, **23**.

Perkins, E. J. (1957) The blackened sulphide-containing layer of marine soils, with special reference to that found at Whitstable, Kent, *Ann. Mag. Nat. Hist*, (12), **10** (109), 23–55.

Perkins, E. J. (1972) Estuaries and industry, *Environment this Month*, **1**, 47–54.

Perkins, E. J. (1974) *The Biology of Estuaries and Coastal Waters*, Academic Press, London and New York.

Perkins, E. J. (1976) The Solway Firth, in: *Problems of a Small Estuary*, ed. Nelson-Smith, A., and Bridges, E. M., University College of Swansea Institute of Marine Studies, pp. 9:1/1–9:1/6.

Perkins, E. J., Bailey, M., and Williams, B. R. H. (1964) The biology of the Solway Firth in relation to the movement and accumulation of radioactive materials, VI. General hydrography, with an appendix on meteorological observations, *U.K. Atom. Energy Auth. Rep.*, **PG604**, 8 pp.

Perkins, E. J., and Williams, B. R. H. (1963) The biology of the Solway Firth in relation to the movement and accumulation of radioactive materials, I. General introduction, *U.K. Atom. Energy. Auth. Rep.*, **PG500**, 8 pp.

Perkins, E. J., and Williams, B. R. H. (1964) The biology of the Solway Firth in relation to the movement and accumulation of radioactive materials, II. Distribution of sediments and benthos, *U.K. Atom. Energy Auth. Rep.*, **PG587**, 37 pp.

Perring, F. H., and Randall, R. E. (1972) An annotated flora of the Monarch Isles National Nature Reserve, Outer Hebrides, *Trans. Proc. bot. Soc. Edinburgh*, **41**, 431–444.

Perring, F. H., and Walters, S. M. (1962) *Atlas of the British Flora*, Nelson, London.

Perring, F. H., and Walters, S. M. (1976) *Atlas of the British Flora*, E. P. Publishing, Wakefield.

Persoone, G., and Pauw, N. de (1968) Pollution in the harbour of Ostend (Belgium), *Helgölander wiss. Meeresunters*, **17**, 302–320.

Petch, C. P. (1945) Reclaimed lands of west Norfolk, *Trans. Norfolk Norwich Nat. Soc.*, **16**, 106–109.

Peterken, G. F., and Hubbard, J. C. E. (1972) The shingle vegetation of southern England: the holly wood on Holmstone Beach, Dungeness, *J. Ecol.*, **60**, 547–571.

Phillips, A. J. (1972) Chemical processes in estuaries, in: *The Estuarine Environment*, ed. Barnes, R. S. K., and Green, J., Applied Science, London, pp. 33–50.

Phillips, P. (1972) *Physiographic processes and planning on the coastline of Christchurch Bay, Hampshire*, unpublished Ph.D. thesis, University of Southampton.

Phillips, P. H. (1972) *Coast protection: physiography and the planning process*, unpublished Ph.D. thesis, University of Southampton.

Pigott, C. D. (1969) Influence of mineral nutrition on the zonation of flowering plants in coastal salt-marshes, in: *Ecological Aspects of the Mineral Nutrition of Plants*, ed. Rorison, I. H., Blackwell, Oxford, pp. 25–35.

Pitt, M. (1971) *Multiple Land Use of Spurn Nature Reserve*, unpublished B.A. thesis, University of York.

Pizzey, J. M. (1975) Assessment of dune stabilization at Camber, Sussex, using air photographs, *Biol. Conserv.*, **7**, 275–288.

Pons, L. Z., and Zonneveld, I. S. (1965) *Soil ripening and soil classification. Initial soil formation in alluvial deposits and a classification of the resulting soils*, Pbn. No. 13, Int. Inst. for Land Reclamation and Improvement, Wageningen.

Porter, E. (1973) *Pollution in Four Industrialized Estuaries*, H.M.S.O., London.

Portmann, J. E., and Connor, P. M. (1968) The toxicity of several oil-spill remains to some species of fish and shellfish, *Mar. Biol.*, **1**, 322–329.

Postma, H. (1961) Transport and accumulation of suspended matter in the Dutch Wadden Sea, *Neth. J. Sea Res.*, **1**, 148–190.

Postma, H., and Rommets, J. W. (1970) Primary production in the Wadden Sea, *Neth. J. Sea Res.*, **4**, 470–493.

Prater, A. J. (1972) The ecology of Morecambe Bay, III. The food and feeding habits of knot (*Calidris canutus* L.) in Morecambe Bay, *J. Appl. Ecol.*, **9**, 179–194.

Pratt, A. (1929) Notes on strand plants, IV. *Arenaria peploides*, *Trans. Proc. bot. Soc. Edinburgh*, **30**, 157–163.

Precheur, C. (1960) *Le littoral de la Manche, de Ste Adresse à Ault*, Poitiers.

Prior, D. B., Stephens, N., and Archer, D. R. (1968) Composite mudflows on the Antrim coast of north-east Ireland, *Geografiska Annaler*, **50A**, 65–78.

Pritchard, D. W. (1967) What is an estuary: physical viewpoint, in: *Estuaries*, ed. Lauff, G., American Association for the Advancement of Science, Washington, D.C., pp. 3–5.

Projects for Environmental Studies (1970) Berkshire, Buckinghamshire and Oxfordshire Naturalists' Trust.

Randall, R. E. (1970) Salt measurement on the coast of Barbados, West Indies, *Oikos*, **21**, 65–70.

Randall, R. E. (1972) *Vegetation in a maritime environment: the Monach Isles*, Ph.D. thesis, University of Cambridge.

Randall, R. E. (1973) Calcium carbonate in dune soils: evidence for geomorphic change, *Area*, **5**, 308–310.

Randall, R. E. (1973) Shingle Street, Suffolk: an analysis of a geomorphic cycle, *Bull. Geol. Soc. Norfolk*, **24**, 15–35.

Randall, R. E. (1974) Airborne salt deposition and its effects upon coastal plant distribution: the Monach Isles National Nature Reserve, Outer Hebrides, *Trans. Proc. bot. Soc. Edinburgh*, **42**, 153–162.

Randall, R. E. (1977) Past and present status and distribution of sea pea (*Lathyrus japonicus* Willd.) in Britain, *Watsonia*, **13**, in press.

Randerson, P. F. (1969) *A quantitative investigation of the effects of public pressure*, M.Sc. thesis, University College, London.

Ranwell, D. S. (1958) Movement of vegetated sand-dunes at Newborough Warren, Anglesey, *J. Ecol.*, **46**, 83–100.

Ranwell, D. S. (1959) Newborough Warren, Anglesey, 1. The dune system and the dune-slack habitat, *J. Ecol.*, **47**, 571–601.

Ranwell, D. S. (1960) Newborough Warren, Anglesey, 2. Plant associes and succession cycles of the sand-dune and dune-slack vegetation, *J. Ecol.*, **48**, 117–141.

Ranwell, D. S. (1961) *Spartina* salt-marshes in Southern England, I. The effect of sheep grazing at the upper limits of *Spartina* marsh in Bridgwater Bay, *J. Ecol.*, **49**, 325–340.

Ranwell, D. S. (1964a) *Spartina* salt marshes in Southern England, II. Rate and seasonal pattern of sediment accretion, *J. Ecol.*, **52**, 79–94.

Ranwell, D. S. (1964b) *Spartina* salt marshes in Southern England, III. Rates of establishment, succession and nutrient supply at Bridgwater Bay, Somerset, *J. Ecol.*, **52**, 95–105.

Ranwell, D. S. (1964c) Conservation and management of estuarine marsh in relation to *Spartina* marsh in the British Isles, *Proc. MAR Conf.*, I.U.C.N. Publications, N.S., No. 3, Part I/C, pp. 281–287.

Ranwell, D. S. (1967) World resources of *Spartina townsendii* (*sensu lato*) and economic use of *Spartina* marshland, *J. Appl. Ecol.*, **4**, 239–256.

Ranwell, D. S. (1968a) *Coastal Marshes in Perspective*, Regional Studies Group Bull. Strathclyde No. 9, pp. 1–26.

Ranwell, D. S. (1968b) Extent of damage to coastal habitats due to the *Torrey Canyon* incident, in: The biological effects of oil pollution on littoral communities, ed. Carthy, J. D., and Arthur, D. R., *Field Studies*, **2** (suppl.), 39–47.

Ranwell, D. S. (1972a) *Ecology of Salt Marshes and Sand Dunes*, Chapman & Hall, London.

Ranwell, D. S. (1972b) *The management of Sea Buckthorn Hippophaë rhamnoides L. on selected sites in Great Britain*, Nature Conservancy, London.

Ranwell, D. S. (1974) *Sand Dune Machair: Report of a Seminar at Coastal Ecology Research Station, Norwich*, Institute of Terrestrial Ecology, Natural Environment Research Council, London.

Ranwell, D. S. (1975a) Management of salt-marsh and coastal-dune vegetation, *Estuarine Research*, **2**, 471–483.

Ranwell, D. S. (1975b) The dunes of St. Ouen's Bay, Jersey: an ecological survey, *Ann. Bull. Soc. Jersiaise*, **21**(3), 381–391.

Ranwell, D. S., Bird, E. C. F., Hubbard, J. C. E., and Stebbings, R. E. (1974) *Spartina* salt marshes in Southern England, V. Tidal submergence and chlorinity in Poole Harbour, *J. Ecol.*, **52**, 627–641.

Ranwell, D. S., and Downing, B. M. (1959) Brent goose (*Branta bernicla* L.) winter feeding pattern and *Zostera* resources at Scolt Head Island, Norfolk, *Animal Behaviour*, **7**, 52–56.

Ranwell, D. S., and Downing, B. M. (1960) The use of Dalapon and substituted urea herbicides for control of seed-bearing *Spartina* (cordgrass) in inter-tidal zones of estuarine marsh, *Weeds*, **8**, 78–88.

Ranwell, D. S., Winn, J. M., and Allen, S. E. (1973) *Road salting effects on soil and plants*, Natural Environment Research Council, 24 pp.

Ratcliffe, D. A. (1971) Criteria for the selection of Nature Reserves, *Adv. Sci.*, **27**, 294–296.

Rattray, J. (1886) The distribution of the marine algae of the Firth of Forth, *Trans. bot. Soc. Edinburgh*, **16**, 420–466.

Raymont, J. E. G. (1972) Some aspects of pollution in Southampton Water, *Proc. Roy. Soc.* (*B*), **180**, 451–468.

Reed, A., and Moisan, G. (1971) The *Spartina* tidal marshes of the St. Lawrence estuary and their importance to aquatic birds, *Naturaliste Canad.*, **98**, 905–922.

Reed, L. (1972) *An Ocean of Waste*, Conservative Political Centre, London.

Reish, D. J. (1960) The use of marine invertebrates as indicators of water quality, in: *Waste Disposal in the Marine Environment*, ed. Pearson, E. A., Pergamon, New York, pp. 92–103.

Reish, D. J. (1972a) The use of marine invertebrates as indicators of varying degrees of marine pollution, in: *Marine Pollution and Sea Life*, ed. Ruivo, M., FAO/Fishing News Books, London, pp. 203–207.

Reish, D. J. (1972b) Literature review: marine and estuarine pollution, *J. Water Poll. Control Fed.*, **44**, 1218–1226.

Remane, A. (1934) Die Brackwasserfauna, *Verh. dt. zool. Ges.*, **36**, 34–74.

Remane, A., and Schlieper, C. (1971) *Biology of Brackish Water. Die Binnengewasser*, Vol. 25, Wiley, New York.

Richards, F. J. (1934) The salt-marshes of the Dovey Estuary, IV, The rates of vertical accretion, horizontal extension and scarp erosion, *Ann. Botan.*, **48**, 225–259.

Riemann, F. (1966) Die Verbreitung der interstiellen Fauna im Elbe-Aestuar, *Ver. Inst. Meeresforsch. Bremerhaven*, **2**, 117–123.

Ritchie, W. (1971) *The Beaches of Barra and the Uists*, Department of Geography, University of Aberdeen.

Robinson, A. H. W. (1961) The hydrography of Start Bay and its relationship to beach changes at Hallsands, *Geogr. J.*, **121**, 63–77.

Roebert, A. J. (1975) Kunstmatige Infiltratie, *Versl. Meded. Comm. Hydrol. Onderz. T.N.O.*, **16**, 66–84.

Rogers, J. A. (1961) *The autecology of Hippophaë rhamnoides L.*, Ph.D. Thesis, University of Nottingham.

Rohlf, F. J., and Sokal, R. R. (1969) *Statistical Tables*, Freeman, San Francisco.

Rompaey, E. van, and Delvosalle, L. (1972) *Atlas de la Flore Belge et Luxembourgeoise*, Jardin Bot. Nat. Belgique, Brussels.

Rooth, J. (1972) Ontwerp inrichtingsplan voor de natuurterreinen Hompelvoet, Veermansplaat, Stampersplaat, *Contactblad Oecol.*, **8**, 71–83.

Rose, F. (1953) *Dungeness*. Internal report (unpublished) prepared for the Nature Conservancy, London.

Royal Commision on Coastal Erosion and Afforestation (1907–1911) *Evidence and Reports*, H.M.S.O., London.

Royal Commission on Environmental Pollution (1972) *Third Report: Pollution in some British estuaries and coastal waters*, H.M.S.O., London.

Royal Society of Edinburgh (1972) Symposium on the Forth–Tay estuaries (various authors), *Proc. Roy. soc. Edinburgh*, ser. B., **71**, 97–226.

Rudberg, S. (1967) The cliff coast of Gotland and the rate of cliff retreat, *Geografiska Annaler*, **49A**, 283–298.

Rudge, P. (1970) The birds of Foulness, *Br. Birds*, **63**, 49–66.

Ruivo, M. (ed.) (1972) *Marine Pollution and Sea Life*, FAO/Fishing News Books, London.

Russell, F. S., and Gilson, H. C. (ed.) (1972) A discussion on freshwater and estuarine studies on the effects of industry, *Proc. Roy. Soc. (B)*, **180**, 363–536.

Russell-Hunter, W. D. (1970) *Aquatic Productivity*, Collier–Macmillan, London.

Ryti, R. (1965) On the determination of soil pH, *Maataloust. Aikakausk*, **37**, 51–59.

Saelan, O. H. (1962) The natural history of the Hardangerfjord, 3. The hydrographical observations 1955–56, *Sarsia*, **6**, 1–25.

Saito, K., Yoshiola, K., and Ishizuka, K. (1965) Ecological studies on the vegetation of dunes near Sarugamori, Aomori Prefecture, *Aomori Pref. Ecol. Rev.*, **16**, 163–180.

Salisbury, E. J. (1952) *Downs and Dunes*, Bell, London.

Sandsborg, J. (1970) The effect of wind on the precipitation distribution over a hillock, *Nordic Hydrology*, **4**, 235–244.

Sawyer, C. N. (1965) The sea lettuce problem in Boston Harbour, *J. Wat. Pollut. Control Fed.*, **37**, 1122–1133.

Schachter, D. (1969) Ecologie des eaux saumâtres, *Verh. Internat. Verein. Limnol.*, **17**, 1052–1068.

Schäfer, W. (1962) *Aktuo-paläontologie nach Studien in der Nordsee*, Kramer, Senkenberg.

Schlieper, C. (1968) *Methoden der Meeresbiologischen Forschung*, VEB Gustav Fisher, Jena.

Schothorst, C. J. (1965) Grassland with low loadbearing capacity, 1. Drawbacks and causes, *Landbouwvoorlichting*, **22**, 492–500.

Schou, A. (1949) Danish coastal cliffs in glacial deposits, *Geografiska Annaler*, **31**, 357–364.

Schreven, D. A. van (1963a) Nitrogen transformations in the former subaqueous soils of polders recently reclaimed from Lake IJssel, I. Water-extractable, exchangeable and fixed ammonium, *Plant Soil*, **18**, 143–162.

Schreven, D. A. van (1963b) Nitrogen transformation in the former subaqueous soils of polders recently reclaimed from Lake IJssel, II. Losses of nitrogen due to denitrification and leaching, *Plant Soil*, **18**, 163–175.

Schreven, D. A. van (1963c) Nitrogen transformations in the former subaqueous soils of polders recently reclaimed from Lake IJssel, III. The uptake of mineral nitrogen by the pioneer vegetation on nitrogen mineralization in recently drained polder soils, *Plant Soil*, **18**, 277–297.

Scott, G. A. M. (1963a) The ecology of shingle beach plants, *J. Ecol.*, **51**, 517–527.

Scott, G. A. M. (1963b) *Glaucium flavum* Crantz, *J. Ecol.*, **51**, 743–754.

Scott, G. A. M. (1963c) *Mertensia maritima* (L.), S. F. Gray., *J. Ecol.*, **51**, 733–742.

Scott, G. A. M. (1965) The shingle succession at Dungeness, *J. Ecol.*, **53**, 21–31.

Scott, G. A. M., and Randall, R. E. (1976) *Crambe maritima* L., *J. Ecol.*, **64**, 1077–1091.

Segal, S. (1960) *Een vegetatiekundig onderzoek in de Kennemereduinen, 1956–1960*, Verslag Univ. van Amsterdam.

Segerstråle, S. G. (1959) Brackishwater classification—a historical survey, *Arch. Oceanogr. Limnol.*, **11** (suppl.), 7–33.

Shepard, F. P. (1937) Revised classification of marine shorelines, *J. Geol.*, **45**, 602–624.

Shimwell, D. W. (1971) *The Description and Classification of Vegetation*, Sidgwick & Jackson, London.

Side, A. G. (1973) Vegetational changes at Egypt Bay, Kent, following the exclusion of tidal waters, 1964–71, *Trans. Kent. Field Club*, **5**.

Silvester, R. (1960) Stabilization of sedimentary coastlines, *Nature*, **188**(4749), 467–469.

Silvester, R. (1970) Growth of crenulate bays to equilibrium, *Proc. Amer. Soc. Civ. Eng.*, *J. Waterways Hbrs. Divn.*, **96**, WW2, 201–218.

Sjöstedt, L. G. (1928) Littoral and supralittoral studies on the Scanian shores, *Lunds Univ. Årsskr.*, **24**, 1–36.

Smith, J. E. (ed.) (1968) *Torrey Canyon Pollution and Marine Life*, Cambridge University Press, London.

Smith, R. J., and Kavanagh, N. J. (1969) The measurements of benefits of trout fishing: preliminary results of a study at Grafham Water, Great Ouse Water Authority, Huntingdonshire, *J. Leisure Res.*, **1**, 316–332.

Smyth, J. C. (1968) The fauna of a polluted shore in the Firth of Forth, *Helgölander wiss. Meeresunters*, **17**, 216–223.

So, C. L. (1965) Coastal platforms of the Isle of Thanet, *Trans. Inst. Br. Geogr.*, **37**, 145–156.

So, C. L. (1967) Some coastal changes between Whitstable and Reculver, Kent, *Proc. Geol. Assoc.*, **77**, 475–490.

Sokal, R. R., and Rohlf, F. J. *Biometry*, Freeman, San Francisco.

Southgate, B. A. (1972) *Report to the Hampshire River Authority on Langstone Harbour.*

South Hampshire Structure Plan (1973) Hampshire County Council, Portsmouth City Council, Southampton City Council.

Southward, A. J. (1965) *Life on the Sea-shore*, Heinemann, London.

Southwood, T. R. E. (1966) *Ecological Methods with Particular Reference to the Study of Insect Populations*, Methuen, London.

Sparks, B. W. (1972) *Geomorphology*, 2nd Edn., Longmans, London.

Spedding, C. R. W., and Large, R. V. (1957) A point-quadrat method for the description of pasture in terms of height and density, *J. Br. Grassland Soc.*, **12**, 229–234.

Spooner, G. M., and Moore, H. B. (1940) The ecology of the Tamar estuary, VI. An account of the macrofauna of the intertidal muds, *J. mar. biol. Ass. U.K.*, **24**, 283–330.

Spurn Management Plan (1976) *Management Plan for Spurn Nature Reserve, January 1977 to December 1981*, Yorkshire Naturalists' Trust.

Stamp, D. L. (1939) Recent coastal changes in south-eastern England, V. Some economic aspects of coastal loss and gain, *Geogr. J.*, **93**, 496–503.

Stamp, D. (1969) *Nature Conservation in Britain*, Collins, London.

Stebbings, R. E. (1970) Recovery of salt-marsh in Britanny sixteen months after heavy pollution by oil, *Environ. Pollution*, **1**, 163–167.

Steeman Nielsen, E. (1958) Experimental methods for measuring organic production in the sea, *Rapp. Cons. Expl. Mar.*, **144**, 38–46.

Steeman Nielsen, E. (1964) Recent advances in measuring and understanding marine primary production, *J. Ecol.*, **52** (suppl.), 119–130.

Steers, J. A. (1926) Orford Ness—a study in coastal physiography, *Proc. Geol. Assoc.*, **37**, 306–325.

Steers, J. A. (1953) *The Sea Coast*, Collins, London.

Steers, J. A. (ed.) (1960) *Scolt Head Island*, 2nd Edn., Heffer, Cambridge.

Steers, J. A. (1962) Coastal cliffs: a symposium, *Geogr. J.*, **128**, 302–320.

Steers, J. A. (1964) *The Coastline of England and Wales*, 2nd Edn., Cambridge University Press, London.

Steers, J. A. (1971) *Blakeney Point and Scolt Head Island*, 3rd Edn., National Trust, London.

Steers, J. A. (1973) *The Coastline of Scotland*, Cambridge University Press, London.

Stege, K. van der (1965) *Verslag onderzoek embryonale duinvorming en aanleg van stuifdijken*, Rapport Rijkswaterstaat, Deltadienst.

Stephenson, T. A., and Stephenson, A. (1972) *Life Between the Tidemarks on Rocky Shores*, Freeman, San Francisco.

Stewart, W. D. P. (1967) Transfer of biologically fixed nitrogen in a sand dune-slack region, *Nature*, **214**, 603–604.

Stewart, W. D. P. (1972) Estuarine and brackish waters—an introduction, in: *The Estuarine Environment*, ed. Barnes, R. S. K., and Green, J., Applied Science, London, pp. 1–9.

Stopford, S. C. (1951) An ecological survey of the Cheshire foreshore of the Dee estuary, *J. anim. Ecol.*, **20**, 103–122.

Straaten, L. M. J. U. van (1961) Directional effects of winds, waves, and currents along the Dutch North Sea coast, *Geol. en Mijnb.*, **40**, 333–346; *ibid.*, 363–391.

Straaten, L. M. J. U. van (1965) Coastal barrier deposits in south and north Holland in particular in the area around Scheveningen and IJmuiden, *Meded. Geol. Stichr. N.S.*, **17**, 41–75.

Streeter, D. T. (1971) The effects of public pressure on the vegetation of chalk downland at Box Hill, Surrey, in: *The Scientific Management of Animal and Plant Communities for Conservation*, ed. Duffey, E., and Watt, A. S., Blackwell, Oxford, pp. 459–468.

Strickland, J. D. M., and Parsons, T. R. (1968) A practical handbook of seawater analysis, *Fish. Res. Bd. Canada*, Bull. 167, 3rd Edn., Fisheries Research Board of Canada, Ottawa.

Sunamura, T., and Horikawa, K. (1971) A quantitative study of the effects of beach deposits on cliff erosion, *Coastal Engineering in Japan*, **14**, 97–106.

Swedmark, B. (1964) The interstitial fauna of marine sand, *Biol. Rev.*, **39**, 1–42.

Symposium on Estuarine Fisheries (1966) Amer. Fish. Soc. Spec. Publ. No. 3, Vol. 95 (suppl.), *Trans. Am. Fis. Soc.*, Washington, D.C.

Tansley, A. G. (1939, 1949) *The British Islands and Their Vegetation*, Cambridge University Press, Cambridge.

Taylor, C., and Gear, A. (1972) The computer as an aid to environmental planning, *J. Environ. Planning*, **1**, 24–32.

Teagle, W. G. (1966) *Public pressure on South Haven Peninsula and its effect on Studland Heath National Nature Reserve*, unpublished, Nature Conservancy, London.

Teal, J. M. (1962) Energy flow in the salt marsh ecosystem of Georgia, *Ecology*, **43**, 614–624.

Teal, J. M. (1969) *Life and Death of the Salt Marsh*, Little, Brown, Boston.

Tentsmuir Management Plan (1963) *Tentsmuir National Nature Reserve Management Plan*, Nature Conservancy.

Terzaghi, K. (1950) Mechanism of landslides, *Bull. Geol. Soc. Amer.*, Berkey Volume, 83–122.

Terzaghi, K., and Peck, R. B. (1948) *Soil Mechanics in Engineering Practice*, Wiley, New York and London.

Thomas, A. S. (1960) Changes in vegetation since the advent of myxomatosis, *J. Ecol.*, **48**, 287–306.

Thomas, A. S. (1963) Further changes in vegetation since the advent of myxomatosis, *J. Ecol.*, **51**, 151–186.

Thomas, D. J. (1972) A review of gull damage and management methods at nature reserves, *Biol. Conserv.*, **4**, 117–127.

Thorn, R. B. (1960) *The Design of Sea Defence Works*, Butterworth, London.

Thorson, G. (1957) Bottom communities (sublittoral or shallow shelf), *Geol. Soc. Am. Mem. 67*, **1**, 461–534.

Tindall, F. P. (1967) The care of a coastline, *J. Town Plan. Inst.*, **53**, 387–392.

Toorn, J. van der, Donougho, B., and Brandsma, M. (1969) Verspreiding van Wegbermplanten in Oostelijk Flevoland, *Gorteria*, **4**, 151–160.

Trew, M. J. (1973) The effects and management of trampling on coastal sand dunes, *J. Environ. Planning Pollution Control.*, **1**, 1–12.

Trice, A. H., and Wood, S. E. (1958) Measurement of recreation benefits, *Land Econ.*, **34**, 195–207.

Tubbs, C. R. (1966) *Langstone Harbour, Hampshire: A study of harbour uses and conservation*, Nature Conservancy Report, London.

Tubbs, C. R. (1977) *Wader and Wildfowl Populations in Langstone Harbour, Hampshire* (in press).

Tucker, B. M. (1955) The determination of exchangeable calcium and magnesium in carbonate soils, *Aust. J. agric. Res.*, **5**, 706–715.

Tulkki, P. (1968) Effects of pollution on the benthos off Gothenburg, *Helgölander wiss. Meeresunters*, **17**, 209–215.

Turrill, W. B. (1959) *British Plant Life*, Collins New Naturalist, London.

Tutin, T. G. (1942) Biological flora of the British Isles, *Zostera* L., *J. Ecol.*, **30**, 217–226.

Tyler, G. (1967) On the effects of phosphorus and nitrogen, supplied to Baltic shore-meadow vegetation, *Botan. Notiser*, **120**, 433–447.

Tyler, G. (1968) Some chemical properties of Baltic shore-meadow clays. Studies in the ecology of Baltic sea-shore meadows, I, *Botan. Notiser*, **121**, 89–113.

Tyler, G. (1969) Flora and vegetation. Studies in the ecology of Baltic sea-shore meadows, II, *Opera Botan.*, **25**, 1–101.

Tyler, G. (1971a) Hydrology and salinity of Baltic sea-shore meadows. Studies in the ecology of Baltic sea-shore meadows, III, *Oikos*, **22**, 1–20.

Tyler, G. (1971b) Distribution and turnover of organic matter and minerals in a shore-meadow ecosystem. Studies in the ecology of Baltic sea-shore meadows, IV, *Oikos*, **22**, 265–291.

Usher, M. B. (1967) *Aberlady Bay Local Nature Reserve: Description and Management Plan*, County Planning Department, East Lothian County Council, Haddington.

Usher, M. B. (1973) *Biological Management and Conservation*, Chapman & Hall, London.

Usher, M. B., Pitt, M., and Boer, G. de (1974) Recreation pressures in the summer months on a nature reserve on the Yorkshire coast, England, *Environ. Conservation*, **1**, 43–49.

Vaas, K. F. (1966) Lakes in Dutch reclamation schemes, *Symp. Inst. Biol.*, **15**, 119–128.

Valentin, H. (1952) *Die Küste der Erde*, Petermanns Geog. Mitt. Ergänzungsheft, 246.

Valentin, H. (1954) Der Landverlust in Holderness, Ostengland von 1852 bis 1952, *Die Erde*, **6**, 296–315.

Vatova, A. (1940) Caratteri di alcune facies bentoniche della laguna Veneta, *Thalassia*, **3** (fasc. 10), 1–28.

Vatova, A. (1961) Primary production in the High Venice lagoon, *J. Conseil int. Explor. Mer.*, **26**.

Venice System (1959) Final resolution of the symposium on the classification of brackish waters, *Arch. Oceanogr. Limnol.*, **11** (suppl.), 243–245.

Visser, A. de (1971) De grote intbreiding van *Puccinellia fasciculata* (Torr.) Bickn. in Zeeland als gevolg van inundatie en overstroming, *Gorteria*, **5**, 231–234.

Voderberg, K., and Fröde, E. (1967) Abschliessende Betrachtung der Vegetationsentwicklung auf der Insel Bock in den Jahren 1946–1966, *Feddes Rep. Berlin*, **74**, 171–176.

Vollenweider, R. (ed.) (1969) *Methods for Measuring Primary Production in Aquatic Environment*, I.P.B. Handbook No. 11, Blackwell, Oxford and Edinburgh.

Vollenweider, R. (ed.) (1973) *Methods for Measuring Primary Production in Aquatic Environment*, 2nd Edn., Blackwell, Oxford and Edinburgh.

Voltolina, D. (1973a) Phytoplankton concentration in the Malamocco channel of the lagoon of Venice, *Arch. Oceanogr. Limnol.*, **18**, 1–18.

Voltolina, D. (1973b) A phytoplankton bloom in the lagoon of Venice, *Arch. Oceanogr. Limnol.*, **18**, 19–37.

Voo, E. E. van der (1964) Danger to scientifically important wetlands in the Netherlands by modification of the surrounding environment, *Proc. MAR Conference*, I.U.C.N. Publication N.S. 3, pp. 247–278.

Wagret, P. (1959) *Les Polders*, Dunod, Paris.

Wagret, P. (1968) *Polderlands*, Methuen, London.

Waisel, Y. (1972) *Biology of Halophytes*, Academic Press, New York and London.

Wallace, T. J. (1953) The plant ecology of Dawlish Warren Pt. 1, *Rep. Trans. Devon. Ass. Advmt. Sci. Lit. Art.*, **85**, 85–94.

Walne, P. R. (1972) The importance of estuaries to commercial fisheries, in: *The Estuarine Environment*, ed. Barnes, R. S. K., and Green, J., Applied Science, London, pp. 107–118.

Warming, E. (1906) *Dansk Plantevaekst, I. Strandbegetationen*, Copenhagen, pp. 36–66.

Warren, A. (1976) Dune trends and the Ekman spiral, *Nature*, **259**, 653–654.

Warren, C. E. (1971) *Biology and Water Pollution Control*, Saunders, Philadelphia.

Wass, M. L., and Wright, T. D. (1969) Coastal wetlands of Virginia: Interim Report of the Governor and General Assembly, *Virginia Inst. Marine Sci. Spec. Rep. Appl. Marine Sci.*, **10**, 1–154.

Wastes Management Concepts for the Coastal Zone (1970) National Academy of Sciences & National Academy of Engineering, Washington, D.C.

Water Resources in England and Wales (1974) Vol. 1, Report; Vol. 2, Appendices, H.M.S.O., London.

Welsh Office (1974) *Report of the working party on possible pollution in Swansea Bay. Vol. 2, Technical reports*, Cardiff.

Wendelberger, G. (1950) Zur Soziologie der kontinentalen Halophytenvegetation Mitteleuropas. Unter besonderer Berücksichtigung der Salzpflanzengesellschaften am Neusiedler See, *Denkschr. Österreich. Akad. Wissensch.*, **108**, 1–180.

Werf, S. van der (1970) Recreatie-invloeden in Meijendel, *Meded. Landbouwhogeschool, Wageningen*, **70**, 1–24.

Westhoff, V. (1952) Gezelschappen met houtige gewassen in de duinen en langs de binnenduinrand, *Dendrol. Jaarb.*, **1952**, 9–49.

Westhoff, V. (1955) Hedendaagse aspecten der natuurbescherming, *Wetenschap en Samenleving*, **9**, 25–34.

Westhoff, V. (1957) Regeneration of dune areas in the Netherlands, which have been biologically devastated by man, *Proceedings and Papers 6th. techn. meeting I.U.C.N. Edinburgh*, I.U.C.N, London, 164–165.

Westhoff, V. (1969) Langjährige Beobachtungen an Aussüngs–Dauerprobeflächen beweideter und unbeweideter Vegetation an der ehemaligen Zuiderzee, in: *Experimentelle Pflanzensoziologie*, ed. Tüxen, R., Junk, The Hague, pp. 246–253.

Westhoff, V. (1970) New criteria for nature reserves, *New Scientist*, 108–113.

Westhoff, V. (1971) The dynamic structure of plant communities in relation to the objectives of conservation, in: *The Scientific Management of Animal and Plant Communities for Conservation*, ed. Duffey, E., and Watt, A. S., Blackwell, Oxford, 3–14.

Westhoff, V., Bakker, P. A., Leeuwen, C. G. van, and Voo, E. E. van der (1970) *Wilde Planten*, deel 1, Ver. tot Behoud van Natuurmonumenten, Amsterdam.

Westhoff, V., and Held, A. J. den (1969) *Planten Gemeenschappen in Nederland*, Thieme, Zutphen.

Westrup, A. W. (1964) Vascular plants, in: *A Survey of Southampton and its Region*, ed. Monkhouse, F. J., British Association for the Advancement of Science, Southampton, pp. 109–111.

White, D. J. B. (1961) Some observations on the vegetation of Blakeney Point, Norfolk, following the disappearance of rabbits in 1954, *J. Ecol.*, **49**, 113–118.

Whittaker, R. H. (1965) Dominance and diversity in land plant communities, *Science*, **157**, 250–260.

Wiegel, R. L. (1959) *Oceanographical Engineering*, Prentice-Hall, Englewood Cliffs, N.J.

Wigham, G. D. (1976) Heavy-metal loads of Bristol Channel biota, in: *Problems of a Small Estuary*, ed. Nelson-Smith, A., and Bridges, A. M., University College of Swansea Institute of Marine Studies.

Williams, B. R. H., Perkins, E. J., and Gorman, J. (1965) The biology of the Solway Firth in relation to the movement and accumulation of radioactive materials, IV. Algae, *U.K. Atom. Energy Auth. Rep.*, **PG650**, 6 pp.

Williams, B. R. H., Perkins, E. J., and Hinde, A. (1965) The biology of the Solway Firth in relation to the movement and accumulation of radioactive material, III. Fisheries and food chains, *U.K. Atom. Energy Auth. Rep.*, **PG611**, 37 pp.

Williams, R. E. (1955) Development and improvement of coastal marsh range, *U.S. Year Book of Agriculture*, **1955**, 444–449.

Williams, R. E. (1959) *Cattle walkways*, U.S. Department of Agric. Leaflet No. 459, pp. 1–8.

Williams, W. W. (1960) *Coastal Changes*, Routledge & Kegan Paul, London.

Willis, A. J. (1965) The influence of mineral nutrients on the growth of *Ammophila arenaria*, *J. Ecol.*, **53**, 735–745.

Willis, A. J. (1973) *Introduction to Plant Ecology*, George Allen & Unwin, Oxford.

Willis, A. J., Folkes, F. B., Hope-Simpson, H. J. F., and Yemm, E. W. (1959) Braunton Burrows: the dune system and its vegetation, *J. Ecol.*, **47**, 1–24; *ibid.*, 249–288.

Wohlenberg, E. (1938) Biologische Kulturmasznahmen mit dem Queller (*Salicornia herbacea* L.) zur Landgewinnung im Wattenmeer, *Westküste*, **1**, 52–104.

Wohlenberg, E. (1939) Die Nutzanwendung biologischer Erkenntnisse im Wattenmeer zu Gunsten der praktischen Landgewinnung an der deutschen Nordseeküste, *Conseil Perm. Int. l'Explor. Mer. Extrait Rapports Procès-Verbaux*, **59**, 125–130.

Wohlenberg, E. (1965) Deichbau und Deichpflege auf biologischer Grundlage, *Die Küste*, **14**, 73–103.

Wood, H. A. (1950) Procedure in studying shore erosion, *Canadian Geographer*, **1**, 31–37.

Wood, P. C. (1969) *The Production of Clean Shellfish*, Min. of Ag. Fish. Food, Lab. Leaflet (N.S.) No. 20.

Woodwell, G. M. (1971) Toxic substances and ecological cycles, in: *Man and the Ecosphere*, Freeman, San Francisco, pp. 128–135.

Woolhouse, A. R. (1974) Further assessment of the effectiveness of a slow release nitrogen fertilizer on sports turf, *J. Sports Turf. Res. Inst.*, **50**, 34–46.

Wright, T. W. (1955) Profile development in the sand dunes of Culbin Forest, Morayshire, 1. Physical properties, *J. Soil Sci.*, **6**, 270–283.

Yonge, C. M. (1949) *The Sea Shore*, Collins, London.

Young, A. (1972) *Slopes*, Oliver & Boyd, Edinburgh.

Zaruba, Q., and Mencl, V. (1969) *Landslides and their Control*, Academia and Elsevier, Prague.

Zeigler, J. M., Tuttle, S. D., Giese, G. S., and Tasha, H. J. (1964) Residence time of sand composing beaches and bars of Outer Cape Cod, *Proc. 9th Conf. Coastal Eng.*, **26**, 401–416.

Zenkovich, V. P. (1967) in: *Processes of Coastal Development*, ed. Steers, J. A., Transl. Fry, D. G., Oliver & Boyd, Edinburgh.

Zijlstra, J. J. (1972) On the importance of the Waddensea as a nursery area in relation to the conservation of the southern North Sea fishery resources, *Symp. Zool. Soc. London*, **29**, 233–258.

Zuur, A. J. (1938) *Over de ontzilting van den bodem in de Wieringermeer. Een studie over de zonten waterbeweging in jonge poldergronden*. The Hague.

Index

337